Self-cleaning Coatings

Structure, Fabrication and Application

RSC Smart Materials

Series Editors:
Professor Hans-Jörg Schneider, *Saarland University, Germany*
Professor Mohsen Shahinpoor, *University of Maine, USA*

Titles in this Series:

1: Janus Particle Synthesis, Self-Assembly and Applications
2: Smart Materials for Drug Delivery: Volume 1
3: Smart Materials for Drug Delivery: Volume 2
4: Materials Design Inspired by Nature
5: Responsive Photonic Nanostructures: Smart Nanoscale Optical Materials
6: Magnetorheology: Advances and Applications
7: Functional Nanometer-Sized Clusters of Transition Metals: Synthesis, Properties and Applications
8: Mechanochromic Fluorescent Materials: Phenomena, Materials and Applications
9: Cell Surface Engineering: Fabrication of Functional Nanoshells
10: Biointerfaces: Where Material Meets Biology
11: Semiconductor Nanowires: From Next-Generation Electronics to Sustainable Energy
12: Supramolecular Materials for Opto-Electronics
13: Photocured Materials
14: Chemoresponsive Materials: Stimulation by Chemical and Biological Signals
15: Functional Metallosupramolecular Materials
16: Bio-Synthetic Hybrid Materials and Bionanoparticles: A Biological Chemical Approach Towards Material Science
17: Ionic Polymer Metal Composites (IPMCs): Smart Multi-Functional Materials and Artificial Muscles Volume 1
18: Ionic Polymer Metal Composites (IPMCs): Smart Multi-Functional Materials and Artificial Muscles Volume 2
19: Conducting Polymers: Bioinspired Intelligent Materials and Devices
20: Smart Materials for Advanced Environmental Applications
21: Self-cleaning Coatings: Structure, Fabrication and Application

How to obtain future titles on publication:
A standing order plan is available for this series. A standing order will bring delivery of each new volume immediately on publication.

For further information please contact:
Book Sales Department, Royal Society of Chemistry, Thomas Graham House, Science Park, Milton Road, Cambridge, CB4 0WF, UK
Telephone: +44 (0)1223 420066, Fax: +44 (0)1223 420247
Email: booksales@rsc.org
Visit our website at www.rsc.org/books

Self-cleaning Coatings
Structure, Fabrication and Application

Edited by

Junhui He
Technical Institute of Physics and Chemistry, Chinese Academy of Sciences, Beijing, China
Email: jhhe@mail.ipc.ac.cn

ROYAL SOCIETY OF CHEMISTRY

THE QUEEN'S AWARDS FOR ENTERPRISE: INTERNATIONAL TRADE 2013

RSC Smart Materials No. 21

Print ISBN: 978-1-78262-286-4
PDF eISBN: 978-1-78262-399-1
EPUB eISBN: 978-1-78262-911-5
ISSN: 2046-0066

A catalogue record for this book is available from the British Library

Published by The Royal Society of Chemistry,
Thomas Graham House, Science Park, Milton Road,
Cambridge CB4 0WF, UK

Registered Charity Number 207890

For further information see our web site at www.rsc.org

Printed in the United Kingdom by CPI Group (UK) Ltd, Croydon, CR0 4YY, UK

Preface

Recent years have seen a fast development in the field of self-cleaning coatings towards varied applications, such as solar cells, flat display panels, smart cellular phones, building windows, oil pipelines, vehicle-related coatings, optical devices, safety products, intelligent devices, oil–water separation, *etc.* The field has been attracting the attention of not only research and teaching scientists and application engineers, but also a huge population of college and graduate students.

This book is aimed at documenting the latest research and development in the hot topic, and entertaining readers with both natural mysteries and artificial miracles. The book contains contributions from world-renowned specialists in the field, and ten reviews that discuss natural counterparts with self-cleaning functions, theoretical aspects of self-cleaning phenomena, fabrication strategies and methods, varied applications, and industrial impacts, *etc.*, under the title of "Self-cleaning Coatings—Structure, Fabrication and Application".

After a year of hard work, the book has eventually been finalized. At this point, I would like to take the opportunity to acknowledge all the authors who spent their precious time preparing their great contributions to the book. I would also like to thank Mr Pu He and Dr Lin Yao, who contributed to the preparation of the book. I am also very grateful to Prof. Dr Hans-Jörg Schneider for inviting me to participate in the publishing program of the RSC Smart Materials Series, and to the Royal Society of Chemistry, UK, for providing the opportunity to publish this book.

Junhui He
Functional Nanomaterials Laboratory
Center for Micro/Nanomaterials and Technology
Technical Institute of Physics and Chemistry
Chinese Academy of Sciences

RSC Smart Materials No. 21
Self-cleaning Coatings: Structure, Fabrication and Application
Edited by Junhui He

Published by the Royal Society of Chemistry, www.rsc.org

Biography

Junhui He is a professor and head of the Functional Nanomaterials Laboratory at the Technical Institute of Physics and Chemistry (TIPC), Chinese Academy of Sciences (CAS). He also serves as the director of the Center for Micro/Nanomaterials and Technology, TIPC, and managing editor of the *International Journal of Nanoscience*. He received his B.Sc. degree from the National University of Defense Technology and his M.Sc. degree and Ph.D. degree under Prof. E. Wang from the Institute of Photographic Chemistry, CAS. He then spent two years as a postdoctoral fellow with Prof. N. S. Allen at Manchester Polytechnic and over four years as a STA fellow and FRS researcher with Prof. T. Kunitake at Frontier Research System (FRS), RIKEN. He returned as part of the Hundred Talents Program. He has so far co-authored over 260 peer-reviewed scientific papers and filed over 60 patents. His research interests involve functional nanomaterials, smart surfaces, thin films, biomimetic materials, sensors, and their energy and environmental applications.

RSC Smart Materials No. 21
Self-cleaning Coatings: Structure, Fabrication and Application
Edited by Junhui He

Published by the Royal Society of Chemistry, www.rsc.org

Contents

RSC Smart Materials No. 21
Self-cleaning Coatings: Structure, Fabrication and Application
Edited by Junhui He

Published by the Royal Society of Chemistry, www.rsc.org

CHAPTER 1

The Inspiration of Nature: Natural Counterparts with Self-cleaning Functions

TONG LI,[a,b,†] TINGTING REN[a,b,†] AND JUNHUI HE*[a]

[a] Functional Nanomaterials Laboratory, Center for Micro/Nanomaterials and Technology and Key Laboratory of Photochemical Conversion and Optoelectronic Materials, Technical Institute of Physics and Chemistry, Chinese Academy of Sciences, Zhongguancundonglu 29, Haidianqu, Beijing 100190, China; [b] University of Chinese Academy of Sciences, Beijing 100864, China

*Email: jhhe@mail.ipc.ac.cn

1.1 Introduction

Generally, a self-cleaning surface is defined as the capability of a material to maintain a contamination-free surface, avoiding the deposition of dirt.[1–3] There are currently two known types of self-cleaning surfaces. One has a special surface structure and chemical composition and is derived from nature, while the other is composed of semiconductor nanocrystals and functions *via* a process called photocatalysis.[3,4] Self-cleaning surfaces have great prospects for vehicle and building glasses, greenhouses, solar cell panels, building-integrated photovoltaics (BIPV), corrosion resistant ships, low resistance oil pipelines, and many daily goods.[2,5–8] As a result, a lot of

[†]These two authors contribute to this chapter equally.

RSC Smart Materials No. 21
Self-cleaning Coatings: Structure, Fabrication and Application
Edited by Junhui He

Published by the Royal Society of Chemistry, www.rsc.org

effort has recently been devoted to both understanding the mechanisms of self-cleaning functions and fabricating self-cleaning surfaces.[9–11] In this chapter, only the first type of self-cleaning surfaces will be discussed, which were found in nature and have been mimicked successfully using artificial surfaces.

Nature is full of mysteries, where human beings can learn, be inspired and thus create biomimicked objects. Successful examples so far include airplanes, radars, sonars, sharkskin swimsuits, jellyfish ear storm prediction instruments, *etc.* Among natural species, the lotus leaf surface shows superhydrophobicity with a high water contact angle and low rolling angle. When it rains, rain droplets can adsorb surface dust and roll off the lotus leaf surface, leaving a clean surface behind. This contamination-free, *i.e.*, self-cleaning, phenomenon had been investigated, and was attributed to the special surface, which is composed of a hydrophobic micro-nano papilla matrix.[1,12] Until now, self-cleaning surfaces have also been found in other plants (*e.g.*, onion skins, rice leaves) and animals (*e.g.*, butterfly wings, cicada wings, waterfowl feathers, gecko feet, shark skins).[2,3,11–19] Many peculiar surface structures with special chemical compositions were observed, and the self-cleaning principles of these organisms were revealed, which may guide the future design and fabrication of bionic self-cleaning surfaces.

1.2 Theories

Natural self-cleaning surfaces generally function *via* either interfacial wettability or interfacial force action, which could be described by the following theories.

1.2.1 Interfacial Wettability

The interfacial wettability of materials' surfaces is an important aspect of surface science. Several models have been proposed to interpret interfacial wetting phenomena. The contact angle and hysteresis are introduced to quantify the wettability.

Young's Model

Young's model was proposed in 1805 to describe the state of a liquid on an ideal rigid surface.[2] The equation is shown as follows:

$$\gamma_{SG} = \gamma_{SL} + \gamma_{LG} \cos \theta_c \tag{1.1}$$

Where θ is the contact angle and γ_{SG}, γ_{SL} and γ_{LG} are the interfacial surface tensions of solid–gas, solid–liquid and liquid–gas, respectively. The equation reveals the contact angle of a liquid droplet on a solid surface from knowledge of the three surface tensions involved. As shown in Figure 1.1a, the three surface tensions reach an equilibrium state, and the contact angle

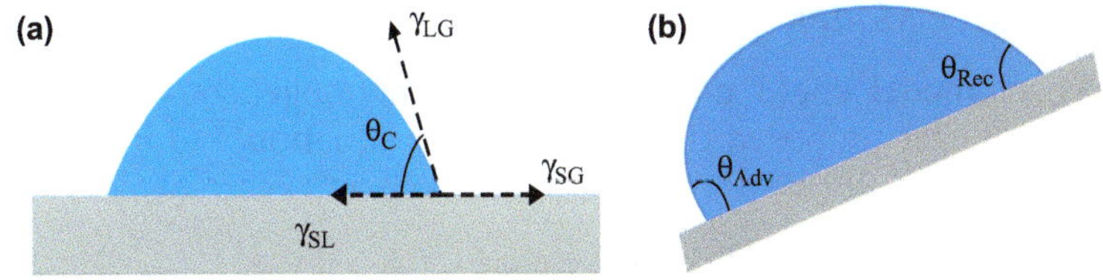

Figure 1.1 (a) Young's model: liquid droplet on a smooth surface, (b) schematic illustration of contact angle hysteresis on a tilted surface.

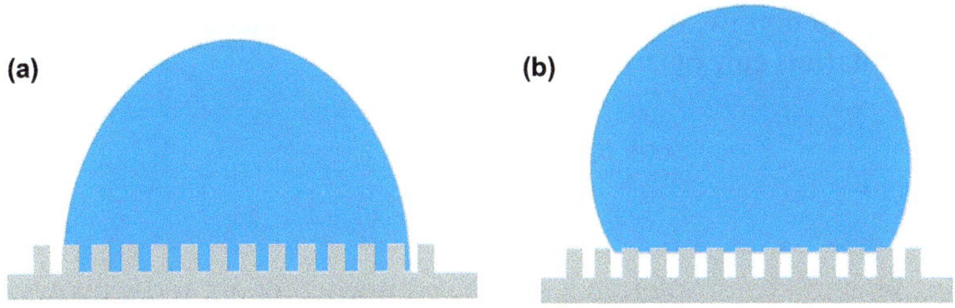

Figure 1.2 (a) A liquid droplet in the Wenzel state, (b) a liquid droplet in the Cassie state.

is the angle between γ_{SL} and γ_{LG}. However, under practical conditions, the structure and chemical heterogeneity of the surface lead to the value of the contact angle varying in a certain range (Figure 1.1b). Consequently, the maximal and the minimal contact angle are defined as the advancing contact angle (θ_a) and the receding contact angle (θ_r), respectively, and the contact angle hysteresis is normally defined as $\theta_a - \theta_r$.[20,21]

Wenzel Model

The Wenzel model's[22] purpose was to describe homogeneous wetting regimes and is defined by the following equation for contact angle on a rough surface:

$$\cos\theta_W = r\cos\theta_Y\ (r \geq 1), \quad (1.2)$$

where θ_W is the apparent contact angle, θ_Y is Young's contact angle for an ideal rigid surface, and r is surface roughness factor, which is defined as the ratio of true area of the solid surface to the apparent area. The Wenzel model reveals the relationship between the surface tension and structure of a homogeneous surface. As shown in Figure 1.2a, the true surface area is larger than the apparent surface area, thus the value of r is greater than 1. Hence, when $\theta_Y < 90°$, the larger the value of r, the more hydrophilic the surface; when $\theta_Y > 90°$, the larger the value of r, the more hydrophobic the surface. The Wenzel model only applies to a thermodynamically stable state and a single chemical component surface, which restrict its applications.

Cassie–Baxter Model

The Wenzel model could not well describe heterogeneous surfaces, and Cassie and Baxter proposed the Cassie–Baxter equation[23] to interpret composite surfaces:

$$\cos\theta_c = f_1\cos\theta_1 + f_2\cos\theta_2,\ (f_1 + f_2 = 1) \tag{1.3}$$

where θ_c is the apparent contact angle in the Cassie–Baxter model, θ_1 and θ_2 are the intrinsic contact angle of component 1 and component 2, and f_1 and f_2 are the apparent area fraction of component 1 and component 2, respectively. Generally, if one component of the surface is air, the $\theta_2 = 180°$, and the equation can be simplified to:

$$\begin{aligned}\cos\theta_c &= f_1\cos\theta_1 - f_2\\ &= f_1(\cos\theta_1 + 1) - 1,\end{aligned} \tag{1.4}$$

Figure 1.2b shows a droplet on a surface composed of air and a hydrophobic component. As a result, the surface shows a hydrophobic property.

1.2.2 Interfacial Force Disequilibrium

In addition to interfacial wettability, interfacial force disequilibrium was also found to lead to a self-cleaning function. When dust particles make contact with two surfaces, the particles are subjected to adhesions from the two surfaces. When one of the surfaces moves, the particles will remain on the surface with the stronger adhesion. As shown in Figure 1.3, the net adhesion force can be calculated by the following equation:

$$F = F_1 - F_2 \tag{1.5}$$

where F_1 and F_2 are the adhesion forces of the substrate surface and self-cleaning surface, respectively. The net force direction is the vector sum of F_1 and F_2. Supposing that the two surfaces have the same compositions, the particles will adhere to the surface with the larger contact areas. The surface with fewer dust particles thus has a self-cleaning property in the interfacial contacting process.

1.3 Self-cleaning Surfaces in Nature

1.3.1 Hydrophobic Self-cleaning Surfaces

1.3.1.1 Isotropic Superhydrophobic Surfaces

Since ancient times, the lotus leaf has won its reputation in China for the characteristic of "Live in the silt but not imbrued". In the 1990s, German botanist Barthlott[12] and his coworkers solved this puzzle. Because of epicuticular wax and a micro-nanoscale papillae structure on the surface of

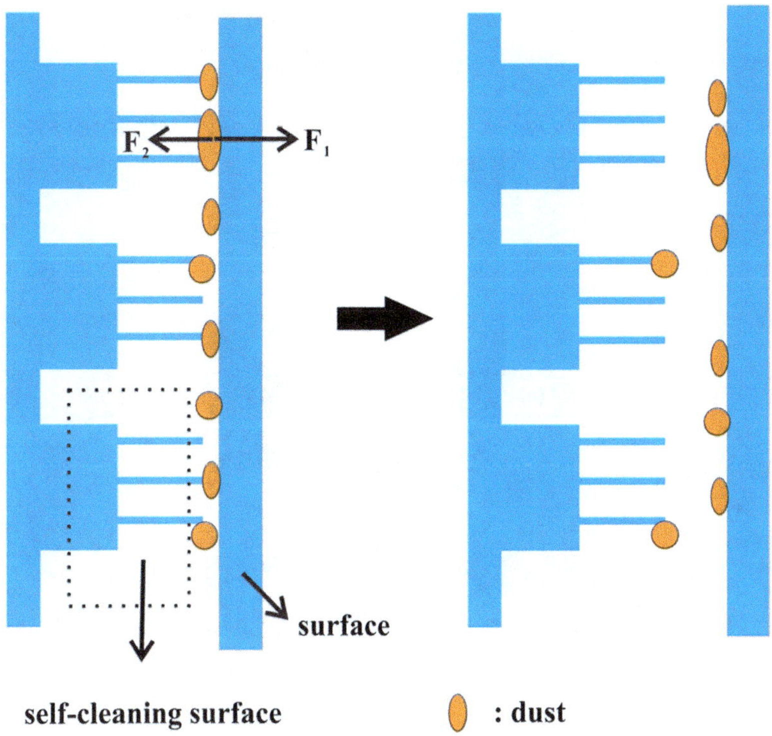

Figure 1.3 Schematic illustration of disequilibrium of interfacial force induced self-cleaning.

lotus leaves, raindrops or water droplets can easily roll off the leaf surface, taking away contamination particles, such as dust or dirt (Figure 1.4a–c).[24] This phenomenon is defined as superhydrophobicity. More specifically, superhydrophobicity is used to describe a surface with a high water contact angle (larger than 150°) and a low rolling angle (smaller than 10°).[25,26] And it is also the core property that leads to "lotus effect"-based self-cleaning. Superhydrophobic surfaces may have wide applications in the coating industry,[27] including building windows,[28] solar cells,[29] textile industry,[30] electronic devices,[31] bioengineering,[32] and drug delivery.[33] Figure 1.4d–f show scanning electron microscopy (SEM) images of the lotus leaf with different magnifications. The surface of the lotus leaf consists of a myriad of papillae with diameters ranging from 5 to 9 μm. And each papilla is composed of cilium-like nanostructures with an average diameter of 124.3 ± 3.2 nm. Furthermore, on the lotus leaf, the micro-nanoscale hierarchical surface structure is built by convex cells and a much smaller superimposed layer of hydrophobic three-dimensional wax tubules (Figure 1.4e,f),[34] which can effectively prevent the lower part of the leaf from wetting. According to the measurements,[35] the water contact angle of the lotus leaf is $161° \pm 2.7°$, and the rolling angle is only about 2°.

Figure 1.4 (a) A flowering plant of lotus. A lotus leaf contaminated with clay (b) and re-movement of the adhering particles by water (c). (d)–(f) The SEM images of the lotus leaf surface in different magnifications: (d) randomly distributed cell papillae; (e) a detail of the cell papillae and (f) the epicuticular wax tubules on the cells.[24]
Reproduced from K. Koch *et al.*, Multifunctional surface structures of plants: An inspiration for biomimetics, *Prog. Mater. Sci.*, **54**, 137–178. Copyright 2009 with permission from Elsevier.

Inspired by the lotus leaf, lotus-leaf-like hierarchical surface morphologies and thus artificial superhydrophobic self-cleaning surfaces have so far been fabricated on different substrates *via* varied methods, including atom transfer radical polymerization,[36] chemical etching,[37] chemical vapor deposition,[38] electrospinning,[39] hydrothermal approach,[40] sol–gel method,[41] *etc.* For example, Jiang *et al.* developed a lotus-leaf-like porous microsphere/nanofiber composite film of polystyrene (PS) with a water contact angle of 160.4° by a simple electrohydrodynamics (EHD) technique (Figure 1.5a) without any modification.[42] Li *et al.* fabricated a micro/nano dual-scale lotus-leaf-like structure by chemically depositing Ag nanonubs on Cu microcones (Figure 1.5b).[43] The surface was tested to be superhydrophobic with a water contact angle (WCA) of 158.9° and highly adhesive with a rolling angle (RA) as high as 90°. Recently, He's group prepared superhydrophobic thin films with high transmittance by using the sol–gel method,[45] layer-by-layer assembly[46] and precursor-derived one-step growth[44] *etc.* And the thin films had excellent mechanical robustness, showing great potential in

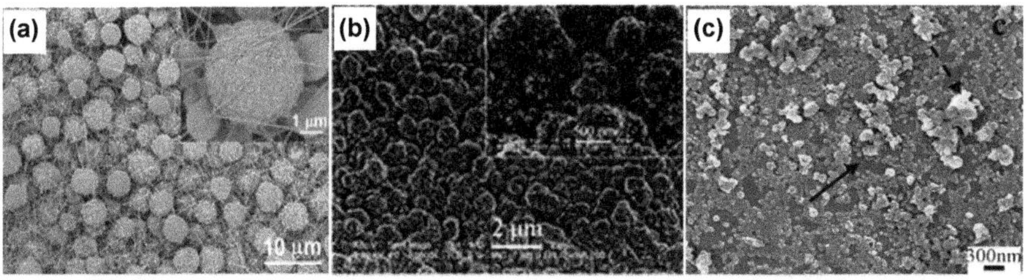

Figure 1.5 (a) SEM image of a superhydrophobic polystyrene film with special microsphere/nanofiber composite structures prepared by the EHD method.[42] Reproduced with permission from John Wiley and Sons. Copyright © 2004 WILEY-VCH Verlag GmbH & Co. KGaA, Weinheim. (b) Lotus-leaf-like silver film the inset is the magnified SEM image.[43] Reproduced with permission from the Royal Society of Chemistry. (c) SEM image of the thin film on a glass substrate. The arrow indicates nano-scale hollow silica nanoparticles and the dashed arrow indicates micro-scale mesoporous silica nanosheets.[44] Reproduced with permission from L. Xu *et al.*, *ACS Appl. Mater. Interfaces*, 2014, **6**, 9029–9035. Copyright (2014) American Chemical Society.

many applications. For instance, they used the precursor-derived one-step growth approach to fabricate hierarchically structured lotus-leaf-like superhydrophobic thin films (Figure 1.5c).[44] The films showed broad-band antireflection (the maximum transmittance, $T_{max} \approx 95\%$), super-hydrophobicity (WCA = 162°) and remarkable durability.

The cicada wing is also a typical example of a self-cleaning function among various insect surfaces. The regularly aligned nanoposts on the cicada wing (Figure 1.6a) endow the cicada wing with superhydrophobicity (WCA ≈ 160°) and thus self-cleaning functionality, which can effectively remove dust particles from the wing surfaces.[47] As shown in Figure 1.6b, the height of the pillar structure is approximately 250 nm, the diameter is about 70 nm.[48] This structure also gives cicadas an antireflection property, which inspires scientists in designing artificial superhydrophobic and antireflective surfaces.

By mimicking cicada wings, Min *et al.* developed a simple and scalable templating technique to fabricate pillar arrays with a high aspect ratio on a Si surface and a glass substrate (Figure 1.6c), which exhibited excellent broadband antireflection and nonwetting properties.[49] A reactive ion etching (RIE) process and fluorosilane modification endowed the surface of the silicon and glass with antireflection and superhydrophobicity (WCA = 172°, Figure 1.6d).

Surface geometrical micro-nanostructure and low surface energy are two significant factors that account for self-cleaning surfaces. For lotus leaves and cicada wings, water droplets can roll off their surfaces in any direction. In other words, these surfaces are isotropic in wettability, where the roughness geometry or chemical heterogeneity is isotropic.

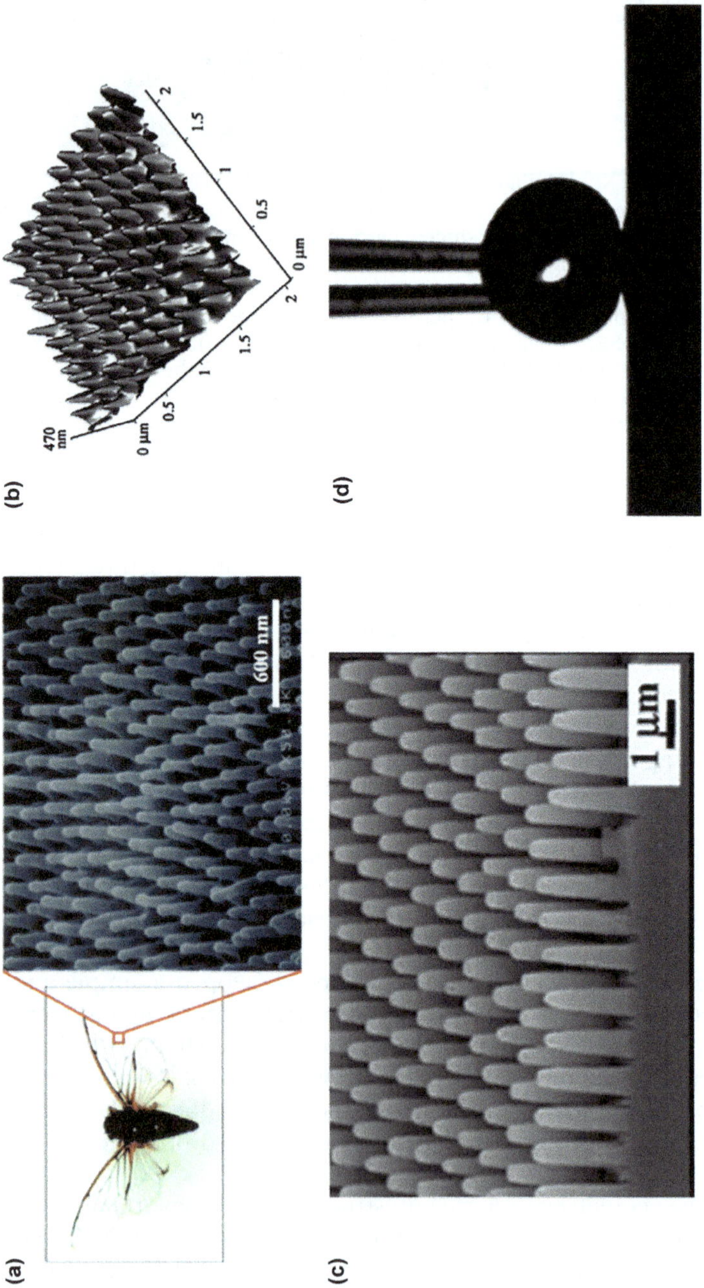
(a)
600 nm
(b)
470 nm
0 µm
0.5
1
1.5
2
(c)
1 µm
(d)

1.3.1.2 Anisotropic Superhydrophobic Surfaces

Apart from lotus leaves and cicada wings, the surfaces of some other plants and insects also show a superhydrophobic self-cleaning property, such as rice leaves, the feathers of ducks and geese, and butterfly wings. However, their surfaces exhibit anisotropic wettability,[14] which is contrary to the surfaces of lotus leaves and cicada wings.

Rice leaves show anisotropic wettability (Figure 1.7a and b). As seen in Figure 1.7c and d, the surface of rice leaves also has a hierarchical micro-nanoscale structure, which is similar to lotus leaves. Nevertheless, papillae arrange in order along the direction parallel to the leaf edges, while the vertical direction is in disorder.

It was measured that the water contact angle along the parallel direction is $157 \pm 2^\circ$, while it is $146 \pm 2^\circ$ in the vertical direction.[50] Accordingly, the rolling angle of the former is 3–5°, and the latter is 9–15°. The observations show that the arrangement of micro-papillae influences the movement of water droplets, which makes rice leaves possess anisotropic super-hydrophobicity. Therefore, rolling water droplets can pick up contaminants on the rice leaf surface. Besides rice leaves, the feathers of ducks and geese also exhibit such anisotropic wettability.[14] Figure 1.8 shows SEM images of duck feathers. Duck feathers are composed of micro-sized backbones, trunks and barbules. And the branch consists of many tomenta, which have nano-sized grooves and protuberances.[51] With a neatly arranged micro and nano-scale structure on the surface of its feathers, a duck can easily keep dry in water.

Gao *et al.* replicated a rice-leaf-like surface by a two-step replication process combining regular replica molding (REM) and temperature-induced phase separation micro-molding (PSμM).[52] First, poly(dimethylsiloxane) (PDMS) was used to replicate the rice-leaf surface structure. Second, they replicated the negative PDMS replica using PSμM of poly(*N*-isopropylacrylamide) (PNIPAAm) aqueous solution, which could be easily carried out just by adjusting the temperature. The artificial rice-leaf-like surface of PNIPAAm showed thermally good responsive and anisotropic wettability (Figure 1.9).

Figure 1.6 (a) FE-SEM image of the cicada wing with regularly aligned nanoposts.[47] Reproduced with permission from W. Lee *et al.*, *Langmuir*, 2004, **20**, 7665–7669. Copyright (2004) American Chemical Society. (b) High-resolution AFM three-dimensional image of the dorsal region of a cicada wing membrane.[48] Reproduced from G. S. Watson *et al.*, Putative function and functional efficiency of ordered cuticular nanoarrays on insect wings, *Biophys. J.*, **94**, 3352–3360. Copyright 2008 with permission from Elsevier. (c) SEM image of silicon pillars after 50 min RIE. (d) Water drop profile on a fluorosilane-modified silicon pillar array (60 min RIE).[49] Reproduced with permission from John Wiley and Sons. Copyright © 2008 WILEY-VCH Verlag GmbH & Co. KGaA, Weinheim.

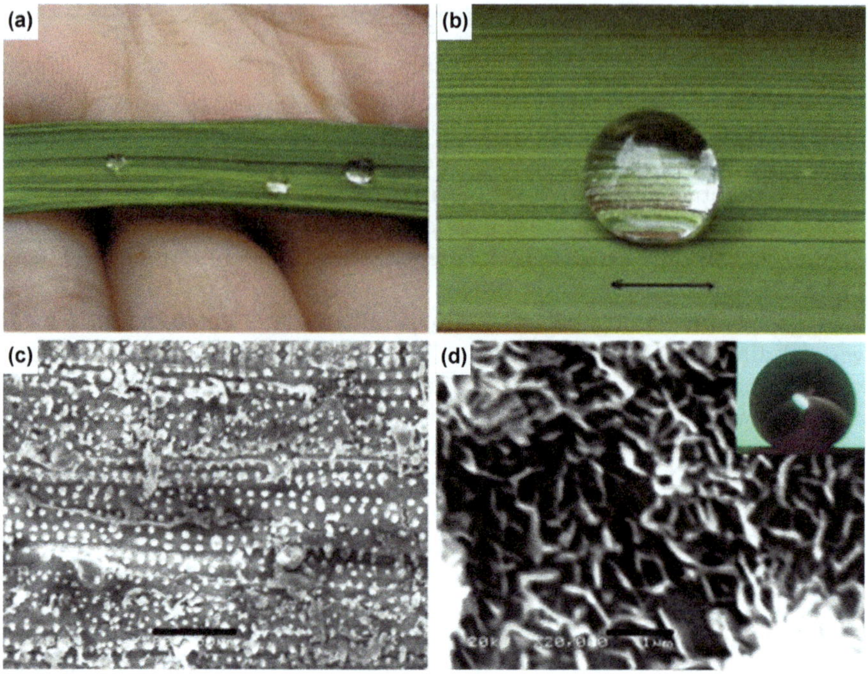

Figure 1.7 (a) Photo of a few water droplets on a rice leaf. (b) A water droplet floating on a rice leaf. (c), (d) SEM images of a rice leaf with different magnifications. The inset of (d) is a water contact angle on a rice leaf, with a value of $157 \pm 2°$, showing a good superhydrophobic property. The bars of (c) and (d) are 50 and 1 μm, respectively.[50] Z. Guo and W. Liu, Biomimic from the superhydrophobic plant leaves in nature: Binary structure and unitary structure, *Plant Sci.*, **172**, 1103–1112. Copyright 2007 with permission from Elsevier.

Another typical example of anisotropic wettability is butterfly wings. It has been reported that the butterfly wing exhibits superhydrophobicity with a water contact angle of $152 \pm 2°$.[53] Figure 1.10 shows the anisotropic wettability on butterfly wings. A water droplet can easily roll off the surface of a butterfly wing along the radial outward direction (Figure 1.10a), while it can be tightly pinned in the opposite direction (Figure 1.10b and c). This unique ability is ascribed to the anisotropic arrangement of flexible nanostrips and microscales overlapping on the wings in one dimension.[14] Figures 1.10d and e show SEM images of butterfly wings, which are covered by a large number of quadrate scales along the RO direction with a length of ~150 μm and a width of ~70 μm. And on the surface of each scale, there are numerous ridging nano-strips of 184.3 ± 9.1 nm in width.[53] Moreover, these nano-strips are composed of cuticle lamellae of apparently different lengths, which are stacked stepwise along the RO direction.

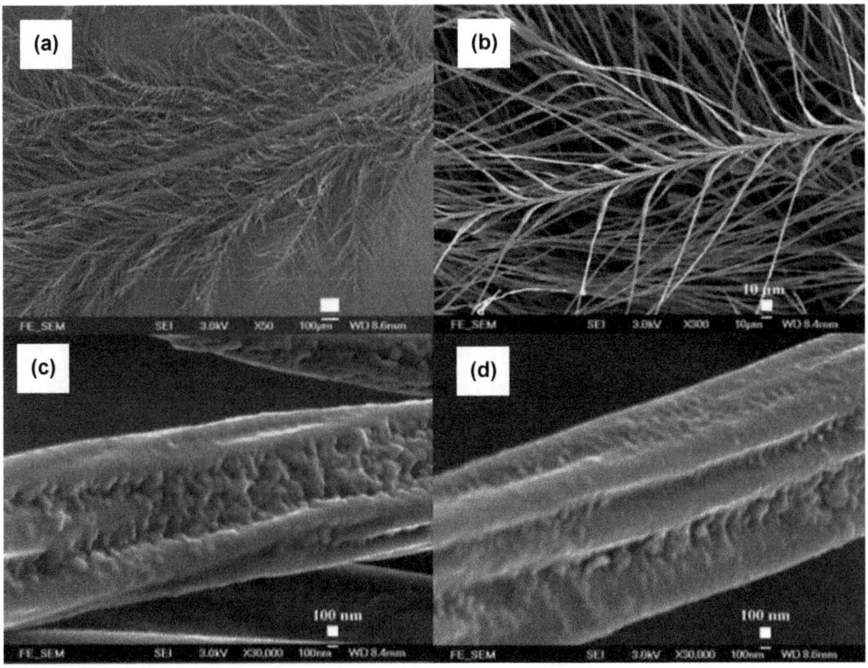

Figure 1.8 SEM images of duck feathers, which show multi-scale structures. (a) bar = 100 μm; (b) bar = 10 μm; (c), (d): bar = 100 nm.[51]

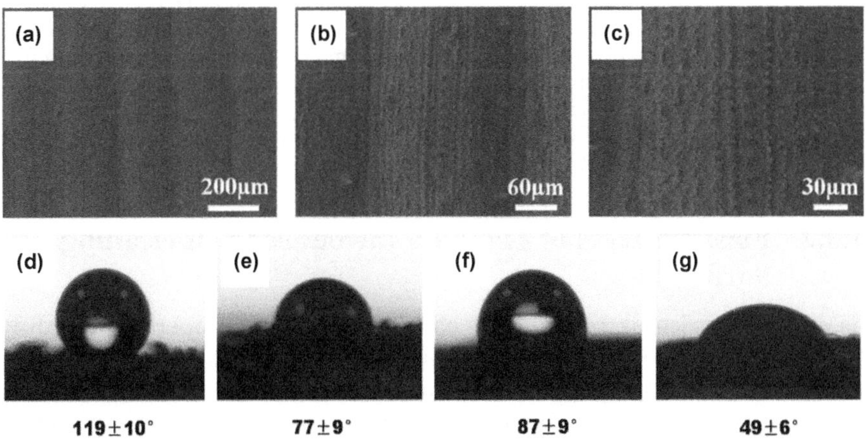

Figure 1.9 (a)–(c) SEM images of the surface structures of the PNIPAAm replica of a rice leaf with different magnifications. (d)–(g) Static contact angles of water droplets on a replicated artificial rice leaf using PNIPAAm, as taken from the direction parallel to the grooves ((d): 50 °C; (e): 20 °C) and from the direction perpendicular to the grooves ((f): 50 °C; (g): 20 °C).[52]

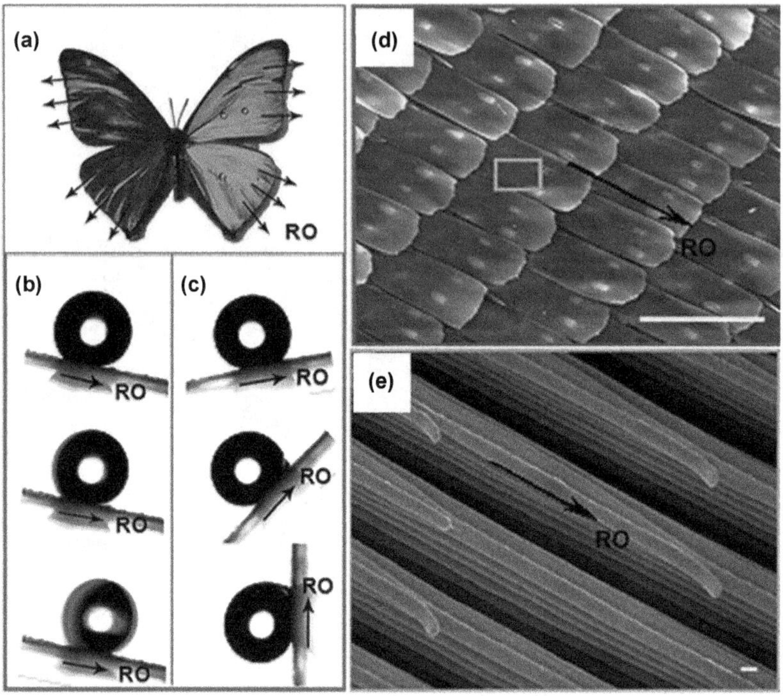

Figure 1.10 Anisotropic wettability on butterfly wings. (a) The black arrows denote the radial outward (RO) direction away from the body's centre axis. (b) The droplet easily rolls along the RO direction when the wing is tilted downwards by 9°. (c) The droplet is firmly pinned on the wing that is tilted upward, even when fully upright. (d), (e) SEM images of the periodic arrangement of overlapping micro-scales on the wings and fine lamella-stacking nano-stripes on the scales. Scale bars: (d) 100 μm; (e) 100 nm.
Reproduced from ref. 53 with permission from the Royal Society of Chemistry.

1.3.2 Elastic Effects of Tomenta Introduced Self-cleaning Surfaces

As well as lotus leaves, other plants have a self-cleaning property.[1] According to Herminghaus,[54] surfaces covered with tomenta, such as Alchemilla leaves, also show superhydrophobicity. As shown in Figure 1.11a and b, a water droplet can easily remain round when it is on the surface of an Alchemilla leaf. Although it seems that the water droplet does not contact the surface of the leaf, the water contact angle of the surface is close to 180°. In Figure 1.11c, the tomenta would play a role perfectly analogous to cuticular wax crystals in the case of microrough surfaces. However, it is found that due to the effect of surface tension, tomenta tend to cluster, which leads to the tomenta curve, and have elastic potential energy.[54] Because of the elastic potential energy, the tomenta prevent water droplets wetting the surface of

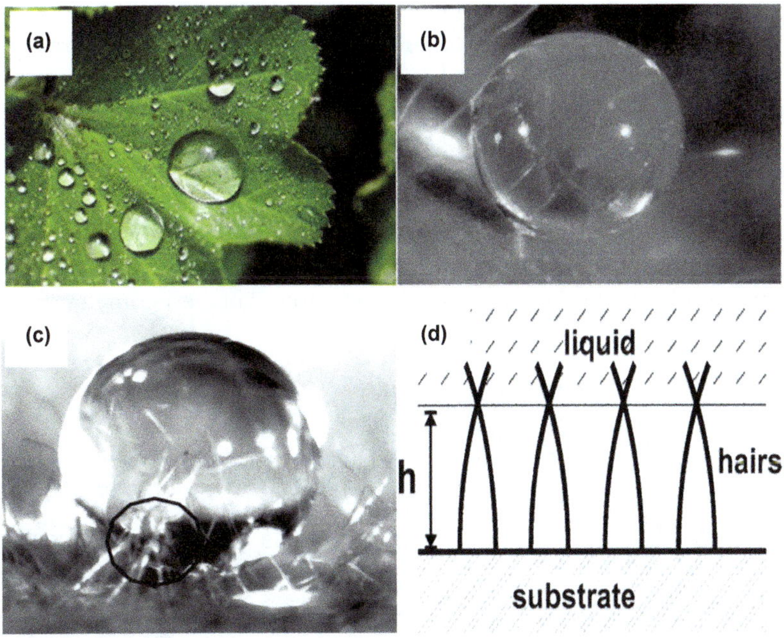

Figure 1.11 (a), (b) A water droplet on an Alchemilla leaf. (c) Condensed hair cluster holding a water droplet (2 mm diameter). (d) Sketch of the elastic deformation of the hairs due to bundle formation at the liquid–air interface.[54]
Reproduced with permission from A. Otten and S. Herminghaus, *Langmuir*, 2004, **20**, 2405–2408. Copyright (2004) American Chemical Society.

the Alchemilla leaf (Figure 1.11d). However, this kind of superhydrophobic structure is just of dynamic stability. The tomenta are hydrophilic and flexible, and therefore the surface will be totally wetted when a higher water pressure makes water droplets contact the bottom of the surface of the Alchemilla leaf. In this case, the wettability will be decided by the thermodynamic stability, which consists of roughness and chemical composition.

As mentioned, for the leaves of plants that are covered with tomenta, elasticity and distribution area are two basic parameters to judge if spherical water droplets can be stably present on the leaf epidermis.

1.3.3 Micro/Nanostructured Slippery Surfaces

In nature, some plants have so-called slippery surfaces, and such surfaces inspired scientists to fabricate slippery self-cleaning surfaces.[55–57] A typical example is pitcher plants (*Nepenthes*), also known as carnivorous plants.[58] Pitcher plants have highly specialized leaves adapted to attract, capture, retain, and digest prey.[58] The capture of insects is based on the presence of lubricating water or the nectar film and the regular microstructure of the peristome surface. Figure 1.12a shows the pitcher plant. Figure 1.12b shows the different sections of the pitcher's inner surface. On top of the plant, there is a lid (operculum), which keeps rain from diluting the digestive fluid in the cup.[59] The distinct regions of the pitcher, illustrated from top to bottom, are the collar-shaped peristome, the waxy zone, the transitional zone and the digestive zone, respectively.[60] The waxy zone constitutes about half the pitcher height (51.3 ± 2.8%), the transitional zone 5.1 ± 0.9%, and the digestive zone constitutes the remaining part. As shown in Figure 1.12c, the underside of the inner margin of the peristome consists of downward-pointing tooth-like projections and nectar pores. The peristome surface has

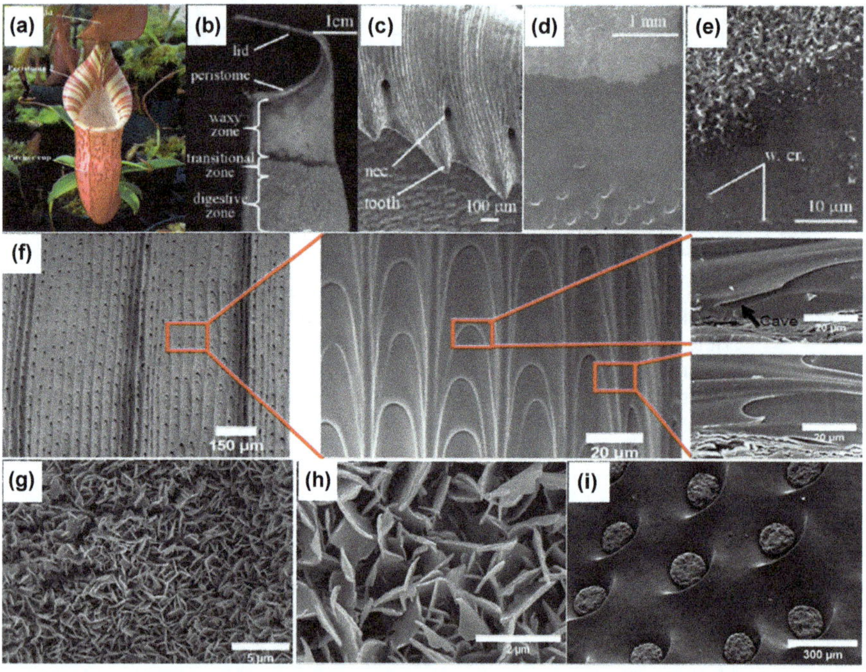

Figure 1.12 (a) *Nepenthes* carnivorous pitcher,[59] copyright 2014 Multidisciplinary Digital Publishing Institute. (b) Longitudinal section of the pitcher showing the different zones. (c) Peristome with downward-pointing teeth above the waxy zone. (d) Smooth transitional zone between waxy zone (upper) and digestive zone (lower). (e) Limit of the waxy zone: wax crystals (w. cr.) on the transitional zone that have detached from the waxy zone.[60] Reproduced with permission from John Wiley and Sons. Copyright © 2002, John Wiley and Sons. (f) SEM micrographs of the peristome surface and longitudinal section. Top view (left and middle) and longitudinal view (right) along the radial direction (the black arrow indicates the cave). (g) SEM image showing the microstructure of the waxy zone. (h) SEM image showing the detailed morphology of the plate-like wax crystals at high magnification. (i) SEM image showing the well-organized glandular cells in the digestive zone.[61] Reproduced with permission of Springer Science + Business Media. Copyright 2015.

a very regular microstructure consisting of first- and second-order radial ridges (Figure 1.12f, left). The second-order radial ridges are formed by straight rows of lunate overlapping epidermal cells, showing an anisotropic topography (Figure 1.12f, middle).[61] And small caves (indicated by the black arrow, Figure 1.12f, right) between two adjacent cells can be observed in the longitudinal SEM image, because of the incomplete overlapping. The zone adjoining the peristome is the waxy zone, which is covered with three-dimensional plate-like wax crystals (Figure 1.12g and h). Figure 1.12d shows the smooth transitional zone that separates the waxy zone and digestive zone. A few isolated wax crystals, which have detached from the waxy zone, are located on this zone (Figure 1.12e). At the bottom of the pitcher, there is a digestive zone, which is composed of orderly arranged glandular cells that release digestive fluid and are located in concave depressions (Figure 1.12i). In an exceptional pitcher species of *Nepenthes bicalcarata*, the digestive zone extends upward to the peristome without having the wax zone.[58]

The hygroscopic nectar and hierarchical microstructure collaborate synergistically and lead to a superhydrophilic peristome, making water droplets spread quickly and effectively form homogeneous thin films.[59,61] And this is in fact the actual trapping mechanism of the pitcher plant in humid conditions. The peristome surface is so slippery that insects are unable to attach to the peristome and thus fall into the pitcher. Consequently, the trapping ability of the pitcher plant will increase on rainy days.

Inspired by pitcher plants, Aizenberg *et al.* first fabricated an amphiphobic and multifunctional slippery liquid-infused porous surface by infusion of fluorinated oil (lubricating fluids) into a porous substrate.[55] Since then, through the use of different infused liquids and porous structures, researchers have designed slippery materials possessing self-cleaning,[56,62,63] adaptive,[64] antifouling,[57] and temperature stimuli-responsive[65] properties. And these materials can facilitate applications in vapor condensation[66] and anti-fouling, anti-icing, self-cleaning windows and optical devices.[67–69]

1.3.4 Hydrophilic and Oleophobic (in Water) Self-cleaning Surfaces

Interfacial self-cleaning mechanisms can in fact be divided into two categories: one is superhydrophobicity, the other is superhydrophilicity. In nature, besides the creatures mentioned earlier, there also exist many underwater creatures that have both superhydrophilic and superoleophobic capabilities. The surfaces of such aquatic creatures may meet a three-phase system involving a solid, water and any organic liquid instead of the solid–water–air system on land. In this system, water takes the place of air, and the organic liquid plays a similar role to water in the solid–water–air system. Those surfaces, including fish scales, shark skins, snail shells and pilot whale skins, may inspire scientists to create underwater self-cleaning and antifouling surfaces.

1.3.4.1 Fish Scales

In the sea, the fish body can be protected from plankton and possess a self-cleaning property even though the sea is polluted by oil. Researchers found that fish scales have a superoleophobic property in water with a contact angle of $156.4 \pm 3.0°$ for an oil droplet (1,2-dichloroethane) and a superoleophilic property in air.[70] Figure 1.13 shows optical and SEM images of fish scales at different magnifications. Fish scales consist of sectorial scales with diameters of 4–5 mm (Figure 1.13a). Each sectorial scale is covered by papillae of 100–300 μm in length and 30–40 μm in width (Figure 1.13b). The surfaces of micropapillae are rough. And nanostructures can be observed on the papillae (Figure 1.13c,d). Furthermore, fish scales are composed of calcium phosphate, protein and a thin layer of mucus, which results in hydrophilicity. Therefore, the superoleophilicity in air and superoleophobic properties in water of fish scales are basically attributed to the hydrophilic mucus and the rough hierarchical structure.

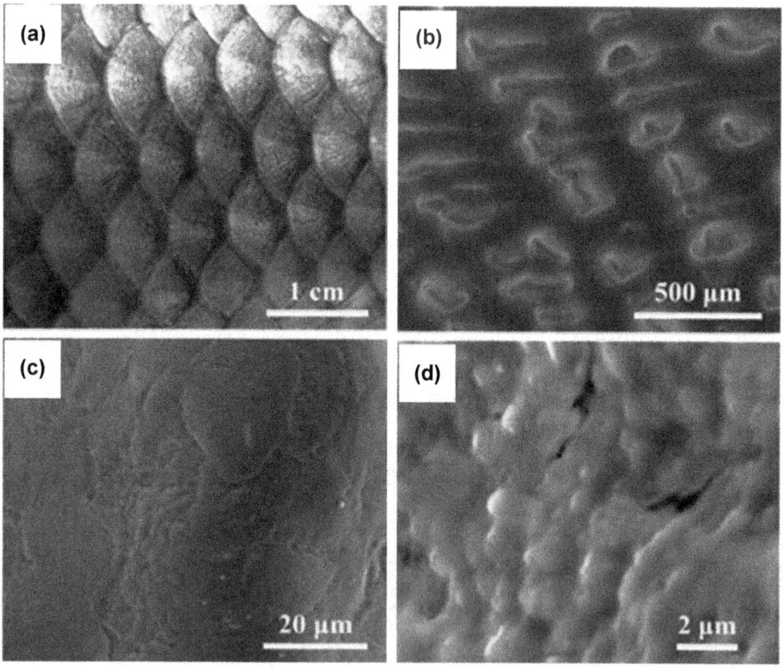

Figure 1.13 Surface structures of fish scales. (a) Optical image of the fish scale. The sector-like scales with diameters of 4–5 mm arrange into an array. (b) SEM image of the fish scale. Oriented papillae 100–300 mm in length and 30–40 mm in width existing on the fish scales arrange in the radial direction. (c) SEM image of the fish scale at high magnification; the papillae surface is rough. (d) SEM image of the papillae.[70]
Reproduced with permission from John Wiley and Sons. Copyright © 2009 WILEY-VCH Verlag GmbH & Co. KGaA, Weinheim.

Similarly, grass carp can always swim freely and resist biofouling and oil pollution in water, illustrating a self-cleaning property.[3] The mechanism of their self-cleaning ability is contributed to the cooperation of surface hydrophilic mucus and multiscale structures.

1.3.4.2 Snail Shells

On rainy days, we can see snails everywhere (Figure 1.14a). The surface of a snail shell is always clean, exhibiting self-cleaning ability in a wet environment. In fact, the water and oil contact angles of a snail shell's surface are about 80° and 10° in air respectively, showing hydrophilicity and oleophilicity.[11] But in water, an oil droplet is difficult to attach to the snail shell (Figure 1.14d–f), indicating superoleophobicity (the oil contact angle is higher than 150°). Figures 1.14b and c show SEM images of a snail shell. The surfacc of a snail shell, which is coated by protein, consists of line grooves with a pitch of 0.5 mm, grooves crossing to the line groove with a pitch of 0.1 mm and micro-grooves with a pitch of 10 μm between the line grooves, forming a hierarchical structure.[11] The key to the self-cleaning property of the snail shell is its superoleophobicity, *i.e.*, the wetted surface of the snail shell is hardly contaminated so that the snail shell can keep clean on rainy days.

1.3.4.3 Shark Skins

As is well known, sharks are fast swimmers, which effectively prevents them from being invaded by bacteria and parasites. Recently, shark skin

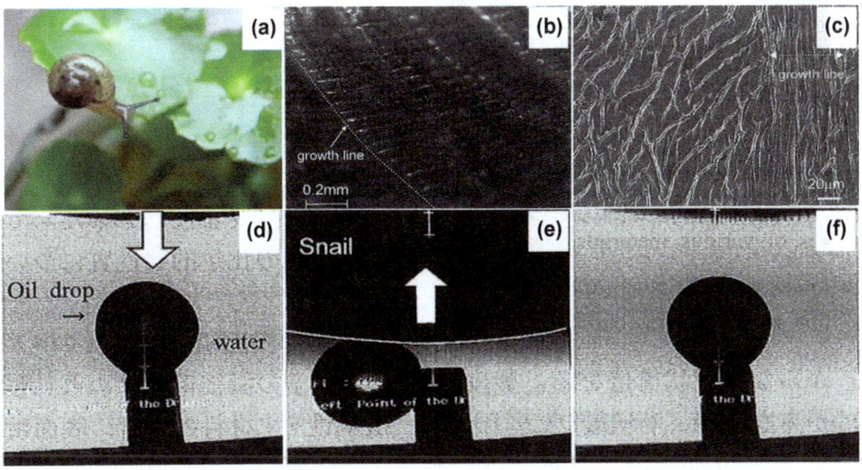

Figure 1.14 (a) A snail on a leaf after it has rained. (b) Optical microscopy image of a snail shell. (c) SEM image of a snail shell. (d)–(f) Measurement of the oil-contact angle in water.[11]
Reproduced from ref. 11 with permission from the Royal Society of Chemistry.

has attracted increasing attention for its self-cleaning and low-drag capabilities.[71,72] Similarly, its surface microstructure leads to the self-cleaning property of shark skin. As shown in Figure 1.15a–c, shark skin is covered by individual tooth-like scales called dermal denticles, on which there are spaced riblets oriented parallel to the swimming direction.[11,73] Figure 1.15d shows the different sizes and shapes of scales on different parts. The scales on the head are larger and saponaceous on the edge; the scales on the back are smaller and have less height; and the scales on the abdomen are sharply pointed on the edge and have deeper micro-grooves.[74]

Due to the shark skin's unique microstructures, sharks show oleophobicity and low-drag properties. In detail, water flow moves quickly past a shark's surface and washes away fouling organisms. Meanwhile, because of these microstructures, it is difficult for microorganisms and contaminants to adhere to the skin. Also, the dermal denticles perpetually realign or flex in response to changes in internal and external pressure as the shark moves through water, creating a 'moving target' for fouling organisms.

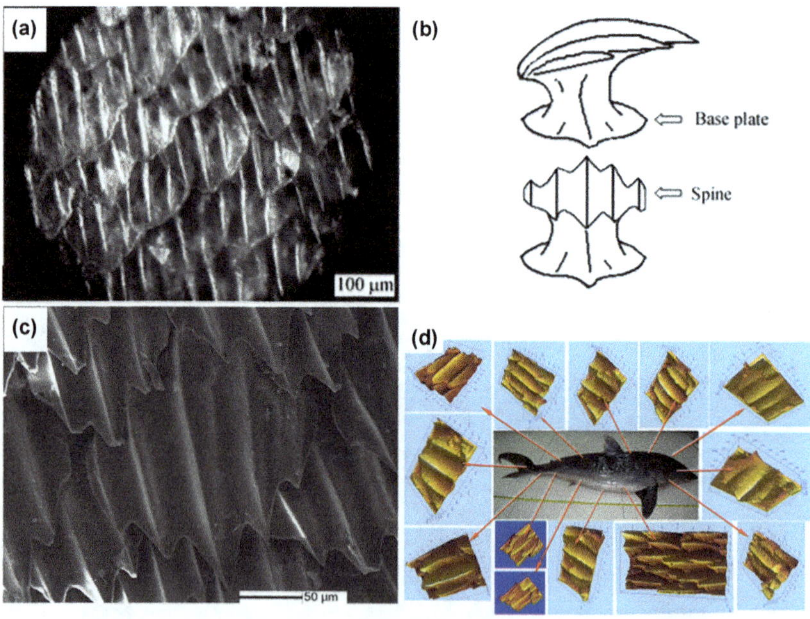

Figure 1.15 (a) Optical microscopy image of shark skin. (b) Illustration of a single shark scale,[73] X. Han, Study on the micro-replication of shark skin, *Sci. China Ser., E: Eng. Mater. Sci.*, 2008, **51**, with permission from Springer. (c) SEM image of shark skin. (d) Sharkskin patterns on different parts,[74] Y. Luo, Improvement of water-repellent and hydrodynamic drag reduction properties on bio-inspired surface and exploring sharkskin effect mechanism, *Appl. Phys. A: Mater. Sci. Process.*, 2015, **120**, with permission from Springer.

1.3.4.4 *Pilot Whale Skins*

In addition to shark skin, pilot whale (*Globicephala melas*) skin also has a clean surface free of contamination, and the mechanism of self-cleaning is shown in Figure 1.16a.[75] This special ability results from nanoridge pores and zymogel on the skin's surface, as seen from the SEM image of pilot whale skin (Figure 1.16b). Baum *et al.* found that there is intercellular gel, containing both polar and nonpolar functional groups, on the surface of pilot whales, which effectively promotes short- and long-term fouling reduction. Moreover, the self-cleaning property, combined with the speed of movement, surfacing, and jumping of pilot whales, may also contribute to removing weakly adhered epibionts.

1.3.5 Disequilibrium of Interfacial Force Induced Self-cleaning Surfaces

Geckos are well-known for the remarkable adhesive property of their toes, which gives them an extraordinary climbing ability. In 2005, Hansen *et al.* revealed how geckos manage to keep their feet clean while walking with sticky toes.[76] Tokay gecko feet contaminated with microspheres could recover their ability to cling to vertical surfaces after only a few steps on clean glass. Studies illustrate that their self-cleaning ability was attributed to the hierarchical structure of their feet.[14] Figure 1.17 shows the structural hierarchy of the gecko feet. Each toe is covered with rows of lamellae (Figure 1.17b) which consist of millions of well-aligned microscale setae (*ca.* 110 μm in length and *ca.* 5 μm in diameter) (Figure 1.17c and d). The setae occur in uniform arrays on overlapping lamellar pads at a density

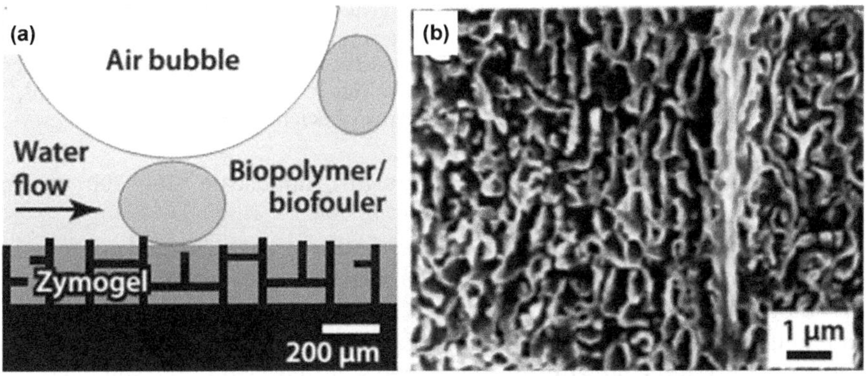

Figure 1.16 (a) Preliminary model of the self-cleaning abilities of pilot whale skin, based on the relief of nanoridges filled with a medium. (b) SEM image of self-cleaning pilot whale skin. The pores enclosed by the nanoridges are displayed very clearly.[75] C. Baum, Average nanorough skin surface of the pilot whale (Globicephala melas, Delphinidae): considerations on the self-cleaning abilities based on nanoroughness, *Mar. Biol.*, 2001, **140**, with permission from Springer.

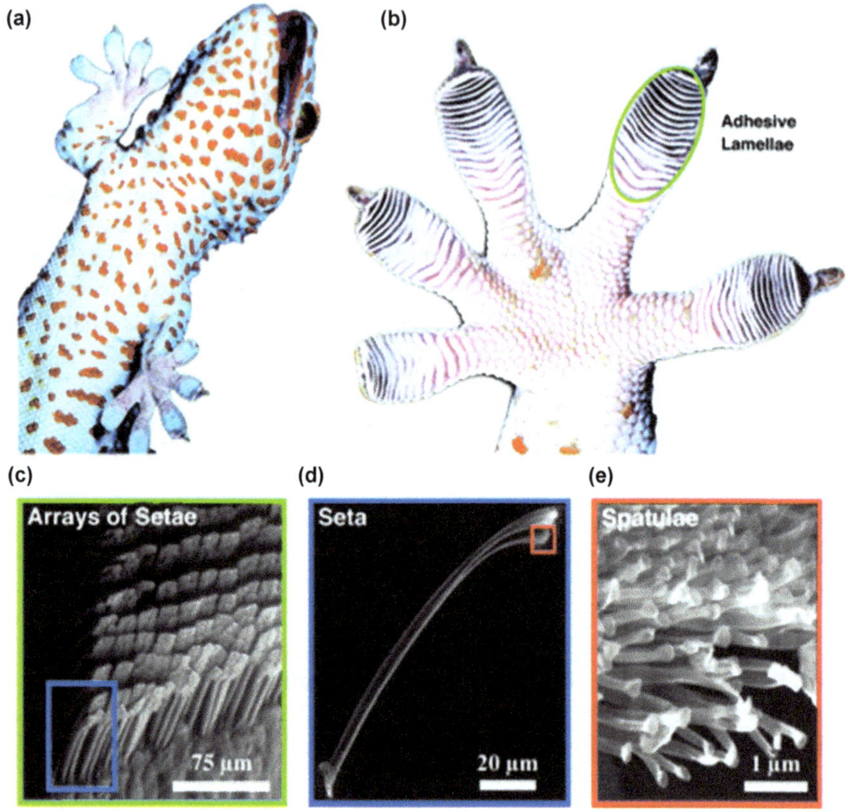

Figure 1.17 Structural hierarchy of gecko feet. (a) Macrostructure: ventral view of a Tokay gecko climbing vertical glass. (b) Mesostructure: ventral view of the foot, with adhesive lamellae visible as overlapping pads. (c) Microstructure: proximal portion of a single lamella, with individual setae in an array visible. (d), (e) Nanostructure: single seta with branched structure at upper right, terminating in hundreds of spatular tips,[76] copyright (2005) National Academy of Sciences, U.S.A.

of 14 400 per mm^2. And each seta splits into nanoscale (200–500 nm in diameter) spatulas (Figure 1.17e).

Nevertheless, early studies did not investigate the gecko's ability to remain clean from contamination, which has become the center of attention for biomimetic self-cleaning surfaces in recent years.[77] Very recently, researchers discovered that disequilibrium in adhesive forces between substrate, dirt and single or multiple gecko spatulas lead to the self-cleaning property of gecko feet. According to the research, dirt particles, which have a small size, can stick to a few spatulas, and the adhesion strength between substrate and dirt particles is higher than between dirt particles and spatulas. As a result, within a few steps, geckos can simply move the setae from the substrate so that it can quickly restore cleanliness in case of contamination.

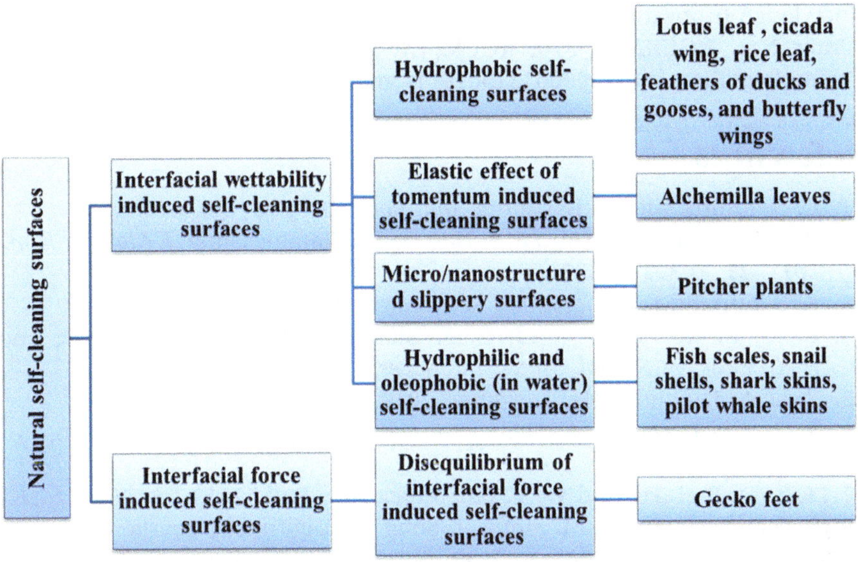

Figure 1.18 Overview of natural self-cleaning surfaces.

1.4 Conclusion and Outlook

Nature is always the source of inspiration for scientists, as already described. The variety of species from terrestrial and aquatic organisms with special wettability (superhydrophobicity, superhydrophilicity and superoleophobicity) provides the answers to puzzles we meet (Figure 1.18).

Terrestrial organisms would inspire us in the design and fabrication of artificial multi-scale self-cleaning materials, while aquatic organisms would inspire us in the design and construction of self-cleaning and antifouling materials for underwater or wet environments. Such materials have great potential in many fields, such as self-cleaning, antifouling, corrosion-resistance, anti-icing, anti-fingerprint and water–oil separation, *etc.* Towards practical applications, however, it will be necessary in the future to consider the robustness and durability of artificial self-cleaning materials by taking into account the surrounding environment in terms of mechanical impacts, water availability (*e.g.*, rain), level and type of pollution, and possibility of erosion (*e.g.*, in sandy environments, in the presence of wind, *etc.*).

Acknowledgements

Financial support from the National Natural Science Foundation of China (Grant Nos. 21571182, 21271177), a Chinese Academy of Sciences Grant (CXJJ-14-M38), the National High Technology Research and Development Program ("863" Program) of China (Grant No. 2011AA050525), the

Knowledge Innovation Program of the Chinese Academy of Sciences (CAS) (Grant Nos. KGCX2-YW-370, KGCX2-EW-304-2), and Beijing Science and Technology Plan (Z151100003315018) is greatly appreciated.

References

1. C. Neinhuis and W. Barthlott, *Ann. Bot.*, 1997, **79**, 667–677.
2. V. A. Ganesh, H. K. Raut, A. S. Nair and S. Ramakrishna, *J. Mater. Chem.*, 2011, **21**, 16304–16322.
3. K. S. Liu and L. Jiang, *Annu. Rev. Mater. Res.*, 2012, **42**, 231–263.
4. B. Xi, L. K. Verma, J. Li, C. S. Bhatia, A. J. Danner, H. Yang and H. C. Zeng, *ACS Appl. Mater. Interfaces*, 2012, **4**, 1093–1102.
5. X. Li, X. Du and J. He, *Langmuir*, 2010, **26**, 13528–13534.
6. S. J. Choi and S. Y. Huh, *Macromol. Rapid Commun.*, 2010, **31**, 539–544.
7. X. Y. Li, J. H. He and W. Y. Liu, *Mater. Res. Bull.*, 2013, **48**, 2522–2528.
8. S. Guldin, P. Kohn, M. Stefik, J. Song, G. Divitini, F. Ecarla, C. Ducati, U. Wiesner and U. Steiner, *Nano Lett.*, 2013, **13**, 5329–5335.
9. T. Li and J. He, *Chin. Sci. Bull.*, 2014, **59**, 715–721.
10. L. Yao and J. He, *Prog. Mater. Sci.*, 2014, **61**, 94–143.
11. S. Nishimoto and B. Bhushan, *Rsc Adv.*, 2013, **3**, 671–690.
12. W. Barthlott and C. Neinhuis, *Planta*, 1997, **202**, 1–8.
13. A. Skarmoutsou, C. A. Charitidis, A. K. Gnanappa, A. Tserepi and E. Gogolides, *Nanotechnology*, 2012, **23**, 505711.
14. K. Liu, X. Yao and L. Jiang, *Chem. Soc. Rev.*, 2010, **39**, 3240–3255.
15. T. S. Wong, S. H. Kang, S. K. Tang, E. J. Smythe, B. D. Hatton, A. Grinthal and J. Aizenberg, *Nature*, 2011, **477**, 443–447.
16. B. Bhushan and Y. C. Jung, *Prog. Mater. Sci.*, 2011, **56**, 1–108.
17. S. Ji, J. Park and H. Lim, *Nanoscale*, 2012, **4**, 4603–4610.
18. N. R. Chiou, C. Lu, J. Guan, L. J. Lee and A. J. Epstein, *Nat. Nanotechnol.*, 2007, **2**, 354–357.
19. M. Toma, G. Loget and R. M. Corn, *ACS Appl. Mater. Interfaces*, 2014, **6**, 11110–11117.
20. J. F. Joanny and P. G. Degennes, *J. Chem. Phys.*, 1984, **81**, 552–562.
21. L. Gao and T. J. McCarthy, *Langmuir*, 2006, **22**, 6234–6237.
22. R. N. Wenzel, *Ind. Eng. Chem.*, 1936, **28**, 988–994.
23. A. B. D. Cassie and S. Baxter, *Trans. Faraday Soc.*, 1944, **40**, 0546–0550.
24. K. Koch, B. Bhushan and W. Barthlott, *Prog. Mater. Sci.*, 2009, **54**, 137–178.
25. M. Callies and D. Quere, *Soft Matter*, 2005, **1**, 55–61.
26. D. Quéré, *Annu. Rev. Mater. Res.*, 2008, **38**, 71–99.
27. Z. Geng and J. He, *J. Mater. Chem. A*, 2014, **2**, 16601–16607.
28. J. Drelich and A. Marmur, *Surf. Innovations*, 2014, **2**, 211–227.
29. X. Li, J. He and W. Liu, *Mater. Res. Bull.*, 2013, **48**, 2522–2528.
30. C. H. Xue, Y. R. Li, P. Zhang, J. Z. Ma and S. T. Jia, *ACS Appl. Mater. Interfaces*, 2014, **6**, 10153–10161.

31. L. Yuan, J. Dai, X. Fan, T. Song, Y. T. Tao, K. Wang, Z. Xu, J. Zhang, X. Bai, P. Lu, J. Chen, J. Zhou and Z. L. Wang, *ACS Nano*, 2011, **5**, 4007–4013.
32. K. L. Song, A. Q. Gao, X. Cheng and K. L. Xie, *Carbohydr. Polym.*, 2015, **130**, 381–387.
33. S. T. Yohe, Y. L. Colson and M. W. Grinstaff, *J. Am. Chem. Soc.*, 2012, **134**, 2016–2019.
34. K. Koch, B. Bhushan and W. Barthlott, *Soft Matter*, 2008, **4**, 1943–1963.
35. K. Koch, B. Bhushan, Y. C. Jung and W. Barthlott, *Soft Matter*, 2009, **5**, 1386–1393.
36. D. Nystrom, J. Lindqvist, E. Ostmark, A. Hult and E. Malmstrom, *Chem. Commun.*, 2006, 3594–3596.
37. T. I. Kim, D. Tahk and H. H. Lee, *Langmuir*, 2009, **25**, 6576–6579.
38. A. Hozumi and T. J. McCarthy, *Langmuir*, 2010, **26**, 2567–2573.
39. R. Asmatulu, M. Ccylan and N. Nurajc, *Langmuir*, 2011, **27**, 504 507.
40. X. Liu and J. He, *Langmuir*, 2009, **25**, 11822–11826.
41. L. Gao and J. He, *J. Colloid Interface Sci.*, 2013, **396**, 152–159.
42. L. Jiang, Y. Zhao and J. Zhai, *Angew. Chem., Int. Ed.*, 2004, **43**, 4338–4341.
43. Y. W. Wu, T. Hang, Z. Y. Yu, L. Xu and M. Li, *Chem. Commun.*, 2014, **50**, 8405–8407.
44. L. Xu, Z. Geng, J. He and G. Zhou, *ACS Appl. Mater. Interfaces*, 2014, **6**, 9029–9035.
45. L. Xu, L. Gao and J. He, *RSC Adv.*, 2012, **2**, 12764.
46. Z. Geng, J. He, L. Xu and L. Yao, *J. Mater. Chem. A*, 2013, **1**, 8721.
47. W. Lee, M. K. Jin, W. C. Yoo and J. K. Lee, *Langmuir*, 2004, **20**, 7665–7669.
48. G. S. Watson, S. Myhra, B. W. Cribb and J. A. Watson, *Biophys. J.*, 2008, **94**, 3352–3360.
49. W. L. Min, B. Jiang and P. Jiang, *Adv. Mater.*, 2008, **20**, 3914–3918.
50. Z. G. Guo and W. M. Liu, *Plant Sci.*, 2007, **172**, 1103–1112.
51. Y. Y. Liu, X. Q. Chen and J. H. Xin, *Bioinspiration Biomimetics*, 2008, **3**, 8.
52. J. A. Gao, Y. L. Liu, H. P. Xu, Z. Q. Wang and X. Zhang, *Langmuir*, 2010, **26**, 9673–9676.
53. Y. M. Zheng, X. F. Gao and L. Jiang, *Soft Matter*, 2007, **3**, 178–182.
54. A. Otten and S. Herminghaus, *Langmuir*, 2004, **20**, 2405–2408.
55. T. S. Wong, S. H. Kang, S. K. Y. Tang, E. J. Smythe, B. D. Hatton, A. Grinthal and J. Aizenberg, *Nature*, 2011, **477**, 443–447.
56. Q. Wei, C. Schlaich, S. Prevost, A. Schulz, C. Bottcher, M. Gradzielski, Z. H. Qi, R. Haag and C. A. Schalley, *Adv. Mater.*, 2014, **26**, 7358–7364.
57. X. Yao, S. S. Dunn, P. Kim, M. Duffy, J. Alvarenga and J. Aizenberg, *Angew. Chem., Int. Ed.*, 2014, **53**, 4418–4422.
58. H. F. Bohn and W. Federle, *Proc. Natl. Acad. Sci. U. S. A.*, 2004, **101**, 14138–14143.
59. M. A. Samaha and M. Gad-el-Hak, *Polymers*, 2014, **6**, 1266–1311.
60. L. Gaume, S. Gorb and N. Rowe, *New Phytol.*, 2002, **156**, 479–489.
61. C. P. Hsu, Y. M. Lin and P. Y. Chen, *JOM*, 2015, **67**, 744–753.

62. H. Liu, P. Zhang, M. Liu, S. Wang and L. Jiang, *Adv. Mater.*, 2013, **25**, 4477–4481.
63. P. F. Zhang, H. W. Chen, L. W. Zhang, T. Ran and D. Y. Zhang, *Appl. Surf. Sci.*, 2015, **355**, 1083–1090.
64. X. Yao, Y. H. Hu, A. Grinthal, T. S. Wong, L. Mahadevan and J. Aizenberg, *Nat. Mater.*, 2013, **12**, 529–534.
65. X. Yao, J. Ju, S. Yang, J. Wang and L. Jiang, *Adv. Mater.*, 2014, **26**, 1895–1900.
66. S. Anand, A. T. Paxson, R. Dhiman, J. D. Smith and K. K. Varanasi, *ACS Nano*, 2012, **6**, 10122–10129.
67. P. Kim, T.-S. Wong, J. Alvarenga, M. J. Kreder, W. E. Adorno-Martinez and J. Aizenberg, *ACS Nano*, 2012, **6**, 6569–6577.
68. K. Manabe, S. Nishizawa, K. H. Kyung and S. Shiratori, *ACS Appl. Mater. Interfaces*, 2014, **6**, 13985–13993.
69. J. Zhang, A. Wang and S. Seeger, *Adv. Funct. Mater.*, 2014, **24**, 1074–1080.
70. M. J. Liu, S. T. Wang, Z. X. Wei, Y. L. Song and L. Jiang, *Adv. Mater.*, 2009, **21**, 665–669.
71. G. D. Bixler and B. Bhushan, *Adv. Funct. Mater.*, 2013, **23**, 4507–4528.
72. Y. H. Liu and G. J. Li, *J. Colloid Interface Sci.*, 2012, **388**, 235–242.
73. X. Han and D. Y. Zhang, *Sci. China, Ser. E: Technol. Sci.*, 2008, **51**, 890–896.
74. Y. H. Luo, Y. F. Liu, J. Anderson, X. Li and Y. Y. Li, *Appl. Phys. A: Mater. Sci. Process.*, 2015, **120**, 369–377.
75. C. Baum, W. Meyer, R. Stelzer, L. G. Fleischer and D. Siebers, *Mar. Biol.*, 2002, **140**, 653–657.
76. W. R. Hansen and K. Autumn, *Proc. Natl. Acad. Sci. U. S. A.*, 2005, **102**, 385–389.
77. S. Sethi, L. Ge, L. Ci, P. M. Ajayan and A. Dhinojwala, *Nano Lett.*, 2008, **8**, 822–825.

CHAPTER 2

Biological Surface: Lotus Leaves and Butterfly Wings

CHENG SONG AND YONGMEI ZHENG*

Key Laboratory of Bio-Inspired Smart Interfacial Science and Technology of Ministry of Education, School of Chemistry and Environment, Beihang University, Xueyuan Road 37, Haidian District, Beijing 100191, China
*Email: zhengym@buaa.edu.cn

2.1 Introduction

The lotus ("He Hua" in Chinese), botanically known as *Nelumbo nucifera*, is a variety of aquatic plant. The lotus is native to China and is now widely cultivated in China, Japan, India, North America, West Asia and Middle Asia. Some seeds found in a dry lake in northeastern China showed that lotuses were planted over 3000 years ago. And some ancient seeds can still blossom today.[1,2] Throughout history, lotuses have been regarded as food (the seeds and the rhizomes), decoration and medicine. Besides, lotuses also play a significant and special role in Chinese cultural activities. The reason for this might be in a rhyme written by a famous Chinese poet, Tao Yuanming, who was living in the Dongjin Dynasty (around 400 A.D.), in which he says "I admire she (lotus) grows and blooms in the mud, yet neatly and clean out of it. I admire she flows in the drops of water the yet graceful seen." It was not until over 1500 years later that Dettre and Johnson found that the surface of the lotus leaf was superhydrophobic,[3] making the lotus "flows in the drops of water the yet graceful seen". And Barthlott and Ehler found that the lotus leaf had a self-cleaning property due to its microstructure,[4] helping the lotus "blooms in the mud, yet neatly and clean out of it".

RSC Smart Materials No. 21
Self-cleaning Coatings: Structure, Fabrication and Application
Edited by Junhui He

Published by the Royal Society of Chemistry, www.rsc.org

Beautiful flowers always attract beautiful fairies. The butterfly, one of the most amazing extant insects on the Earth, dances among the blossoms gently and elegantly with a pair of sparkling wings, making this beautiful picture even more attractive and dynamic. Due to its great beauty, a butterfly called Camberwell Beauty, living in Britain, has been given the names Grand Surprise, the Willow Beauty and the White Petticoat in the past. It has similar names in several different languages for the same reason. In ancient China, there was a famous legend of love named "Liang Shanbo and Zhu Yingtai", a story about a young couple that died for love and turned into two beautiful butterflies so that they could live together freely despite strong opposition from their parents, just like Romeo and Juliet in Shakespeare's play. In addition, due to it coming from an ugly worm, the butterfly is often used in novel imagery to symbolize the transformation of a person.

No matter which – lotus or butterfly – they both leave a deep impression with their beautiful appearance. However, have you wondered how they prevent their beauty from becoming dirty? Do they always need to take a bath to keep themselves clean, just like human beings? The answer is no. They do not need to clean themselves but use a more intelligent and efficient way to maintain their beauty: lotus leaves and butterfly wings are both self-cleaning. In other words, they cannot be polluted so that they have no need to clean themselves.[5,6]

In this chapter, we will begin with the evolution of lotuses and butterflies and then figure out the self-cleaning property of the lotus leaves and butterfly wings, along with their mechanism. Then, we will focus on some artificial surfaces with a similar self-cleaning property, including their morphologies, fabrications and applications.

2.2 The Evolution and Taxonomy of the Lotus and Butterflies

2.2.1 The Lotus

Figure 2.1 shows the flowers, seeds and bodies of a progeny lotus of old Xipaozi fruit (over 1200 years old) collected by I. Ohga in the 1920s and planted by H. V. Wester in 1951.[1] The characteristics of the ancient lotus were very similar to the extant ones. The flower, on the second day of bloom, has a diameter of about 26 cm. There is also a yellow receptacle located in the middle that consists of multiple stigmas. Surrounding the stigmas are over 200 saffron stamens and about 15 pink petals (Figure 2.1a). Different from the extant species, the body of the ancient lotus (seedlings) has some variegation on one of its leaves, together with a large wedge (at the right side of the leaf) encompassing six red veins and specks. Besides, the small plumule (at the upper right) is much thicker, and another plumule (lower right) has some yellow and white specks (Figure 2.1c).[1]

Figure 2.1 Lotus developed from ancient seeds. (a) Flower, ~26 cm in diameter, showing the multiple stigmas and saffron stamens. (b) Lotus seeds. (c) Body of a 400-year-old Xipaozi fruit. (d) 464-year-old fruit on the second day of germination, having a girth of ~1 cm.

Figure 2.2 shows the taxonomical study of the lotus. The lotus (*Nelumbo nucifera*) belongs to the family Nelumbonaceae in plant taxonomy, along with the Brasenia, Euryale, Nuphar and Nymphaea.

Despite the similarity between the ancient lotus and living ones today, there was a clear evolution lineage for the lotus over millions of years (Figure 2.3). According to the morphology of fossils and DNA (gene) sequences discovered, there could be two parallel lineages during the evolution of the extant lotus. *Nelumbo provinciale* living in the Cretaceous period in Fuveau, Southern France,[7] was believed to be the ancestor of *Nelumbo protospeciosa*.[8–11] And *N. protospeciosa* seems to lead to the extant *N. nucifera*, which is now widely seen in Europe and Asia. The other parallel lineage of evolution was for the living lotus in North America. *Nelumbo kempii*, living in the Magothy Formation of Long Island and New Jersey, USA,[12,13] is now believed to be the ancestor of *Nelumbo protolutea*,[9,10] known in the early Eocene Grenada Formation of Meridian, Mississippi, USA.[9] And *N. protolutea* then evolved into the widely seen species seen in North America today, *Nelumbo lutea*.[9]

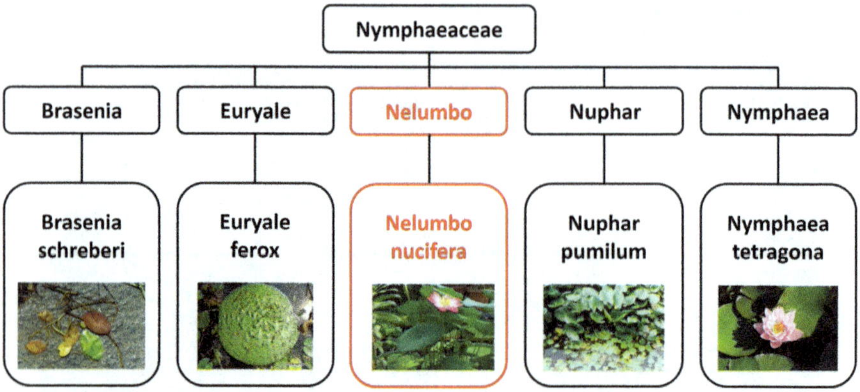

Figure 2.2 Taxonomical diagram of Nelumbo. J. Shen-Miller, Sacred lotus, the long-living fruits of China Antique, *Seed Sci. Res.*, **12**(03), 131–143, reproduced with permission.[1]

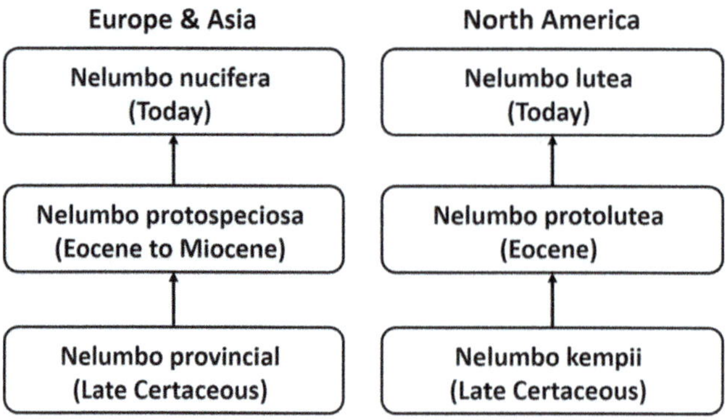

Figure 2.3 The evolution of the lotus.

2.2.2 The Butterflies

Unlike the lotus, which has the botanical name *Nelumbo nucifera*, butterflies are too different to be categorized, so there were some interesting mistakes when distinguishing the butterflies. For example, two butterflies found by early scientists and named the Selvedged Heath Eye and the Golden Heath Eye were then proved to be the male and female of a single species, which was then initially named the Gatekeeper but is now named the Small Heath. What is more interesting, the original name Gatekeeper is now the name of an entirely different species, which was known as the Hedge Brown, Hedge Eye and Large Heath. The name Large Heath is now applied to an entirely different species that was previously known as the Manchester Argus, or Marsh Ringlet!

From the view of modern taxonomy, the word butterfly is not a name for a specific species but is a common name for over 14 000 species in three superfamilies within the Lepidoptera – the Hedyloidea, the Hesperioidea and the Papilionoidea – which evolved from a common ancestor Rhopalocera (Figure 2.4).[14–17]

Today in England, we can still see the ancient insect called the longhorn caddis fly *Mystacides azurea*, which is now believed to be the ancestor of butterflies and moths (Figure 2.5a). In order to hold and grasp the female in front of it during copulation, the male *Mystacides azurea* has a pair of highly modified pincers at the front of its head. Like another old insect, the dragonfly, the larvae of *Mystacides azurea* are aquatic and live in portable cases built in sand grains or fragments of stems bonded to a silk tube out of their body. Modern day bagworm moths (Psychidae) still have larvae that live inside cases.

The fossil records show that insects began to live on the Earth about 300 million years ago. After about 100 to 150 million years, the very first Lepidoptera evolved from the Trichoptera (caddis flies), together with the appearance of the earliest flowering plants. Figure 2.5b shows a butterfly fossil found in the Florissant beds, Colorado, USA, which is named *Prodryas persephone* (Nymphalidae), which lived about 30 million years ago. And it

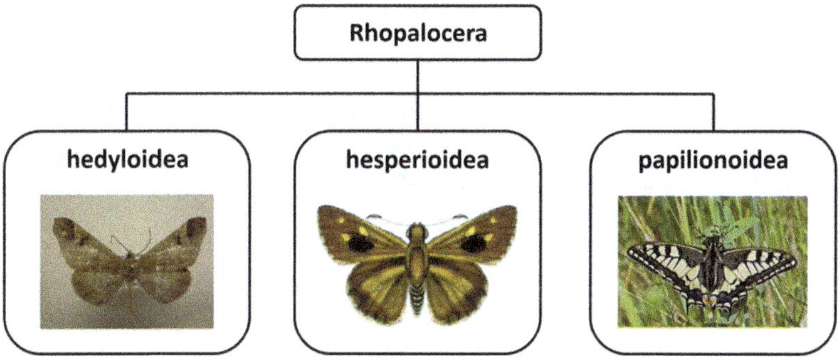

Figure 2.4 Three different species of butterfly.

Figure 2.5 (a) *Mystacides azurea*. (b) Fossil of an ancient butterfly. W. Barthott, Purity of the sacred lotus, or escape from contamination in biological surfaces, *Planta*, **202**, 1997, figure 5, with permission of Springer.[20]

shows some similarities between the ancient butterflies and the extant ones: the double vibrissa, wide and transparent wings and the arthropodic body with an exoskeleton.[18] However, different from the vertebrate which has hard bones, the bodies of the butterflies were so soft that few of them could turn into fossils. And due to the lack of fossil records, there is still no clear evolutionary trace for the butterflies, which still needs to be studied.[18]

2.3 Self-cleaning Properties of Lotus Leaves and Butterfly Wings

2.3.1 The Lotus Leaves

Lotus leaves may have been the first biological surface to inspire scientists to investigate the self-cleaning phenomenon. The self-cleaning property of lotus leaves is so significant that the word "lotus effect" was created to describe all surfaces with self-cleaning properties. In the late 1970s, a group of German biologists, led by Barthlott and Neihuis, studied over 300 plant leaves[19,20] and they found that it was the synergy of the rough structure and the hydrophobic wax cover on the surface that helped with the self-cleaning property (Figure 2.6a).[21] If the wax was removed by cleaning with acetone, the water contact angle would be dramatically reduced and the self-cleaning property would also disappear. Therefore, the superhydrophobicity and self-cleaning property are considered to be caused by the synergy of a rough surface structure that imprisons air beneath water drops and the water repellency of the wax coated on the surface. And the self-cleaning property was then defined with two key criteria:

(1) Superhydrophobicity. The water contact angle (WCA) must be over 150°;
(2) Strong anti-pollution. Pollutants such as dusts and microparticles on the surface can be taken away by water when rolling off with no traces (Figure 2.6b).[21]

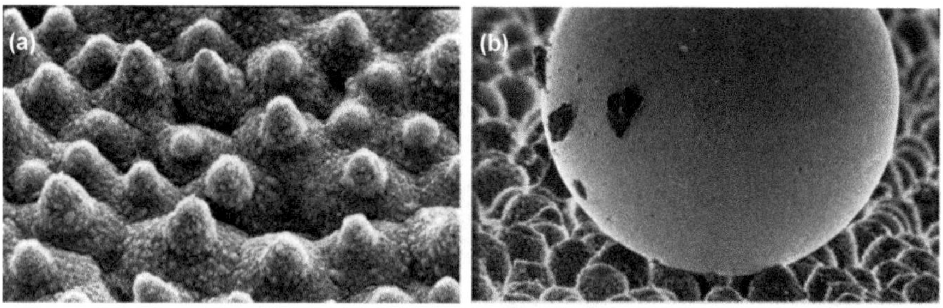

Figure 2.6 (a) The SEM image of a lotus leaf. (b) A drop with dust standing on the lotus leaf.
Reproduced from ref. 22 with permission from John Wiley and Sons.

However, after about 30 years, a new discovery by Jiang Lei's group in China indicated that a nanoscale structure may occur on the microscale sturcture[22] and the combination of such a micro/nano structure is the real reason that the lotus leaf is self-cleaning. In addition, such surfaces have large WCAs and small sliding angles. Thus, water drops can easily roll off the surface, taking away pollutants.[22] Figure 2.7a is an SEM image of a large area of a lotus leaf. Many small papillas in diameters ranging from 5 to 9 μm are spread randomly on the surface. A magnified view of one of the papillas (Figure 2.7b) shows that each papilla is covered by numerous nanobranches with an average diameter of 124.3 nm ± 3.2 nm. In addition, the nanobranches also appear in the spaces among the papillas (Figure 2.7c), which can prevent the lower part of the leaf from being permeated.[23] Jiang Lei's group figured out that this micro/nano structure could lead to a large WCA (161.0° ± 2.7°) and a small sliding angle (about 2°).

When put into moist air, tiny water drops will condense on the surface. The whole process can be observed *in situ* with an environmental scanning electronic microscope (ESEM). At the beginning, a small microdroplet appears in the basin among the papillars (Figure 2.8a). When the pressure keeps increasing, the tiny microdroplet grows to bead in contact with the papillars (Figure 2.8b). With an increase in pressure, the drop keeps growing

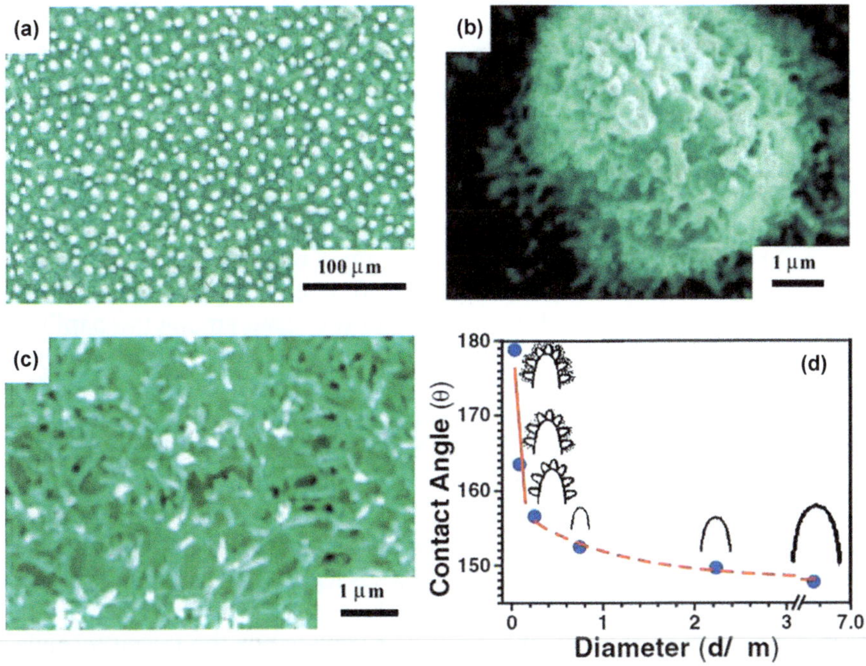

Figure 2.7 SEM images of the lotus leaf surface and the relationship between the nanostructure and the contact angle.
Reprinted with permission from Y. Zheng, D. Han, J. Zhai and L. Jiang, *Appl. Phys. Lett.*, 2008, **92**, 084106. Copyright 2008, AIP Publishing LLC.[24]

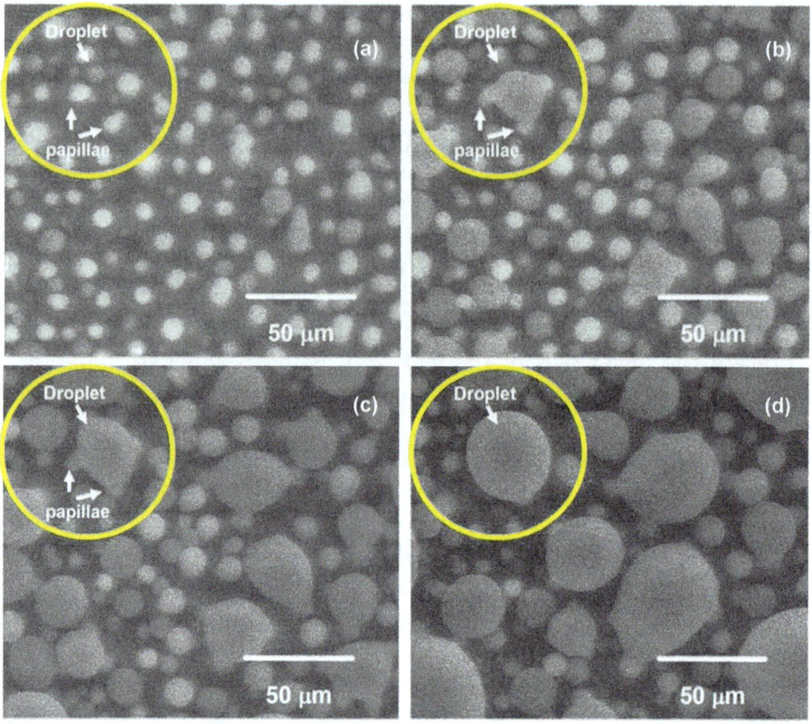

Figure 2.8 ESEM images indicating the water condensation process on the lotus leaf surface.
Reprinted with permission from Y. Zheng, D. Han, J. Zhai and L. Jiang, *Appl. Phys. Lett.*, 2008, **92**, 084106. Copyright 2008, AIP Publishing LLC.[24]

until it is so big, the basin cannot hold it. So it tends to escape the basin (Figure 2.8c). Finally, the drop will stand on top of the papillars, bridging over more than four papillars and leaving the air trapped beneath (Figure 2.8d).[24]

The magnification of the drop standing in the basin among the papillars is shown in Figure 2.9. Different from the condensation process on a flat surface, the water drop does not collapse during the condensation (Figure 2.9a). This may be due to the wet contact on top of the papillars where a thin layer covers them (Figure 2.9) and the thin layer can be clearly seen in the detail of the papillar.[24]

According to Adamson and Gast, a model of the relationship between the superhydrophobicity and the multi-scale structure can be founded. The multi-scale structure on the surface of the lotus leaves (nano structure spreads over the microstructure) can be roughly taken as a fractal described by the Koch curve.[25] Then, the roughness can be calculated by the fractal formula. Considering Wenzl's law, the relationship between the contact angle on the rough surface (θ_f) and the smooth surface (θ) can be written by eqn (2.1)

$$\cos\theta_f = f_s\left(\frac{L}{l}\right)^{D-2}\cos\theta - f_v \tag{2.1}$$

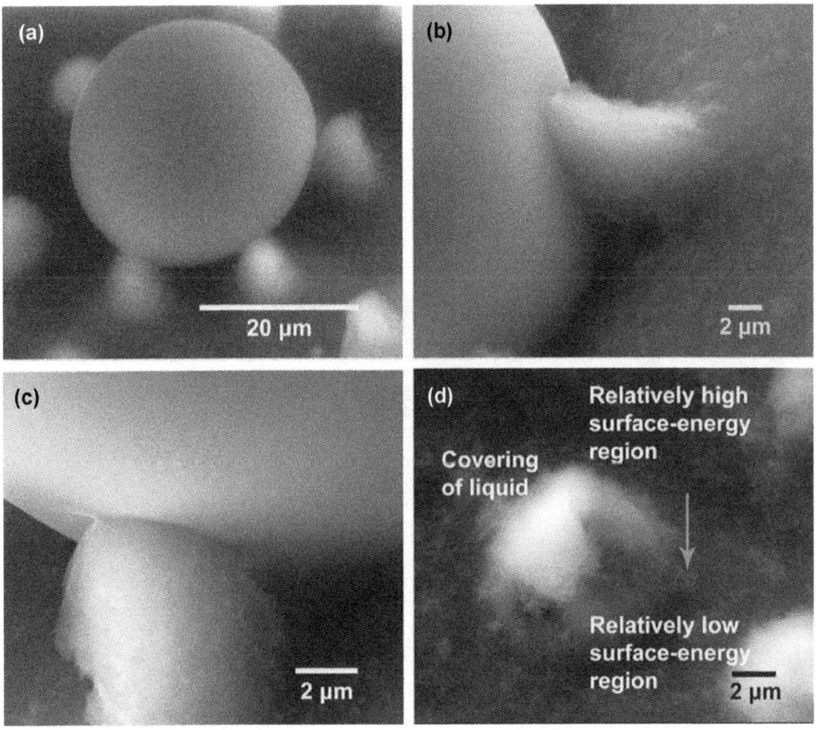

Figure 2.9 ESEM images of water drops in contact with the papillars.

where $(L/l)^{D-2}$ is the roughness, L, l are the upper and lower limits for a fractal surface, respectively and D is the fractal dimension, f_s and f_v are the fraction of area occupied by surface and air ($f_s + f_v = 1$), respectively. As for the surface of lotus leaves, L and l can be roughly approximated by the size of the microscale papillas and the nanoscale branches. In the Koch curve, D is $\sim$2.27 in three-dimensional space, and (L/l) is $\sim 3^n$, where n can be determined be the specific fractal structure. It is easy to know that as n increases, the roughness increases. Thus, if we set L, the upper limit, a constant, with n increasing, l, the lower limit, will decrease. When we put in the diameter of the papillas and branches measured in Figure 2.7a, f_s and f_v can be calculated as 0.2056 and 0.7944, respectively. If we note θ as $104.6° \pm 0.5°$ as reported,[26] according to eqn (2.1), θ_f can be roughly calculated as shown in Table 2.1.

Figure 2.7d gives the relationship between the diameter of the papillas and the appearent contact angles based on the calculation results. With the shrinking of the papillas, the contact angle climbs up very slowly. However, when the nanobranches begin to appear on the papillas, the value of the contact angle rockets from less than 160° to over 170°, which gives a much steeper curve. This illustrates that the nanobranches have played an important role in generating the superhydrophobicity.

Table 2.1 Relationship between θ_f and n.

n	θ_f
0	147.8°
1	149.7°
2	152.4°
3	156.5°
4	163.4°

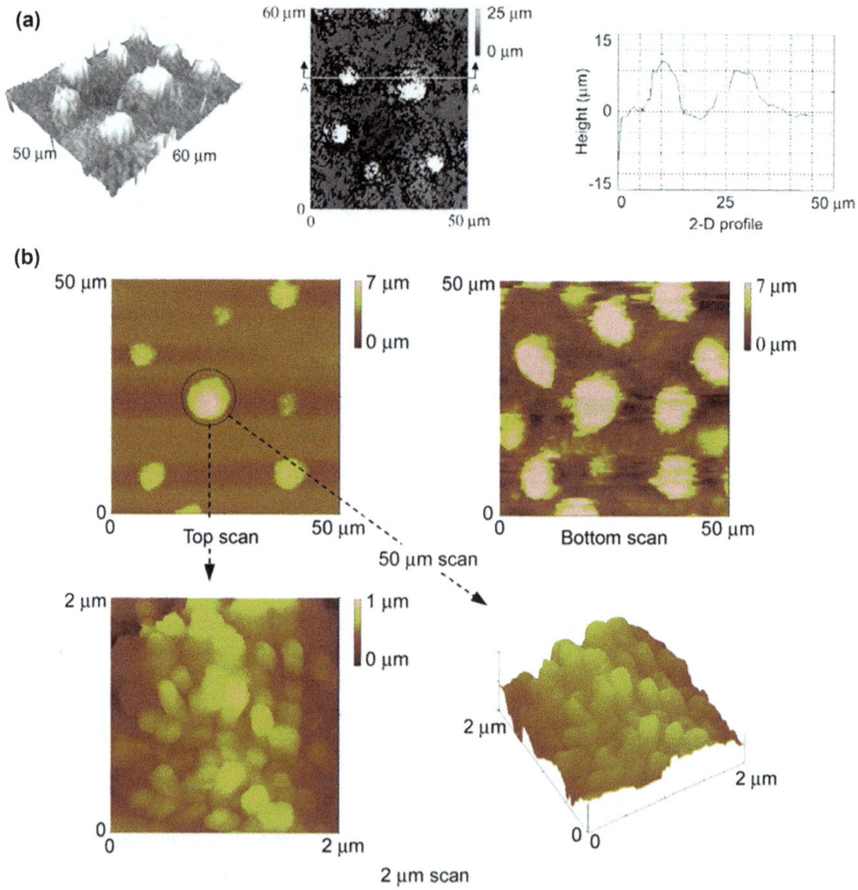

Figure 2.10 AFM images of natural lotus leaves.
Reproduced from ref. 22 with permission from John Wiley and Sons. Copyright © 2002 WILEY-VCH Verlag GmbH & Co. KGaA, Weinheim.

Observed under an atomic force microscope, the micro/nano structure can be clearly revealed, which is made up of micro papillas and nanobranches (Figure 2.10).[27–30]

The second criterion of self-cleaning is strong anti-pollution, which indicates that the suface must be slightly adhesive. The wax coated on the papillas results in a low coefficient. Thus the interfacial area between the surface and the dust (or particles) is very small, causing a rather reduced adhesion. Figure 2.11 is a cartoon illustration demonstrating the differences between the smooth surface and the lotus leaf with a self-cleaning property when water rolls down the surfaces. When water drops, such as rain, fall onto the leaf, it is very easy for them to roll down, catch the adhered dust and take it away, resulting in the self-cleaning property, or the so-called "lotus effect", which protects leaves from pollutants. A considerable amount of dust still leaves non-superhydrophobic smooth surfaces with lower WCAs.[31,32]

In addition to the superhydrophobicity on the upper surface of the lotus leaves, which is the interface between air and soild, superoleophobicity has

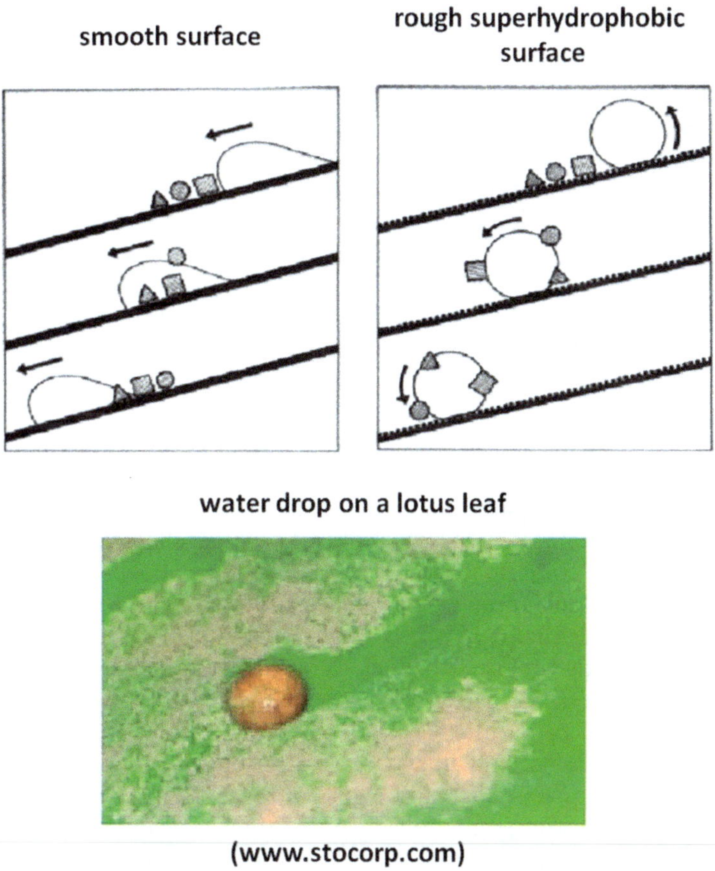

Figure 2.11 Behaviour of water drops rolling off different surfaces and a natural lotus leaf.
Reprinted from ref. 33 with permission from the Royal Society of Chemistry.

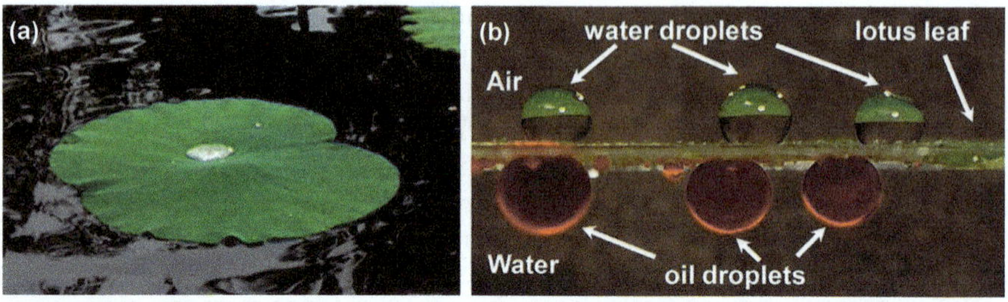

Figure 2.12 Oil drops beneath the lotus leaf.
Reprinted from ref. 33 with permission from the Royal Society of Chemistry.

also been found on the lotus leaves, which is the interface between water and solid.[33] Figure 2.12 shows a photo of a lotus leaf on the water surface. When water drops are dispensed onto the upper surface, they retain a spherical shape due to the superhydrophobicity. However, when oil drops (*n*-hexane dyed red) are dispensed onto the lower surface, they also retain the same shape, which means the lower surface of the lotus leaf is superleophobic.

The ESEM image and AFM scanning results give a possible explanation for this contrast. Different from the upper surface, the lower surface has no three-dimensional multi-scale structure.[19,34] Instead, it consists of randomly ordered tabular and slightly convex papillars with a size of about 40 mm × 20 mm × 4 mm (L × W × H) (Figure 2.13). And each papillar contains numerous nanogroove structures with a length of 200–500 nm. This structure can secrete some hydrophilic compounds, such as in ferns,[35] which can form a water layer covering its surface and make it superleophobic.

2.3.2 Butterfly Wings

Different from lotus leaves, the pretty colors of various butterfly wings are the result of a complex pattern of scales, bristles or cilia covering the surface of the wings.[36] From a biological point of view, each scale (or bristle) is formed by a single epidermal cell. Those cells, forming the basic architecture layer by layer and stripe by stripe, are considered to be the units for the pattern and its details. And since even the simplest scales are perfectly structured, butterfly wings can be seen as one of the most skilful works of art in biological pattern formation at not only the cellular level but also the macro level carved by nature.[36]

Figure 2.14 contains SEM images of several butterfly wings. Despite the difference in the apperenance and colors of the butterflies, their wings are all formed by well-ordered micro structures: the scales or bristles cover the surface of the wings, with the latter partly covered by the former. This structure is very much like tiles (Figure 2.15), which were common structures on the roofs of the ancient Chinese dating back to the Zhou Dynasty (700 B.C.) and were used to prevent leaks from rain.

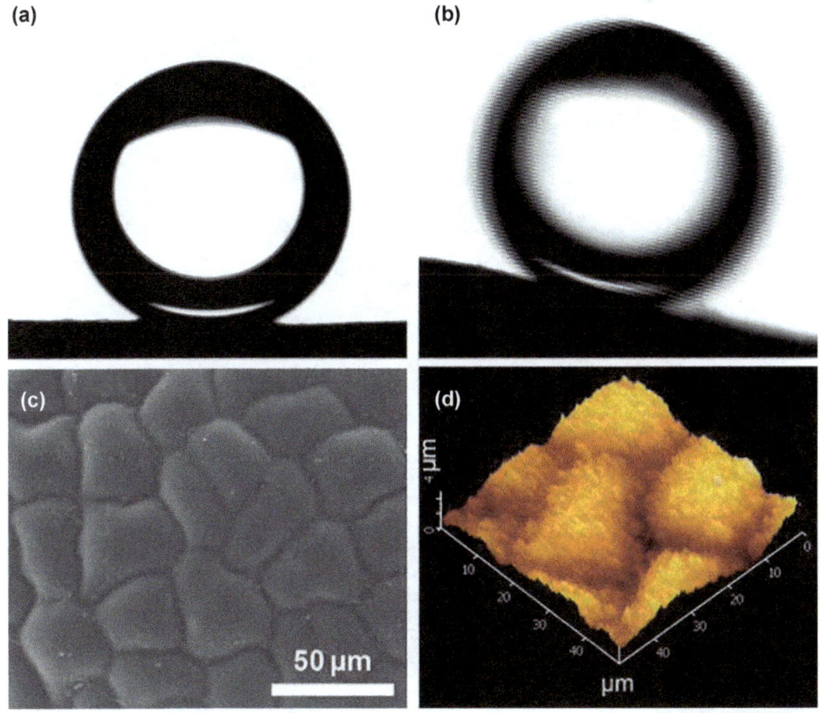

Figure 2.13 Oil sliding off the lower lotus leaf surface and the SEM and AFM images of the surface.
Reproduced from ref. 36 with permission from John Wiley and Sons. Copyright © 1994 Wiley-Liss, Inc.

It is not only in shape but also in function that the scales and tiles are the same. Researchers have found that although the structure and color of different butterfly wings differ from one another, the water repellency properties of them are quite similar, which is also extremely interesting. Figure 2.16 shows two butterflies with quite different wings: one is colourful, thick and non-transparent while the other is thin and transparent. Although their wings are made by highly ordered large scales and randomly arranged small cilia, respectively, the water contact angles on the wings are both higher than 150°, resulting in superhydrophobicity.[37]

The self-cleaning property of butterfly wings is strongly related to their superhydrophobicity.[38,39] The water contact angle on the wings covered by scales is higher than 150° and the contamination particles can be washed away easily with a small tilt angle (~3°), while the contact angle on the wings after the scales are removed is only 100° and the particles cannot be easily washed away.[40] And there is no evidence to indicate that the area, length, width or gaps of the scales affect the contact angle as well as the self-cleaning property of the wings.[41,42]

Figure 2.14 SEM images of several butterfly wings. (a) *Marpho menelaus*, top view of scales, scale bar = 50 μm; (b) *Caligo martia*, part of the upper lamina of a non-specialized scale, scale bar = 1 μm; (c) *Astraptes azul*, longitudinal view of a fractured bristle, scale bar = 2 μm; (d) *Parides sesostris*, longitudinal view of fractured reflective scale, scale bar = 1 μm; (e) *Morpho menelaus*, edge of a cover scale, scale bar = 2 μm; (f) *Caligo beltruo*, part of a longitudinally fractured scale, scale bar = 0.5 μm. P. Perez Goodwyn, Waterproof and translucent wings at the same time: problems and solutions in butterflies, *Naturwissenschaften*, **96**, 2009, figure 1, with permission of Springer.[37]

When put into moist air, water drops begin to condense on the overlapped scales. Figure 2.17 is a combination of the ESEM images indicating water condensation on the wing surfaces of two different butterflies, *Morpho rhotcnor* (left) and *Morpho aega* (right).[43]

Scales on butterfly wings Tiles on an ancient Chinese house roof

Figure 2.15 Comparison of the scales on butterfly wings and the tiles on a Chinese house roof.
Left: Copyright (2015) from Bio-Inspired Wettability Surfaces: Developments in Micro- and Nanostructures by Yongmei Zheng. Reproduced by permission of Taylor and Francis Group, LLC, a division of Informa plc. Right: www.tonghuashiji.net.

Different from the self-cleaning property of lotus leaves where dust can be washed away from any direction, water drops falling onto butterfly wings can only move from a single direction away from the body but they can get pinned in the opposite direction (Figure 2.18).[44,45] This is because the micro scales and the nano rear ends are both arranged along the direction away from the body, forming an overlapping structure (Figure 2.19a and b). And the AFM scanning results show that the height of each step along the rolling direction is about 10 nm for the micro scales and 500 nm for the nano rear ends.

When the butterfly wings slope downward, the scales and rear ends separate from each other, fragmenting the water contact line, making the water drops roll off easily. However, when the wings slope upwards, the scales and rear ends get close to each other, forming a continuous water contact line and preventing the water drops from rolling away (Figure 2.20).

Recently, researchers have found that the flapping of the wings can be another factor driving the tiny water droplets from the surface of the wings.

For butterflies living in habitats with moist air, *e.g. Morpho deidamia*, it is not enough to wait for the water drops to leave their wings before they can fly. They need a much more efficient and active way to get rid of the drops, especially when they are relatively small. As with most butterflies investigated, an overlapped structure and directional adhesion have been discovered on their wings,[44] which can help the drops roll off the wings along the radial outside (RO) direction (Figure 2.21a–d). However, unlike the butterflies with flat scales covering their wings, the scales on the wings of *Morpho deidamia* bend toward the surface. And the tilt angle against the RO direction (ω_2) is larger than that along the RO direction (ω_1) (Figure 2.21b$_2$). In addition, the surface of each scale is covered with parallel ridges with nanowalls, which also tilt toward the RO direction (Figure 2.21c$_2$).[46]

Figure 2.16 Differences between non-transparent and transparent wings. Reproduced from ref. 44 with permission from the Royal Society of Chemistry.

Figure 2.17 Water drops on butterfly wings.
Reproduced from ref. 44 with permission from the Royal Society of Chemistry.

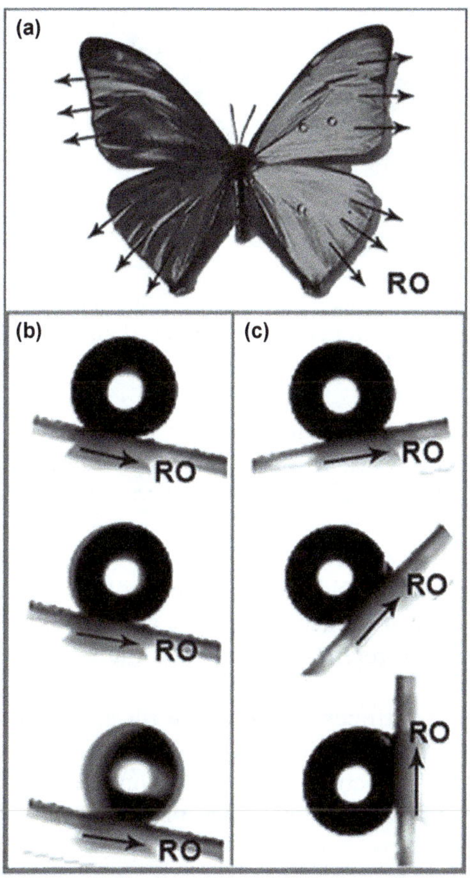

Figure 2.18 The rolling and pinned status on butterfly wings.
Reproduced from ref. 44 with permission from the Royal Society of Chemistry.

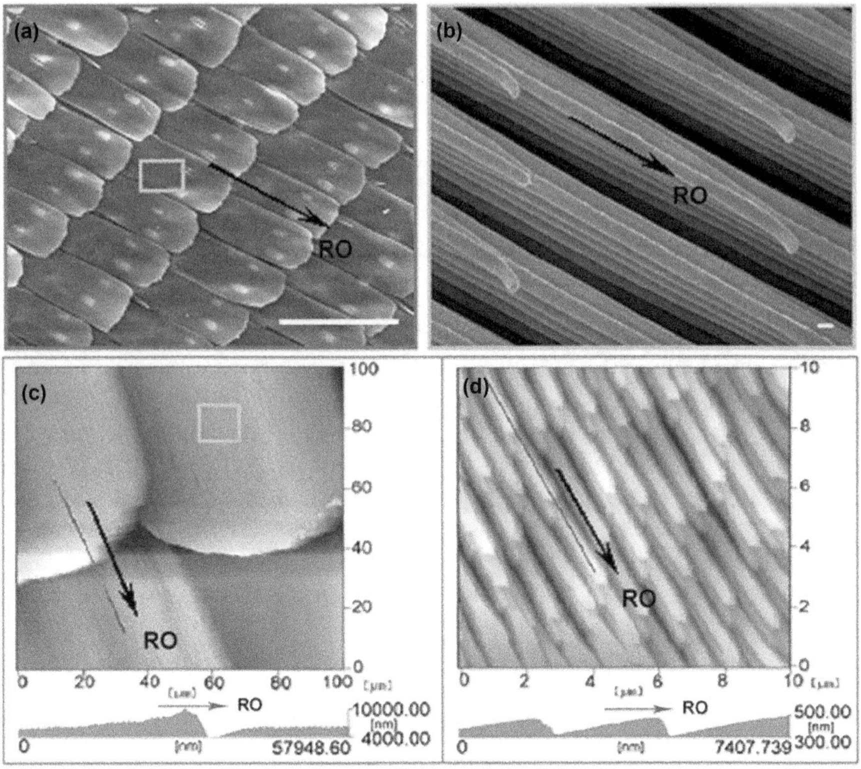

Figure 2.19 SEM and AFM images of scales and nano ridges. Reproduced with permission from C. Liu, J. Ju, Y. Zheng and L. Jiang, *ACS Nano*, 2014, **8**, 1321. Copyright (2014) American Chemical Society.[46]

Due to the asymmetric scales with a micro-/nano-structure, *Morpho deidamia* can get rid of water drops in both static and dynamic situations.

On static wings, tiny water droplets start to generate between the ridges due to their small size. As the surface of the scale is superhydrophobic,[47,48] the drops will gather on the ridges as their volume increases. Because drop 1 penetrates into the ridges more deeply than drop 2, it has a larger fraction of liquid/solid interface, leading to a larger retention force.[49] Thus, drop 2 with a smaller retention force will then be absorbed by drop 1, making drop 1 + 2.

When drop 1 + 2 continues growing, the tilt angle and shapes makes its shape asymmetric, where the side along the RO direction spreads more than that against the RO direction,[46,50] causing an imbalance in the surface tension forces acting on the two sides of the drop.[51] This force can be written as:

$$F_{dl} = \int_{L_{left}}^{L_{right}} \gamma(\cos\theta_{a1} - \cos\theta_{r1})\mathrm{d}l \tag{2.2}$$

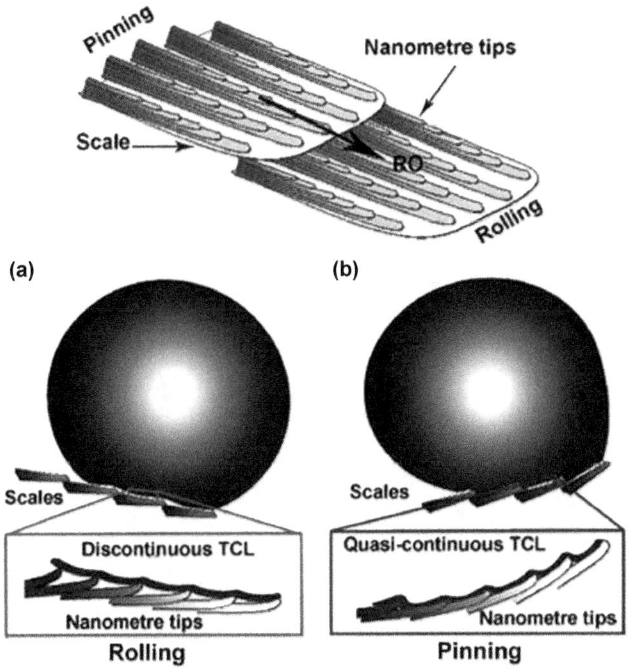

Figure 2.20 Different TCL phase along and against RO direction. Reproduced with permission from C. Liu, J. Ju, Y. Zheng and L. Jiang, *ACS Nano*, 2014, **8**, 1321. Copyright (2014) American Chemical Society.[46]

where γ and dl represent the surface tension of water and the integrating variable along the RO direction.[50,52]

When drop $1+2$ meets its neighbor on the scale (drop 3), the capillary force produced by the water band between the drops makes the initial one move towards the new one.[53] And this fast inside attachment of liquid causes a combined drop $(1+2+3)$, which will then dewet from the surface. During the dewetting process, the part belonging to drop $1+2$ (left side) dewets along the RO direction, whilst the drop 3 part (right side) dewets along the opposite direction (against the RO direction). The asymmetry of drop $1+2+3$ indicates that the adhesive force in the interface between the drop and the surface is relatively smaller when dewetting facing the direction of RO but larger when dewetting against it. This discrepancy in the adhesive force between the two directions may be the result of the different types of three-phase contact lines (TCLs) along the two opposite directions: fragmented along the RO direction while relatively continuous or quasi-continuous against it (Figure 2.22).

After coalescence, the surface energy released during the coalescence process is already at the maximum. And it can convert to the kinetic energy of drop $1+2+3$.[54,55] We note E_k the kinetic energy obtained by drop $1+2+3$, E_k can be defined as:

$$E_k = E_{s1} + E_{s2} - E_{s(1+2)} - E_v \tag{2.3}$$

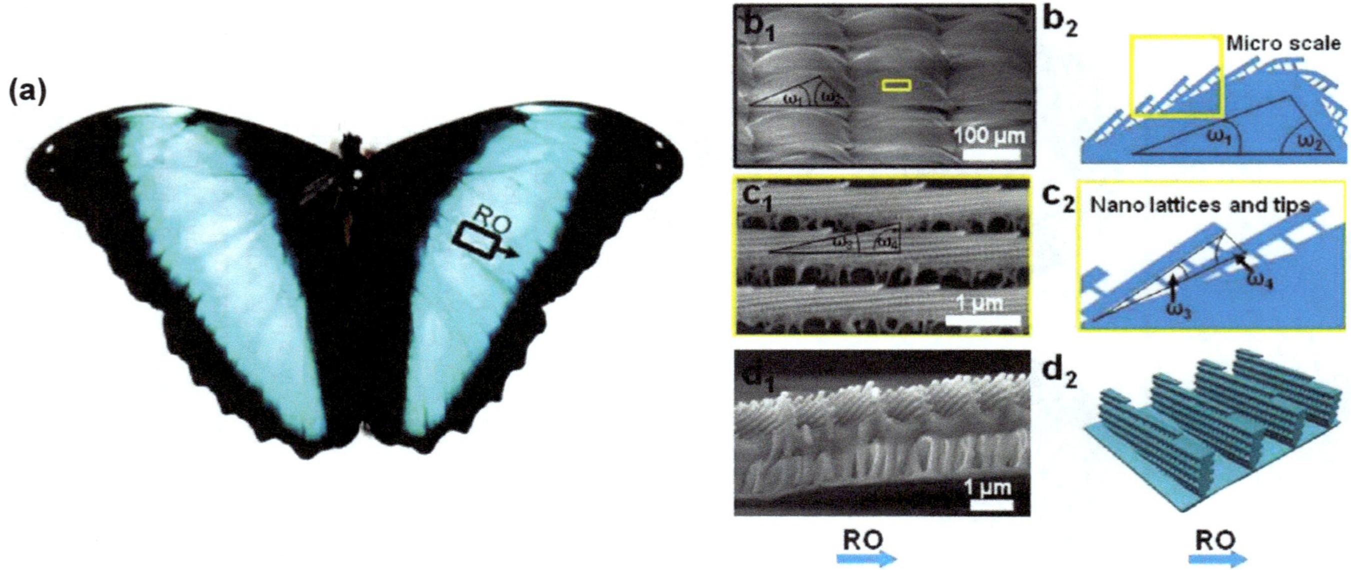

Figure 2.21 Morphology of the scales and ridges of butterfly wings.
Reproduced with permission from C. Liu, J. Ju, Y. Zheng and L. Jiang, *ACS Nano*, 2014, **8**, 1321. Copyright (2014) American Chemical Society.[46]

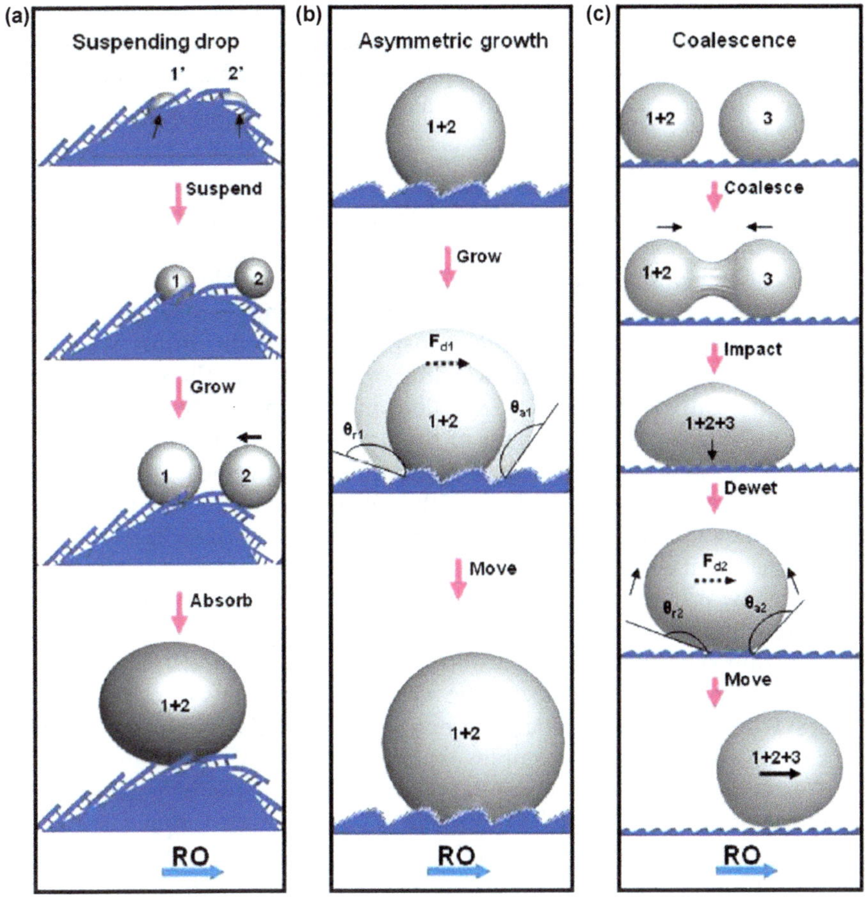

Figure 2.22 A drop's behavior on butterfly wings.
Reproduced from ref. 70 with permission from John Wiley and Sons.

where E_{s1}, E_{s2} and $E_{s(1+2)}$ are the surface energy of the initial drops 1, 2 and the coalescenced drop $1+2$, respectively, and E_v is the viscous dissipation during the coalescence (as the diameter of the drops is smaller than 2.7 mm, there is no need to consider the impact of gravity). If the radius of drops $1+2$ and 3 are assumed to be the same, E_k can be further written as:

$$E_k = \gamma\pi r^2\left\{(2-2^{\frac{2}{3}})(2-2\cos\theta) + \left[2^{\frac{2}{3}}\Psi_b(\Phi) - 2\Psi_a(\Phi)\right]\sin^2\theta\right\} - 64\pi\mu\sqrt{\frac{\gamma r^3}{\rho}} \tag{2.4}$$

where γ, ρ, θ and μ are the surface tension of water, the density of water, the contact angle of water drops on the wings and viscosity coefficient of water, respectively. And $f_a(\varphi)$ and $f_b(\varphi)$ represent the $f(\varphi)$ after and before the coalescence, where $f(\varphi)$ can be noted as:

$$f(\varphi) = \varepsilon\varphi \cos\theta_Y + \varphi - 1 \tag{2.5}$$

In eqn (2.3–2.5), ε is the roughness parameter of the wetted surface area, φ is the ratio of the soild area wetted by water and θ_Y represents the Young's contact angle on a smooth surface.[55] Due to the multi-scale structure formed by the micro scales and nano ridges, the butterfly wings get a large ε, leading to a lower energy barrier for the transportation of the drops with high coalescence speed. Thus when two drops become one, there is a synergy between the unbalanced surface tension and the surface energy released, leading to an apparent movement of the drops along the RO direction.

Unlike flies and bees, who fly with humming, butterflies always fly from flower to flower softly and gently, as if they were dancing a waltz. On average, the frequency of a butterfly vibrating its wings is less than 20 Hz with an amplitude about several centimeters.[56–58]

The behavior of drops attaching to the vibrating wings is similar to those on static wings. With the vibration of the wings, fog drops move along the RO direction (Figure 2.23a). The side view images (Figure 2.23b) show that transportation of the drops on the vibrating wings also contains three phases: contacting, coalescence and bouncing. Analysis of a drop's behavior on static wings gives a possible explanation for this phenomenon. As the TCL is fragmented along the RO direction but continous against it, the dewetting of the drops is easier along the RO direction than opposite the RO direction. Thus an unbalanced surface tension occurs to the fog drops on the asymmetric wings, pointing to the RO direction.[59] With the help of the vibration and released surface energy, the drops can bounce along the RO direction.[60] The relationship between the frequency and the bouncing distance is shown in Figure 2.23c.

2.4 Artificial Self-cleaning Surfaces Inspired by Lotus Leaves and Butterfly Wings

Nature is always the school for scientists and engineers. Inspired by specific multi-scale surfaces with an excellent self-cleaning property, researchers have developed a series of artificial surfaces with a similar structure and similar property.[61–63] The secret of the self-cleaning on lotus leaves and butterfly wings lies in the micro-/nano-structure covering their surfaces. In order to fabricate such structures, lots of methods have been introduced to form the multi-scale surfaces, including self-assembly,[64] CVD,[65–67] bottom-up,[68] spin coating,[69] electrospinning[70] and so on.

Electrospinning is a simple and efficient method to fabricate large-scale functional fibers and membranes, including membranes with a self-cleaning

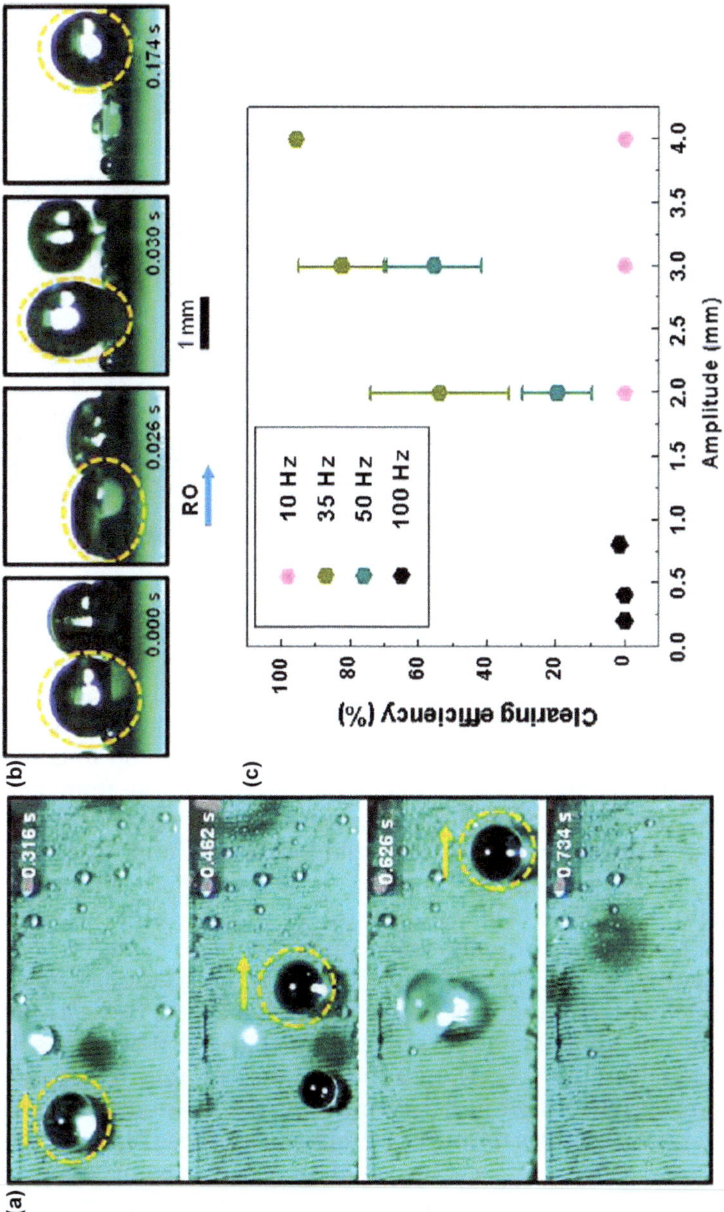

Figure 2.23 A water drop's behavior on vibrating butterfly wings.

Figure 2.24 Self-cleaning membrane fabricated by electrospinning.

property.[70–76] Combining a porous micro sphere and three-dimensional fibers, Jiang *et al.* fabricated a polystyrene (PS) membrane (Figure 2.24). The hydrophobicity mainly depends on the micro spheres while the fiber scaffold is used to enhance mechanical properties such as stability and toughness and grasping the micro spheres in order to prevent them from being washed away.[70] The water contact angle on the membrane is about 160.4°. In addition, they electrospun the mixture solution of polyanion (PAN) and PS and got a similar "fiber-sphere" membrane. On the surface of the sphere are some nano papillars, which can increase the roughness of the surface and lead to a better superhydrophobicity.

A hydrothermal method can be used not only to synthesise various nanocomposites with different amazing functions but also to fabricate nano arrays with superhydrophobicity and self-cleaning. Liu *et al.* fabricated a film made of randomly arranged Sb_2O_3 arrays by taking antimony isopropoxide as the reaction precursor and hexadecylamine as the orientation-directing agent (Figure 2.25).[77] Besides the superhydrophobicity, these films also gain a small sliding angle, which meets the standard of self-cleaning.

If we take the electrospinning and hydrothemal method as the way to fabricate structures inspired by lotus leaves, replication is the way to copy the exact structure. Utilizing natural lotus leaves as a replication template and poly(dimethylsiloxane) (PDMS) as the replication material, the morphology of the upper surface can be reproduced onto the PDMS

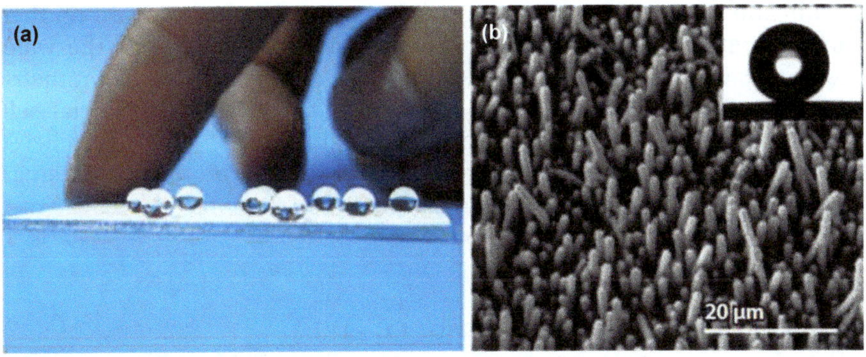

Figure 2.25 Sb_2O_3 arrays and their superhydrophobicity.

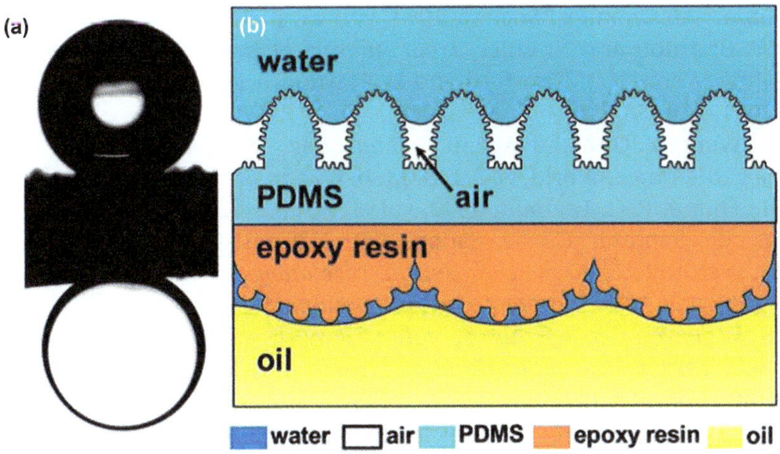

Figure 2.26 Janus artificial membranes.

surface.[33] Through a two-step replication, the lower surface can also be duplicated with epoxy resin. When water and oil are dispensed onto the replicated upper and lower surfaces respectively, they represent the same superhydrophobicity and superleophobicity as the natural lotus leaves (Figure 2.26).[33]

2.5 Conclusion

Lotus leaves and butterfly wings are two common natural self-cleaning surfaces. Due to the micro/nanostructure on the surfaces of lotus leaves, water has a large contact angle and low adhesion force on their surfaces, making it easily to roll off and wash away dust.[47] Different from lotus leaves with a static and isotropic self-cleaning property, the roll-off behavior of water drops on butterfly wings is anisotropic, which means the water drops can only roll off along the RO direction but will get pinned against the

RO direction. Butterflies can also get rid of water drops condensed on their wings by vibrating them. With the growth and coalescence of the water drops, they will bounce along the RO direction as a result of the asymmetric structure of the scale. Inspired by natural surfaces, various artificial self-cleaning surfaces have been designed and fabricated.[61,78,79] It is believed that these surfaces will play a significant role in the production of smart devices, water transportation and new generation sensors.

References

1. J. Shen-Miller, *Seed Sci. Res.*, 2002, **12**, 131.
2. J. Shen-Miller, M. B. Mudgett, J. W. Schopf, S. Clarke and R. Berger, *Am. J. Bot.*, 1995, 1367.
3. E. J. Rulon and H. D. Robert, *Contact Angle, Wettability, and Adhesion*, American Chemical Society, 1964, vol. 43, p. 112.
4. W. Barthlott and N. Ehler, *Trop. Subtrop. Pflanzenwelt*, 1977, **19**, 367.
5. H. K. Webb, R. J. Crawford and E. P. Ivanova, *Adv. Colloid Interface Sci.*, 2014, **210**, 58.
6. G. Wang, Z. Guo and W. Liu, *J. Bionic Eng.*, 2014, **11**, 325.
7. S. g. d. France, World, C. f. t. G. M. o. t. *Mémoires de la Société géologique de France*, F.-G. Levrault, 1842, vol. 5.
8. G. de Saporta, G. de Saporta, F. Naturalist, G. de Saporta and F. Naturaliste, *Recherches sur la végétation du niveau aquitanien de Manosque*, Librairie Polytechnique Baudry, 1891.
9. E. W. Berry, *Geologic history indicated by the fossiliferous deposits of the Wilcox group (Eocene) at Meridian*, Mississippi, Washington Government Printing Office, 1917.
10. F. H. Knowlton, The flora of the Denver and associated formations of Colorado, US Gov't. Print. Off., 1930.
11. Y. Li, J. Yao and C. Li, *Acta Geol. Sin. (Engl. Ed.)*, 2014, **88**, 1252.
12. A. Hollick, *Bull. Torrey Bot. Club*, 1894, 49.
13. C. A. Hollick, *The Cretaceous Flora of Southern New York, and New England*, US Government Printing Office, 1906, vol. 50.
14. P. Ackery, Symposia of the Royal entomological Society of London 1984.
15. P. R. Ackery, R. de Jong and R. I. Vane-Wright, *Volume 1: Evolution, Systematics, and Biogeography*, 1998, vol. 4, p. 263.
16. D. Carter and N. Kristensen, *Lepidoptera, Moths and Butterflies*, 1999, vol. 1, 26.
17. M. J. Scoble, *The Lepidoptera. Form, Function and Diversity*, Oxford University Press, 1992.
18. J.-C. Sohn, C. Labandeira, D. Davis and C. Mitter, *Zootaxa*, 2012, **3286**, 1.
19. C. Neinhuis and W. Barthlott, *Ann. Bot.*, 1997, **79**, 667.
20. W. Barthlott and C. Neinhuis, *Planta*, 1997, **202**, 1.
21. N. A. Patankar, *Langmuir*, 2004, **20**, 8209.
22. L. Feng, S. Li, Y. Li, H. Li, L. Zhang, J. Zhai, Y. Song, B. Liu, L. Jiang and D. Zhu, *Adv. Mater.*, 2002, **14**, 1857.

23. S. Pechook and B. Pokroy, *Adv. Funct. Mater.*, 2012, **22**, 745.
24. Y. Zheng, D. Han, J. Zhai and L. Jiang, *Appl. Phys. Lett.*, 2008, **92**, 084106.
25. B. B. Mandelbrot, *The Fractal Geometry of Nature*, Macmillan, 1983, vol. 173.
26. W. Barthlott, C. Neinhuis, D. Cutler, F. Ditsch, I. Meusel, I. Theisen and H. Wilhelmi, *Bot. J. Linn. Soc.*, 1998, **126**, 237.
27. B. Bhushan and Y. C. Jung, *Nanotechnology*, 2006, **17**, 2758.
28. B. Bhushan and Y. C. Jung, *J. Phys.: Condens. Matter*, 2008, **20**, 225010.
29. Z. Burton and B. Bhushan, *Ultramicroscopy*, 2006, **106**, 709.
30. B. Bhushan, Y. Wang and A. Maali, *Langmuir*, 2009, **25**, 8117.
31. W. Zhao, L. Wang and Q. Xue, *J. Phys. Chem. C*, 2010, **114**, 11509.
32. J. Gao, Y. Liu, H. Xu, Z. Wang and X. Zhang, *Langmuir*, 2010, **26**, 9673.
33. Q. Cheng, M. Li, Y. Zheng, B. Su, S. Wang and L. Jiang, *Soft Matter*, 2011, **7**, 5948.
34. K. Koch, B. Bhushan and W. Barthlott, *Prog. Mater. Sci.*, 2009, **54**, 137.
35. J. M. Osborn and E. L. Schneider, *Ann. Mo. Bot. Gard.*, 1988, 778.
36. H. Ghiradella, *Microsc. Res. Tech.*, 1994, **27**, 429.
37. P. P. Goodwyn, Y. Maezono, N. Hosoda and K. Fujisaki, *Naturwissenschaften*, 2009, **96**, 781.
38. G. D. Bixler and B. Bhushan, *Soft Matter*, 2012, **8**, 11271.
39. S. Niu, B. Li, Z. Mu, M. Yang, J. Zhang, Z. Han and L. Ren, *J. Bio. Eng.*, 2015, **12**, 170.
40. Y. Fang, T. Wang, G. Sun and Q. Cong, http://europepmc.org/.
41. C. Qian, C. Guang-hua, F. Yan and R. Lu-quan, *J. Bio. Eng.*, 2004, **1**, 249.
42. L.-y. Wu, Z.-w. Han, Z.-m. Qiu, H.-y. Guan and L.-q. Ren, *J. Bio. Eng.*, 2007, **4**, 47.
43. Y. Zheng, Q. Cheng, Y. Hou and Y. Chen, *Bio-Inspired Wettability Surfaces: Developments in Micro- and Nanostructures*, CRC Press, Boca Raton, 2015.
44. Y. Zheng, X. Gao and L. Jiang, *Soft Matter*, 2007, **3**, 178.
45. H. Mei, D. Luo, P. Guo, C. Song, C. Liu, Y. Zheng and L. Jiang, *Soft Matter*, 2011, **7**, 10569.
46. C. Liu, J. Ju, Y. Zheng and L. Jiang, *ACS Nano*, 2014, **8**, 1321.
47. A. Lafuma and D. Quéré, *Nat. Mater.*, 2003, **2**, 457.
48. X. Gao and L. Jiang, *Nature*, 2004, **432**, 36.
49. M. Callies and D. Quéré, *Soft Matter*, 2005, **1**, 55.
50. C. Extrand, *Langmuir*, 2007, **23**, 1867.
51. M. K. Chaudhury and G. M. Whitesides, *Science*, 1992, **256**, 1539.
52. P. S. Yadav, P. Bahadur, R. Tadmor, K. Chaurasia and A. Leh, *Langmuir*, 2008, **24**, 3181.
53. S. Thoroddsen and K. Takehara, *Phys. Fluids*, 2000, **12**, 1265.
54. J. B. Boreyko and C.-H. Chen, *Phys. Rev. Lett.*, 2009, **103**, 184501.
55. F.-C. Wang, F. Yang and Y.-P. Zhao, *Appl. Phys. Lett.*, 2011, **98**, 053112.
56. H. Tanaka, K. Matsumoto and I. Shimoyama, Intelligent Robots and Systems, 2008. IROS 2008. IEEE/RSJ International Conference on, IEEE: 2008, p. 3095.

57. H. Tanaka, K. Hoshino, K. Matsumoto and I. Shimoyama, Intelligent Robots and Systems, 2005. (IROS 2005). 2005 IEEE/RSJ International Conference on, IEEE: 2005, p. 2706.
58. A. Brodsky, *J. Exp. Biol.*, 1991, **161**, 77.
59. M. Reyssat, F. Pardo and D. Quéré, *EPL*, 2009, **87**, 36003.
60. S. Daniel and M. K. Chaudhury, *Langmuir*, 2002, **18**, 3404.
61. T. Jiang, Z. G. Guo and W. M. Liu, *J. Mater. Chem. A*, 2015, **3**, 1811.
62. X. D. Shi, R. M. Dou, T. Z. Ma, W. Y. Liu, X. H. Lu, K. J. Shea, Y. L. Song and L. Jiang, *ACS Appl. Mater. Interfaces*, 2015, **7**, 18424.
63. X. Jin, B. R. Shi, L. C. Zheng, X. H. Pei, X. Y. Zhang, Z. Q. Sun, Y. Du, J. H. Kim, X. L. Wang, S. X. Dou, K. S. Liu and L. Jiang, *Adv. Funct. Mater.*, 2014, **24**, 2721.
64. W. K. Cho, S. Park, S. Jon and I. S. Choi, *Nanotechnology*, 2007, **18**, 395602.
65. A. Pakdel, C. Zhi, Y. Bando, T. Nakayama and D. Golberg, *ACS Nano*, 2011, **5**, 6507.
66. C.-T. Hsieh and W.-Y. Chen, *Carbon*, 2010, **48**, 612.
67. Y. Zhang, L. Stan, P. Xu, H.-L. Wang, S. K. Doorn, H. Htoon, Y. Zhu and Q. Jia, *Carbon*, 2009, **47**, 3332.
68. H. Yang and P. Jiang, *Langmuir*, 2010, **26**, 12598.
69. C.-F. Wang, W.-Y. Chen, H.-Z. Cheng and S.-L. Fu, *J. Phys. Chem. C*, 2010, **114**, 15607.
70. L. Jiang, Y. Zhao and J. Zhai, *Angew. Chem.*, 2004, **116**, 4438.
71. Y. Zhu, J. Zhang, Y. Zheng, Z. Huang, L. Feng and L. Jiang, *Adv. Funct. Mater.*, 2006, **16**, 568.
72. Y. Zhu, J. C. Zhang, J. Zhai, Y. M. Zheng, L. Feng and L. Jiang, *ChemPhysChem*, 2006, **7**, 336.
73. X. Li, B. Ding, J. Lin, J. Yu and G. Sun, *J. Phys. Chem. C*, 2009, **113**, 20452.
74. X. Wang, B. Ding, J. Yu and M. Wang, *Nano Today*, 2011, **6**, 510.
75. J. Lin, Y. Cai, X. Wang, B. Ding, J. Yu and M. Wang, *Nanoscale*, 2011, **3**, 1258.
76. L. Valtola, A. Koponen, M. Karesoja, S. Hietala, A. Laukkanen, H. Tenhu and P. Denifl, *Polymer*, 2009, **50**, 3103.
77. K. Liu, J. Zhai and L. Jiang, *Nanotechnology*, 2008, **19**, 165604.
78. C. L. C. Tan, K. Sapiha, Y. F. H. Leong, S. Choi, F. Anariba and B. J. R. Thio, *Soft Matter*, 2015, **11**, 5400.
79. S. H. Liu, X. J. Liu, S. S. Latthe, L. Gao, S. An, S. S. Yoon, B. S. Liu and R. M. Xing, *Appl. Surf. Sci.*, 2015, **351**, 897.

CHAPTER 3

Transparent Superamphiphobic Coatings: Structure, Fabrication and Perspective

ZHI GENG[a,b] AND JUNHUI HE*[a]

[a] Functional Nanomaterials Laboratory, Center for Micro/Nanomaterials and Technology and Key Laboratory of Photochemical Conversion and Optoelectronic Materials, Technical Institute of Physics and Chemistry, Chinese Academy of Sciences, Zhongguancundonglu 29, Haidianqu, Beijing 100190, China; [b] University of Chinese Academy of Sciences, Beijing 100864, China

*Email: jhhe@mail.ipc.ac.cn

3.1 Introduction

Self-cleaning materials, including superhydrophilic coatings and superhydrophobic coatings, have gained much attention in society. The wettability of such coatings can keep their substrates clean with the help of rain. Maintenance can be reduced, and thus time and cost may be saved if substrates have a self-cleaning property. For superhydrophobic coatings, two models have been proposed, including the Wenzel model and the Cassie–Baxter model. In the Wenzel model,[1] a water droplet is assumed to make direct contact with the surface, accounting for surfaces with a large contact angle but high contact angle hysteresis. Such coatings usually do not have a self-cleaning property as water droplets stay on their surfaces. In the Cassie–Baxter model,[2] a microscopic air chamber exists between the water droplet and the surface texture. The special structure and low surface

RSC Smart Materials No. 21
Self-cleaning Coatings: Structure, Fabrication and Application
Edited by Junhui He

Published by the Royal Society of Chemistry, www.rsc.org

energy of superhydrophobic coatings can make water droplets suspended on the surface stable and roll off easily, which better accounts for superhydrophobicity. In recent years, we have witnessed significant advances in building superhydrophobic surfaces.[3–10]

Superhydrophobic coatings cannot, however, avoid contact with oils or other organic liquids in practical applications. Such low energy liquids may destroy the air chamber in the surface texture and lead to transformation from the Cassie–Baxter model to the Wenzel model. As a result, the coatings will lose superhydrophobicity. Fabricating superamphiphobic coatings with both superhydrophobicity and superoleophobicity may solve the problem effectively. As compared to fabricating superhydrophobic coatings, many more challenges must be met in building superamphiphobic coatings. Two methods would be helpful for building superamphiphobic coatings. One is to create a large surface roughness and low energy surface with surface tension less than a quarter of the surface tension of oils.[11–13] The other is to construct a rough surface with a "re-entrant" or "overhanging" structure, which could fabricate oleophobic surfaces with oleophilic materials.[14,15] As superamphiphobic coatings have usually been fabricated by combining the advantages of the two methods, they will be discussed comprehensively in the following sections without distinguishing between the two methods.

Materials with high transmittance are widely used in different fields. Coatings combining high transmittance and superamphiphobicity may be applied on varied transparent substrates, such as lenses, goggles, windshields, and the protecting panels of solar cells, which would largely expand their range of applications. Although superamphiphobic coatings possess a large surface roughness (commonly larger than 1 μm)[14–16] and thickness, which is contradictory to high transmittance coatings (smooth surface and proper film thickness),[3] several groups have succeeded in fabricating superamphiphobic coatings with high transmittance *via* different methods.[13,17–30]

This review aims to provide an account of the recent development in fabricating superamphiphobic coatings with high transmittance. In the second section, we illustrate two principle surface textures of superamphiphobic coatings. In the third section, we summarize the fabrication methods of superamphiphobic coatings with high transmittance in recent research. Finally, we conclude by pointing out challenges that remain in this field and the research direction to meet the challenges. An outlook is also made toward their future prospects.

3.2 Superamphiphobic Coatings: Re-entrant and Overhang Surface Structures and Their Effects on Light Transmittance

Special surface textures are required to build superamphiphobic coatings. Tuteja *et al.* first put forward the concept of re-entrant surface texture, which could achieve superamphiphobic surfaces with oleophilic materials.[31]

Fluorodecyl polyhedral oligomeric silsesquioxane (POSS)-poly(methyl methacrylate) (PMMA) was used to fabricate a superoleophobic surface by electrospinning. According to the Cassie equation, $\cos\theta^* = -1 + \varphi_s(1 + \cos\theta)$, where θ^* is the apparent contact angle on the textured surface, θ is the equilibrium contact angle on a smooth surface of the same material, and φ_s is the fraction of the solid in contact with the liquid. The electrospun fibers exhibited excellent oleophobicity ($\theta^* \gg 90°$). However, the corresponding spin-coated surfaces were oleophilic ($\theta < 90°$) (Figure 3.1a). Another re-entrant structure was fabricated *via* silica deposition and a two-step etching process, which was called a micro-hoodoo structure (Figure 3.1b). These surfaces are considered to be in a metastable Cassie state.[14,15] Two important parameters could be used to design these re-entrant surface textures. For the electrospun fibers, the first one is spacing ratio $D^* = (R + D)/R$, where R is the radius of the nanofibers and $2D$ is the average edge-to-edge spacing of the nanofibers. The second is the robustness parameter $H^* = 2(1 - \cos\theta)Rl_{cap}/D^2$ $[l_{cap} = (\gamma_{lv}/\rho g)^{1/2}]$, which measures the robustness of the metastable Cassie state with respect to the fluid properties, equilibrium contact angle, and surface geometry. In the formula, γ_{lv} refers to the interfacial tension of liquid and vapor phases, ρ is the liquid density, g is the acceleration due to gravity (Figure 3.1a). For the micro-hoodoo geometry, $D^* = (W + D)/W$ and $H^* = 2[(1 - \cos\theta)R + H]l_{cap}/D^2$ (Figure 3.1b). The interface is thought to be metastable when the parameters $D^* \gg 1$ and $H^* \gg 1$.[14,31]

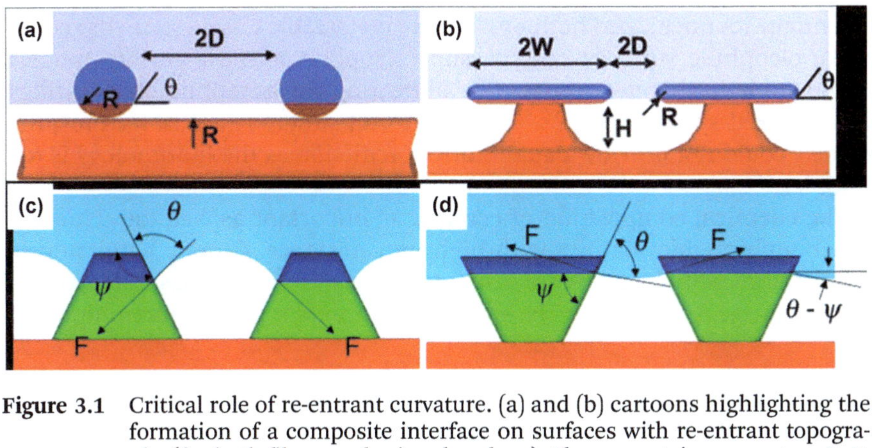

Figure 3.1 Critical role of re-entrant curvature. (a) and (b) cartoons highlighting the formation of a composite interface on surfaces with re-entrant topography (for both fibers and micro-hoodoos). The geometric parameters R, D, H, and W characterizing these surfaces are also shown. The blue surface is wetted while the red surface remains non-wetted when in contact with a liquid whose equilibrium contact angle is θ (<90°).[14,22] Reproduced from ref. 22 with permission from John Wiley and Sons. Copyright © 2015 WILEY-VCH Verlag GmbH & Co. KGaA, Weinheim. (c) and (d) Schematic diagrams illustrating possible liquid–vapor interfaces on two different surfaces with the same solid surface energy and the same equilibrium contact angle (θ), but different geometric angles (ψ).[31,32] Copyright 2008 Proceedings of the National Academy of Sciences U.S.A.

For the metastable Cassie state, by the time the liquid makes contact with the surface, the net force on the interface of three phases (air, liquid and solid) is toward the outside of the structure due to a special surface texture (inverted trapezoidal structure), which could prevent the liquid from penetrating into the surface structure. As a result, the surface made by oleophilic materials can provide superoleophobicity (Figure 3.1d). In contrast, the net force of the coating with the structure of a trapezoidal surface texture is toward the inside on the interface of three phases (Figure 3.1c), which would lead to a superoleophilic surface.[31,32]

Gao's group fabricated an overhanging surface structure, which could also make a hydrophilic surface become superhydrophobic.[15] The capillary force would provide an energy barrier that prevents water from entering into the surface texture and allow it to stay in the metastable Cassie state. Three cross-sectional profiles of a liquid in contact with a solid indent consisting of overhanging structures are shown in Figure 3.2, which could illustrate the mechanism of the special surface structure. When $\theta_{\text{overhang}} > \theta_{\text{flat}}$, the interface of the water and air phases is concave, the net force generated by the meniscus (F_s) is toward the inside of the indent, causing water to penetrate into the structure. As a result, the surface turns to the Wenzel state. When $\theta_{\text{overhang}} = \theta_{\text{flat}}$, the interface of the water and air phases is flat, F_s could be ignored. When $\theta_{\text{overhang}} < \theta_{\text{flat}}$, the interface of the water and air phases is convex, and F_s is toward the outside of the structure, which could prevent the liquid from making contact with the solid surface.[15,32,33]

It must be noted that the re-entrant and overhang surface models are essentially identical, and both provide the metastable Cassie state that could make oleophilic surfaces become superoleophobic. These surface models would guide further development in fabricating superamphiphobic coatings with ordered surface textures. However, for disordered surface textures, the models could not provide a quantitative prediction as the morphology is too complex.

The chemical composition of coatings is important as well for achieving superamphiphobic coatings with high transmittance. On one hand, appropriate fluorocarbon silanes are usually used to modify coatings to create low energy surfaces,[14,33] which is a basic requirement of superamphiphobic

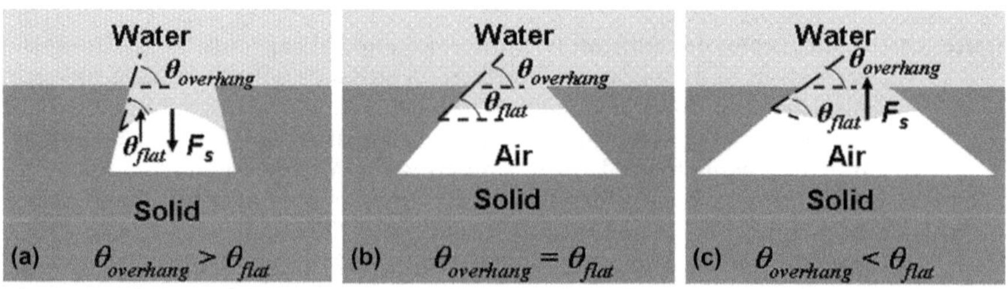

Figure 3.2 Three cross-sectional profiles of water in contact with a solid indent consisting of overhanging structures.
Reproduced with permission from L. Cao, H. H. Hu and D. Gao, *Langmuir*, 2007, **23**, 4310. Copyright (2007) American Chemical Society.

coatings.[11–13] On the other hand, matrices of high refractive index should generally be avoided with respect to coatings of high transmittance.

For coatings with high transmittance, three factors would affect light transmittance. The first is the light absorption of coatings, which is determined by the compositions of both coating and substrate. Tailoring these compositions, *e.g.*, removal of an iron impurity, could decrease the light absorption of coating and substrate.[34] The second factor that affects light transmittance is the light reflection of the substrate surface. Based on recent research, coatings with a proper film thickness and refraction index can significantly suppress light reflection from the substrate surface and present an antireflective property.[35–37] The light reflection could even be negligible because of antireflective coatings.[38,39] Last but not least, light scattering may seriously affect the light transmittance of coatings, especially for superamphiphobic coatings. The particle sizes in superamphiphobic coatings contain rc entrant or overhang surface structures that are commonly much larger than 1 μm.[14,31,33] Based on recent research, the Rayleigh and Mie scattering in the visible region could not be negligible when the size of particles in coatings are larger than 100 nm.[40,41] For Rayleigh scattering, the intensity of scattered light (I) could be estimated by the formula: $I/I_0 = \dfrac{1+\cos^2\theta}{2S^2}\left(\dfrac{2\pi}{\lambda}\right)^4\left(\dfrac{n^2-1}{n^2+2}\right)^2\left(\dfrac{d}{2}\right)^6$, where I_0 is the intensity of incident light, S is the distance between particle and detector (typically a few centimeters), n is the refractive index of the particle, d is the diameter of the particle, and λ is the wavelength of light. When the particle sizes are larger than the wavelength λ, the Mie scattering is dominant in light scattering. The total scattering cross section of Mie scattering could be expressed as $\sigma_M = \dfrac{\lambda^2}{2\pi}\sum_{m=1}^{\infty}(2m+1)\left(|a_m|^2+|b_m|^2\right)$, where a_m and b_m are the Mie coefficients, representing the magnetic and electric multipoles of order m, respectively.[40] They are functions of d and n. Figure 3.3a and b show the contribution of particle diameters to the degree of Rayleigh scattering and Mie scattering cross sections from materials with different refractive indices. It is noted that light scattering becomes much more serious with an increase in particle sizes. Therefore, to fabricate superamphiphobic coatings with high transmittance, the surface roughness of coatings must be largely reduced, which would pose significant challenges to their fabrication. Research has recently proved that light scattering caused by pillar edges could be reduced by increasing the D^* in the re-entrant surface structure.[16]

3.3 Approaches to Building Superamphiphobic Coatings with High Transmittance

3.3.1 Colloidal Lithography and Plasma Nanotexturing

Colloidal lithography, which is high throughput and cost-effective, has been successfully used to fabricate thin films.[42,43] For example, Gogolides's group

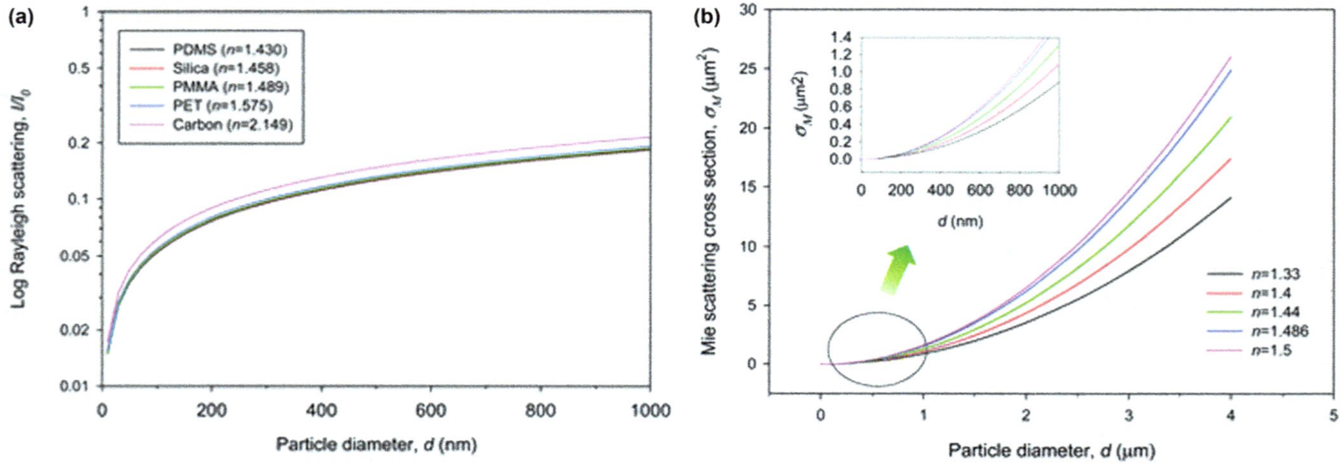

Figure 3.3 The contribution of particle diameters to the degree of Rayleigh scattering (a) and the Mie scattering cross sections (b) from materials with different refractive indices.
Reproduced from ref. 40 with permission from the Royal Society of Chemistry.

fabricated superamphiphobic to amphiphilic surfaces with colloidal lithography and plasma nanotexturing.[18] Polystyrene (PS) colloidal particles were deposited on poly (methyl methacrylate) (PMMA) substrates *via* spin-coating. O_2 plasma ion etching was used in a high-density plasma reactor, and the conditions were as follows: 1900 W, 100 sccm O_2, 0.75 Pa, and a moderate bias voltage (usually −80 V). After O_2 plasma etching, the structured surface was exposed to C_4F_8 plasma in the same reactor for the conformal deposition of a thin fluorocarbon (FC) film (900 W, 0 V, 5.33 Pa C_4F_8, deposition rate 30 nm min^{-1}). The surface morphology is shown in Figure 3.4a and b.

The contact angle of the coating was 168° for water and 153° for diiodomethane; the contact angle hysteresis was 2° for water and 9° for diiodomethane. The surface became superhydrophilic and superoleophilic after treatment with O_2-plasma for even as short as 10 s. The coating was transparent, which is shown in Figure 3.4c. However, a large surface

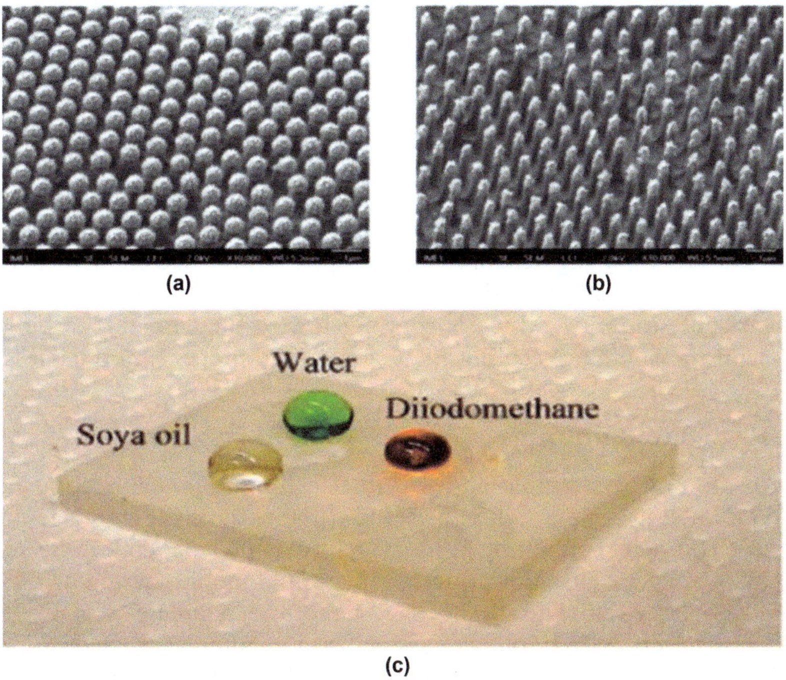

Figure 3.4 Ordered micropillars produced after colloidal lithography of 1 μm PS particles on a PMMA substrate and one etching step in high-density oxygen plasma. (a) 1 min etch, −80 V bias; (b) 2 min etch, −80 V bias. (Other conditions: 1900 W, 100 sccm O_2, 0.75 Pa.); (c) fluorocarbon-coated, oxygen plasma treated superamphiphobic surface under optimal conditions.[18]
Reproduced with permission from K. Ellinas, A. Tserepi and E. Gogolides, *Langmuir*, 2011, **27**(7), 3960. Copyright (2010) American Chemical Society.

roughness results in high light scattering, which thus reduces the transmittance of the coating. Furthermore, as limited by the materials in the fabrication, the robustness of the coating needs to be improved.

3.3.2 3-D Diffuser Lithography

3-D diffuser lithography had been used to make large superhydrophobic coatings, which had a perfectly ordered microbowl and microlens array.[44] Im *et al.* fabricated a superamphiphobic surface with an inverse trapezoidal microstructure on a large transparent flexible substrate.[17] 3-D diffuser lithography was used to fabricate a photoresist mold for duplicating inverse-trapezoidal microstructures by using the soft lithography of PDMS. Spin-coating was used to fabricate a negative photoresist on a metal-patterned glass substrate. A self-assembled monolayer was formed on the coating with a silanizing agent (tridecafluoro-1,1,2,2-tetrahydrooctyl trichlorosilane, CF_3-$(CF_2)_5(CH_2)_2$-$SiCl_3$). Figure 3.5a–d shows the schematic diagram of the backside 3-D diffuser lithography and the surface morphology of the coating. The contact angle of the coating was 153° for water and 135° for methanol (Figure 3.5e–f). The light transmittance of the coating was approximately 85% in the range of 400 nm to 700 nm (Figure 3.5g). The surface morphology fits the re-entrant surface texture, which contributes to the superamphiphobicity of the coating. The superhydrophobicity of the surface is stable, which could withstand the impinging test of a water droplet at high speed. However, the robustness of the surface, which could merely withstand the impinging test of a water droplet, is not good enough to satisfy the needs in practical applications. Moreover, the large surface structure leads to high light scattering, which is contradictory to the demands of building high light transmittance coatings (flat surface and proper film thickness).

3.3.3 Electrospinning

Electrospinning is a simple and versatile technique for fabricating nanofibers on a large scale.[45,46] It consists of three parts: a high voltage power supply, a spinneret containing a sol–gel solution and a flat collector. Ganesh *et al.* fabricated a transparent robust superamphiphobic coating using SiO_2 nanofibers as a template.[21] The SiO_2 nanofiber template was produced by electrospinning SiO_2 on a glass substrate. After that, deposition of triethoxysilane (TS) was carried out inside a desiccator with two bottles of 2 mL ammonia and 2 mL TS solution, respectively. A hybrid silica network containing SiO_2 nanofibers enclosed by an ultrathin porous silica membrane after coating was heated at 600 °C for 2 h. The amphiphobic modification of the coating was carried out by placing the calcinated specimen into a desiccator with a glass bottle containing 40 mL of fluorinated silane for 2 h. Figure 3.6a–d shows the surface morphology of the as-spun SiO_2 nanofibers and the hybrid silica network (SiO_2 nanofibers–silica membrane). The contact angles and contact angle hysteresis (given in parentheses) of the

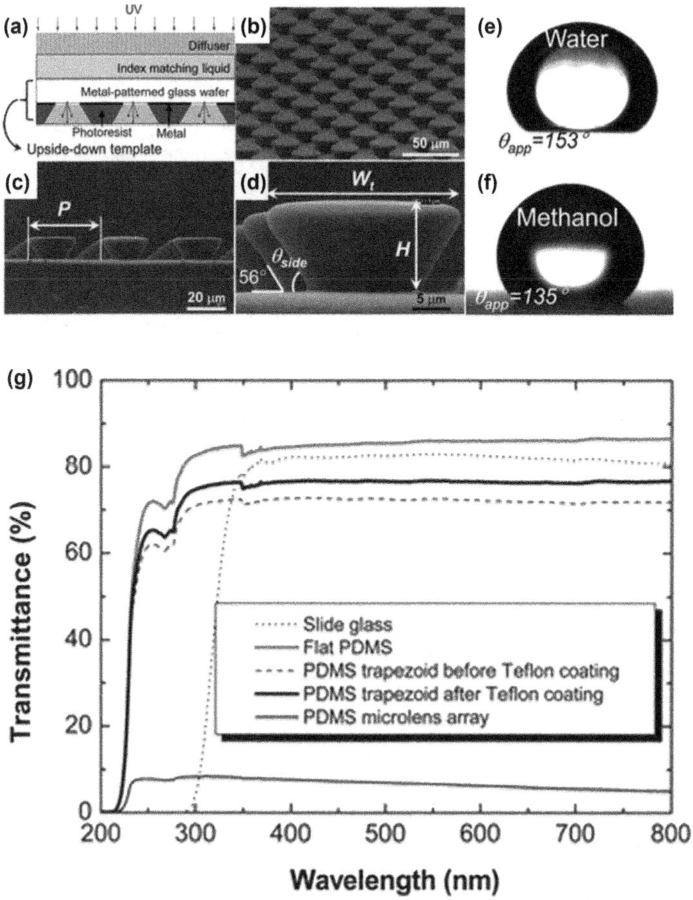

Figure 3.5 (a) A schematic diagram of backside 3-D diffuser lithography. (b)–(d) SEM images of the fabricated PDMS trapezoids with various magnifications and view angles. The dimensions of a trapezoid are $H = 11.9$ μm, $\theta_{side} = 56°$, $P = 40$ μm, and $W_t = 26$ μm. (e)–(f) Contact angles of deionized water and the methanol of the coating. (g) Transmittance characteristics of various samples as a function of the wavelength of the incident light. Reproduced from ref. 17 with permission from the Royal Society of Chemistry.

coating for water and hexadecane were $161° \pm 0.8°$ (2°) and $146.5° \pm 0.7°$ (6°), respectively. The hardness and modulus of the coating were measured to be 1.6 GPa and 24.7 GPa, respectively. However, the transmittance of the transparent superamphiphobic film coated glass substrate is 85%, which has fallen significantly compared to a blank glass substrate (Figure 3.6e). The deposition of the silica membrane leads to the low porosity of coating, which is opposite to the demands of high transmittance coatings, *i.e.*, high porosity. A photograph of water, hexadecane and ethylene glycol droplets on the superamphiphobic surface is shown in Figure 3.6f.

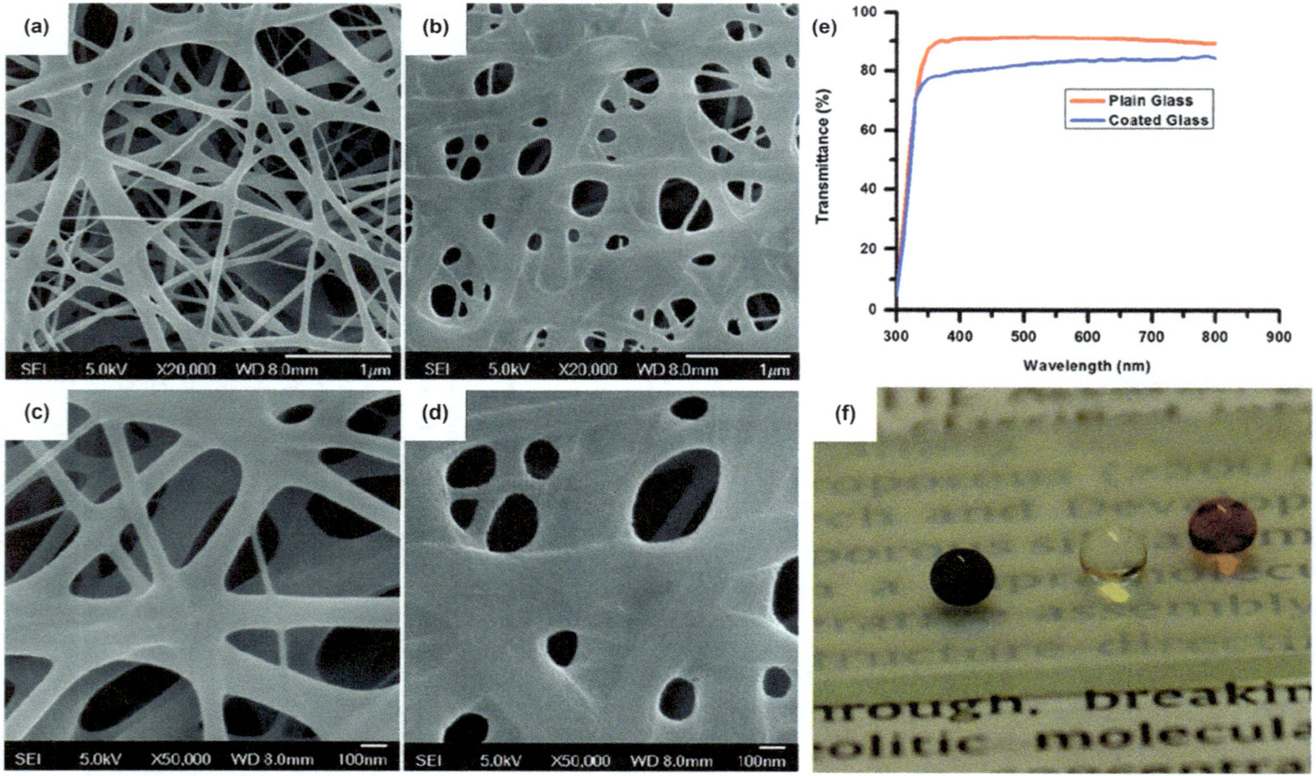

Figure 3.6 (a) and (c) SEM images (low and high magnifications) of the as-spun SiO_2 nanofibers, (b) and (d) SEM images (low and high magnifications) of the hybrid silica network (SiO_2 nanofibers–silica membrane). (e) Comparison of the transmittance of blank and superamphiphobically coated glass substrates. (f) Photograph of water (dyes blue with trypan blue dye), hexadecane (dyed red with alizarin red dye) and ethylene glycol (colourless) droplets on the superamphiphobic surface. Reproduced from ref. 21 with permission from the Royal Society of Chemistry.

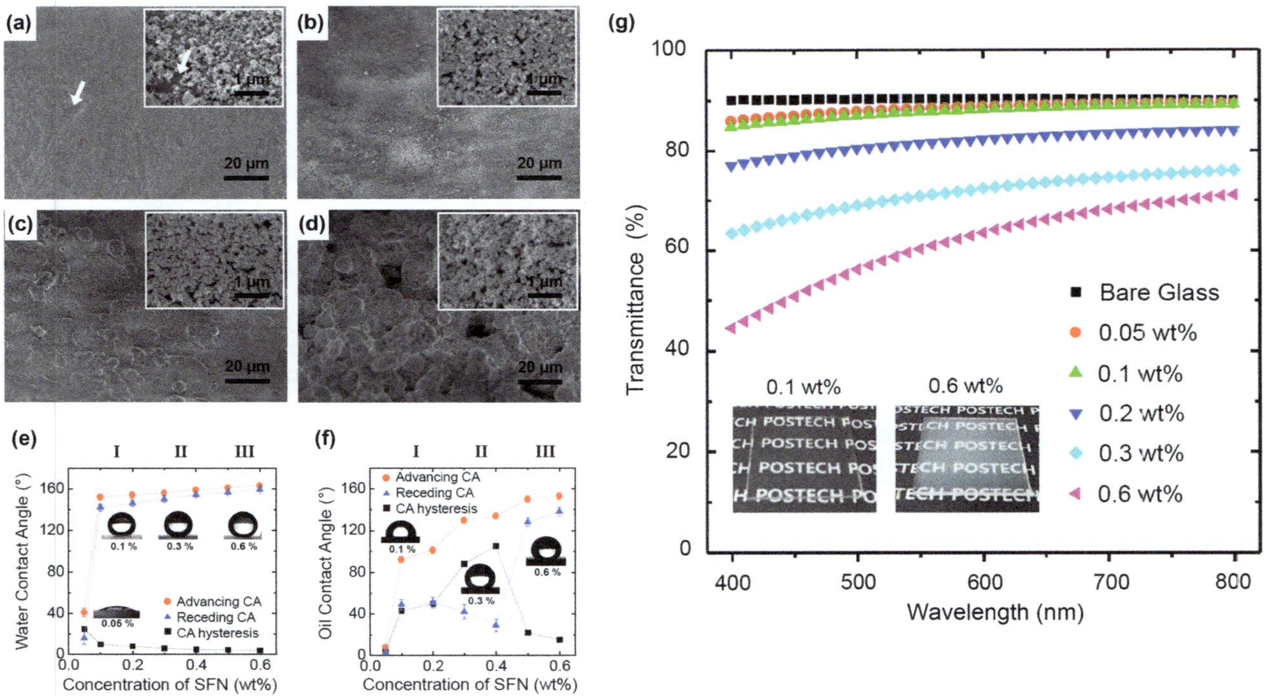

Figure 3.7 SEM images of spray-deposited SFNs on Si wafers with various sol concentrations: (a) 0.05, (b) 0.1, (c) 0.3, and (d) 0.6 wt%. Contact angles of water (e) and hexadecane (f) on surfaces prepared from sols with various SFN concentrations. (g) Transmittance spectra of coatings on glass substrates with SFN sols of various concentrations. The insets show optical images of the coated glass substrates overlaying printed paper.[24]
Reproduced with permission from S. G. Lee, D. S. Ham, D. Y. Lee, H. Bong and K. Cho, *Langmuir* 2013, **29**(48), 15051. Copyright (2013) American Chemical Society.

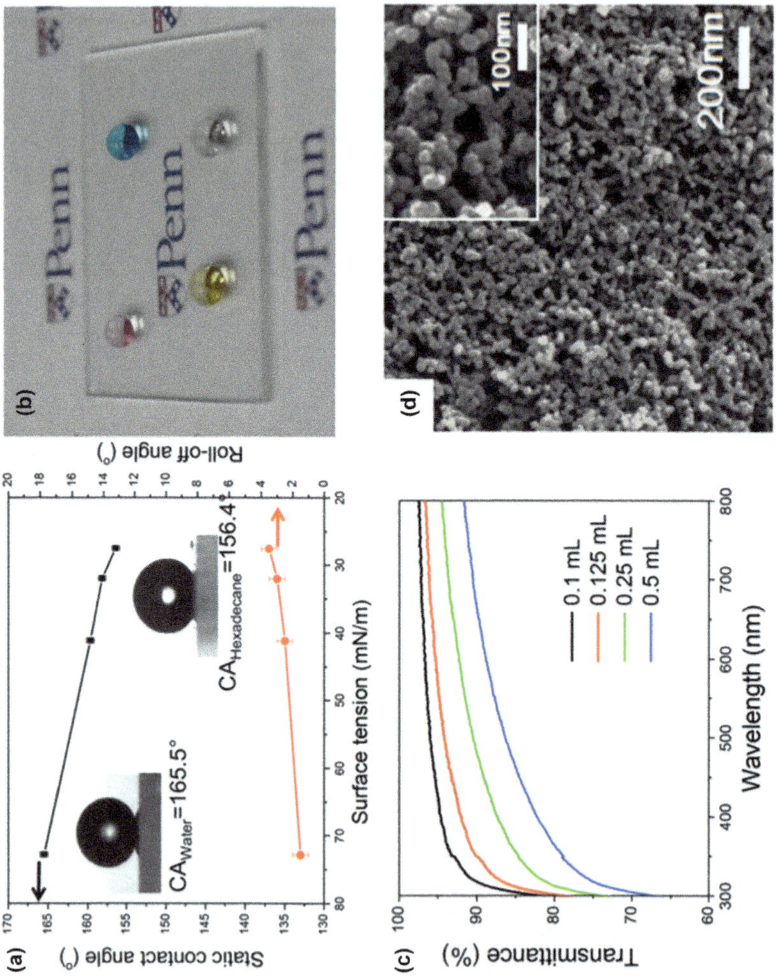
(a)
Static contact angle (°)
Roll-off angle (°)
Surface tension (mN/m)
CA$_{Water}$=165.5°
CA$_{Hexadecane}$=156.4°
(b)
Penn
(c)
Transmittance (%)
Wavelength (nm)
0.1 mL
0.125 mL
0.25 mL
0.5 mL
(d)
100nm
200nm

3.3.4 Spray-coating

Spray-coating is a facile method to fabricate coatings on a large scale.[47] Recently, the fabrication of superamphiphobic coatings has been reported by spraying fluoropolymer blends with nanoparticles (or carbon nanotubes) using cosolvents.[24,48–51] The group of Cho used this method to fabricate superamphiphobic transparent coatings *via* spraying silica–fluoropolymer hybrid nanoparticles (SFNs) without any pre- or post-treatment.[24] The assembly of these nanoparticles could create both microscale and nanoscale roughness, and the fluoropolymer decorated on the silica nanoparticles could provide a low surface energy. The surface morphologies of coatings prepared using varied sol concentrations are shown in Figure 3.7a–d, which lead to varied surface wettability and transmittance. With an increase of the SFN concentration, the contact angles of water and hexadecane on the coating increase, while the contact angle hysteresis of water and the transmittance of the coating decrease (Figure 3.7e–g). This approach of building transparent superamphiphobic coatings is easy to implement, and can be applied to fabricate large scale coatings. However, external forces may easily destroy the surface texture of the superamphiphobic coatings. With the help of PDMS solutions, the robustness of the coating increases. However, the transmittances of the coatings are lower than the substrates. The coatings are not mechanically robust as the adhesions between the nanoparticles are not strong enough to resist outer abrasion forces.

Ge *et al.* sprayed strings of amphiphilic silica nanoparticles dispersed in an amphiphilic sol solution to fabricate transparent superamphiphobic coatings.[51] The contact angles and roll-of angles (given in parentheses) of the water and hexadecane of the coating were 165.6° (1.5°) and 156.4° (3.5°) (Figure 3.8a), respectively. The coating also possessed high transparency. Figure 3.8b and c show a photograph and light transmittance of the coating. Figure 3.8d shows the coating's surface morphology.

Figure 3.8 (a) SCAs and RAs of water ($\gamma = 72.8$ mN m^{-1}), propylene carbonate ($\gamma = 41.1$ mN m^{-1}), olive oil ($\gamma = 32.0$ mN m^{-1}), and hexadecane ($\gamma = 27.5$ mN m^{-1}) droplets on the 5 μm thick amphiphobic coating. Insets: optical images of a 5 μL water droplet and hexadecane droplet on the coating surface. (b) Photograph of four droplets, including dyed water (red), dyed propylene carbonate (blue), olive oil (yellow), and hexadecane (colorless), deposited on a superamphiphobic glass slide. (c) UV-Vis transmittance of the coatings sprayed from different amount of solutions on glass slides. Bare glass was used as a reference. (d) SEM images of the superamphiphobic coating with network of stringed silica NPs, which was taken after the sputtering of Au. The as-prepared silica NP/sol ethanol solution (0.25 mL, TEOS–HDFTES = 2 : 1, v/v) was sprayed on a 1×1 in.^{2} glass slide with a film thickness of ≈ 5 μm. Inset: high-resolution SEM image showing the fractal, nanoporous structure. Reproduced from ref. 51 with permission from John Wiley and Sons.

Our group fabricated robust superamphiphobic coatings with high transmittance *via* dip-coating solid silica nanoparticles, spray-coating silica hollow spheres, followed by spray-coating a binder solution composed of silica nanosheets and acid-catalyzed silica sol. The water contact angle and ethylene glycol contact angle of the coatings reached as high as 171° and 150°, respectively. The maximum transmittance of coated glass reached 96.2% at a wavelength of 550 nm, which is much higher than that of blank glass (91.2%). With the help of the binder solution, the coatings demonstrated excellent mechanical robustness. They fell at the 5A level in the tape adhesion test based on ASTM D3359-93, and could endure the 5H pencil scratching test. They also retained their superhydrophobicity after a tape peeling test, sand abrasion test, water drop impact test, and outdoor durability test for five months.[52]

3.3.5 Micromolding

Micromolding is a cost-effective and low-expertise approach to mass fabrication of patterned polymer surfaces, which could generate polymer replicas without limitation from the same master pattern. Recently, researchers have used the micromolding method to fabricate transparent superamphiphobic coatings.[20,29] Kang *et al.* fabricated a robust superomniphobic coating with high transmittance and durability, using a micromolding treatment to fabricate mushroom-like micropillars on polymeric substrates.[20] Figure 3.9a shows the process of micromolding. Octafluorocyclobutane (C_4F_8) was deposited on the coating to reduce the surface energy by chemical plasma deposition with an ICP etching chamber. The surface morphologies of coatings fabricated on PDMS and PFPE are shown in Figure 3.9b–g. The surface had a re-entrant structure, which contributed to the superamphiphobicity of the coatings. The contact angles of deionized water and ethanol were near 160°, the contact angle hysteresis were less than 10°. The light transmittance of the coating was slightly lower than 90% in the visible spectral range from 380 to 780 nm, while that of bare substrate was 93–94% (Figure 3.9h). The micromolding approach can fabricate accurate surface texture. However, the process of fabrication should be simplified. Furthermore, limited by the materials used in the fabrication, the robustness of surface structures needs to be improved.

3.3.6 Micromolding and Spray-coating

Golovin *et al.* created a transparent superomniphobic surface *via* spraying 1*H*,1*H*,2*H*,2*H*-heptadecafluorodecyl polyhedral oligomeric silsesquioxane (F-POSS) on master-molded polydimethyl siloxane (PDMS) pillars.[16] Silicon master-molds with different square arrays of holes were fabricated by standard photolithography. PDMS pillars were constructed using silicon master molds as templates. Figure 3.10a illustrates the fabrication method

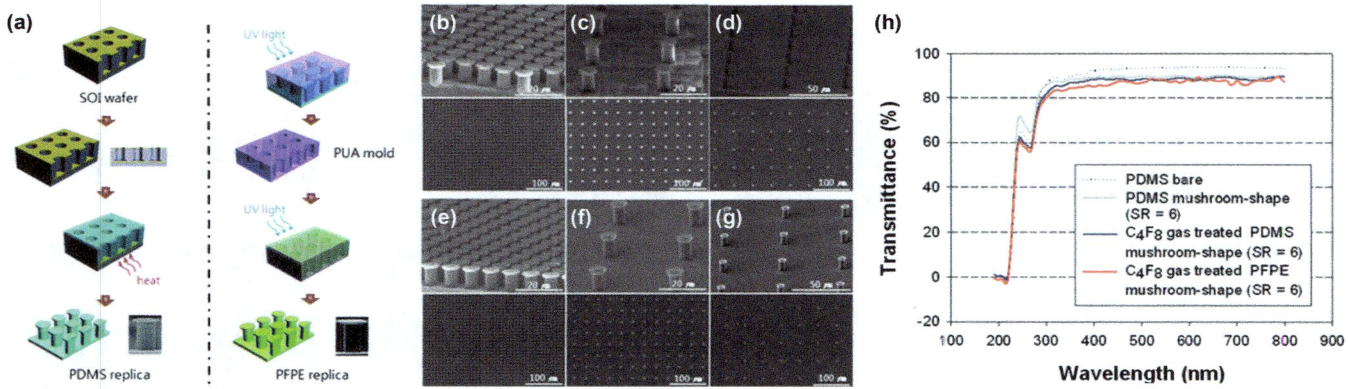

Figure 3.9 (a) Fabrication scheme of PDMS and PFPE mushroom-like micropillar arrays. The bottom SEM images show that the two structures have identical geometry after dual duplication processes. SEM images of the fabricated PDMS (b)–(d) and PFPE micropillars (e)–(g) with side (upper panels) and top views (lower panels), respectively. The micropillars have a 5 μm pillar diameter, 8.5 μm top disk diameter, 10 μm height with different space/diameter (SR): SR = 1 for (b) and (e), 6 for (c) and (f), and 10 for (d) and (g). (h) Transmittance spectra of various samples over a UV-VIS wavelength regime.
Reproduced from ref. 20 with permission from the Royal Society of Chemistry.

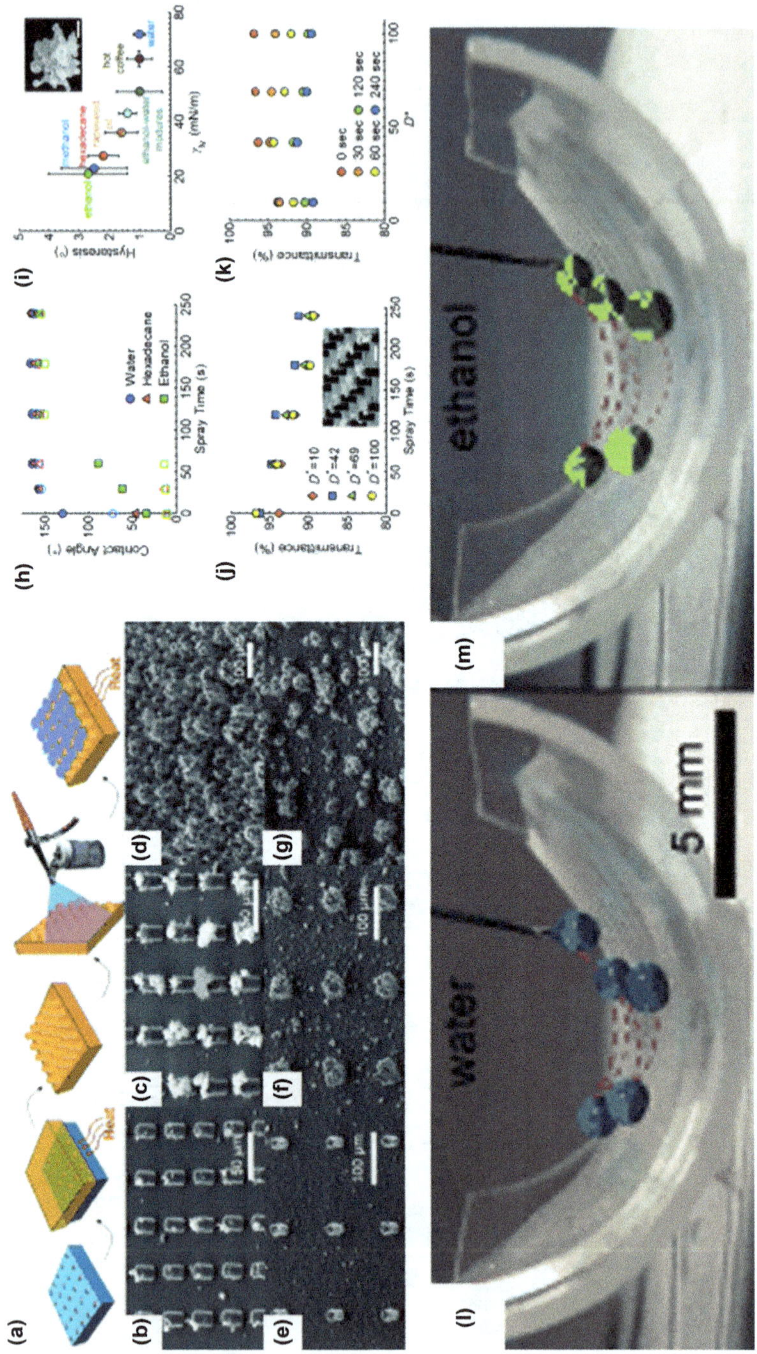
(a)
(b)
(c)
(d)
(e)
(f)
(g)
(h)
(i)
(j)
(k)
(l)
(m)
Contact Angle (°)
Spray Time (s)
Water
Hexadecane
Ethanol
Hysteresis (°)
Transmittance (%)
water
ethanol
5 mm

of surface structures, while Figure 3.10b–g shows their surface morphologies obtained under varied experimental conditions. The surfaces showed high transparency and superomniphobicity. By calculating the surface robustness factors D^* and A^*, the surfaces were proven to be in a metastable Cassie–Baxter state. The transmittances, contact angles and contact angle hysteresis of the surfaces are shown in Figure 3.10h–k. The specimens were flexible, and the contact angle hysteresis was still ultra-low even when the specimens were greatly bended. Although the surface had excellent wetting, optical and bending properties, its surface robustness, anti-abrasion capability and adhesion-to-substrate were not provided in the article, which would play an important role in practical applications of transparent superamphiphobic surfaces.

3.3.7 Template-based Method

The template-based method is an imprint-related way to prepare superhydrophobic surfaces, which involves the use of a matrix with desired features, replication of the features and subsequent lifting off the replica or dissolution of the templates.[53] Deng *et al.* built a transparent superamphiphobic coating using candle soot as a template.[19] A glass slide was coated with candle soot by holding it above the flame of a paraffin candle. Chemical vapor deposition (CVD) of tetraethoxysilane (TEOS) catalyzed by ammonia was used to coat the soot layer with a silica shell. After 24 h TEOS CVD treatment, the coating was calcined at 600 °C for 2 hours to remove the carbon core. The surface was coated with a semifluorinated silane by CVD to reduce the surface energy. The surface morphology of the porous structure is

Figure 3.10 (a) Fabrication method. First, the desired pattern is etched into a silicon master mold using photolithography. PDMS is poured into the master mold and cured in an oven. PDMS/F-POSS is sprayed onto the pillars, which are subsequently cured again. (b)–(d) PDMS pillars ($D^* = 10$) spray-coated with PDMS/F-POSS for 30, 120, and 240 seconds. (e)–(g) PDMS pillars with $D^* = 100$ spray-coated for 30, 120, and 180 seconds. (h)–(i) Contact angle measurements with water, hexadecane, and ethanol for pillars with $D^* = 100$. The filled and open symbols represent advancing and receding contact angles, respectively. The surface in (i) was sprayed for 120 seconds. The inset shows the re-entrant PDMS/F-POSS structure; scale bar: 5 mm. (j)–(k) Transparency of the surfaces coated with the PDMS/F-POSS spray for different spray times and D^* values. For all spray times greater than 0 seconds, a maximum in transparency persists at $D^* = 42$. (l)–(m) Movie frames taken with 10 ms intervals are superimposed for droplets of water and ethanol, respectively. The droplets roll back and forth several times before reaching the edge of the substrate, which highlights the ultralow contact angle hysteresis (CAH).
Reproduced from ref. 16 with permission from John Wiley and Sons. Copyright © 2013 WILEY-VCH Verlag GmbH & Co. KGaA, Weinheim.

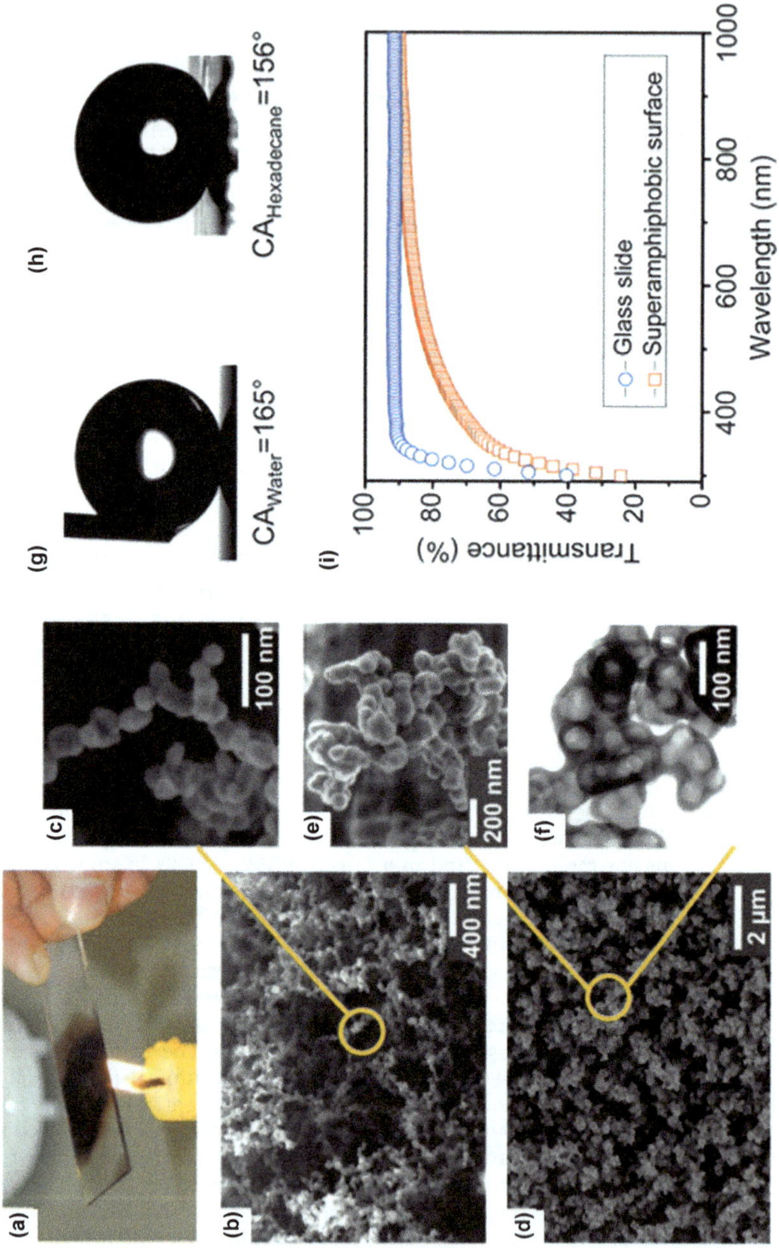
(a)
(b)
400 nm
(c)
100 nm
(d)
2 μm
(e)
200 nm
(f)
100 nm
(g)
CA_{Water}=165°
(h)
$CA_{Hexadecane}$=156°
(i)
Transmittance (%)
Wavelength (nm)
Glass slide
Superamphiphobic surface
0
20
40
60
80
100
400
600
800
1000

shown in Figure 3.11a–f. The coating had a contact angle and roll-off angles (given in parentheses) of $165 \pm 0.8°$ ($1 \pm 1°$) for water and $156 \pm 0.8°$ ($5 \pm 1°$) for hexadecane (Figure 3.11g–h), respectively. The transmittance of the coating was reduced by less than 10% compared to that of pristine glass for wavelengths above 500 nm (Figure 3.11i). The coating could withstand the sand abrasion test. Although the coating possesses superamphiphobicity and good robustness, the light transmittance of the coating is unfortunately lower than that of the substrate due to the large roughness of the coating.

Zhu *et al.* fabricated transparent superamphiphobic coatings using carbon nanotubes (CNTs) as templates.[26] CNTs were entrapped by silica nanoparticles and the compounds were sprayed onto the glass substrate. The coating possessed superamphiphobicity; water, dodecane and other organic liquids could remain as typical spherical shapes on the coating and could roll off it easily. The coating was transparent but the light transmittance of the coating was much lower than the glass substrate.

3.3.8 Spin-coating

Spin-coating is widely used to fabricate thin films, which could deposit uniform coatings of organic materials, or uniformly distribute particulate matter on flat surfaces.[54] Fu and coworkers fabricated a superhydrophobic and near-superoleophobic transparent coating *via* spin-coating.[13] The coating was first prepared with polydimethylsiloxane (PDMS) and hydrophobic silicon dioxide (SiO_2) nanoparticles by spun assembly. Then a sintering treatment was used to degrade the PDMS into SiO_2. After the sintering treatment, the coating was immersed in piranha solution in 50 °C for 30 min. Finally, the coating was immersed in 1*H*,1*H*,2*H*,2*H*-perfluorooctyltrichlorosilane ($CF_3(CF_2)_5CH_2CH_2SiCl_3$, PFTS, 1.0 v% in toluene) for about one minute and then put into a drying cabinet at 80 °C for 5 h.

Figure 3.11 (a) Photograph depicting sample preparation. A glass slide is held in the flame of a candle until a soot layer a few micrometers thick is deposited. (b) Scanning electron microscope (SEM) image of the soot deposit. (c) High-resolution SEM image showing a single particle chain made up of almost spherical carbon beads 40 ± 10 nm in diameter. (d) SEM image of the deposit after being coated with a silica shell. (e) High-resolution SEM image of a cluster after the carbon core was removed by heating for 2 hours at 600 °C. (f) High-resolution TEM image of a cluster after calcination, revealing the silica coating with holes that were previously filled with carbon particles. The silica shell is 20 ± 5 nm thick. (g) and (h), a 2 μl water drop and 5 μl hexadecane drop deposited on the surface possess static contact angles of $165° \pm 1°$ and $156° \pm 1°$, respectively. (i) Ultraviolet-visible transmittance spectra of a 3 μm-thick superamphiphobic surface compared to pristine glass.[19,22] Reproduced from ref. 22 with permission from John Wiley and Sons. Copyright © 2015 WILEY-VCH Verlag GmbH & Co. KGaA, Weinheim.

The contact angle of the coating was 152.7 ± 2.1° for water and 140.7 ± 3.2° for diiodomethane. The contact angle hysteresis was below zero for water and 12.4 ± 5.2° for diiodomethane. Figure 3.12 shows the light transmittance, roughness average and height average of the coatings with different treatments. The robustness of the coating should be improved as the adhesion strength between the nanoparticles of the coating is weak. Furthermore, the contact angle of diiodomethane is merely 140°, which does not meet the criteria of superamphiphobic coatings.

3.3.9 Layer-by-layer Assembly

In 1966, Iler demonstrated a novel method for creating multilayers of inorganic colloidal particles, which is now known as a layer-by-layer (LbL) assembly.[55] It is a simple and accurate way to fabricate coatings with varied compositions and structures. Cao and Gao fabricated transparent superhydrophobic and highly oleophobic coatings through layer-by-layer assembly of silica nanoparticles and sacrificial polystyrene nanoparticles.[56] After the assembly, the coatings were calcined at 550 °C for 4 h to remove the sacrificial polystyrene nanoparticles. Then the coatings were immersed in (tridecafluoro-1,1,2,2-tetrahydrooctyl) trichlorosilane (FTS) solution for 15 min to reduce the surface energy. Figure 3.13a and b show the surface morphologies of the coatings with 25 bilayers before and after removing the polystyrene particles. The contact angles of the coating were 160° for water (Figure 3.13c) and 140° for hexadecane (Figure 3.13d). The contact angle hysteresis was less than 5° for water, but that for hexadecane was not mentioned. The coating had similar transparency to the glass slide (Figure 3.13e). However, the fabrication process may be simplified to satisfy the needs in practical applications. Furthermore, the robustness of the coating may be improved as the adhesion strength is weak between the nanoparticles of the coating.

Our group designed coatings that are simultaneously highly antireflective and superamphiphobic.[57] Three different nanoparticles were used to fabricate the coatings *via* layer-by-layer assembly. After the assembly, the coating was calcined at 550 °C for 3 h. 1*H*,1*H*,2*H*,2*H*-Perfluorooctyltrichlorosilane (PFTS) was deposited on the coating *via* CVD at 120 °C for 2 h. The contact angles and sliding angles for different kinds of liquids are shown in Figure 3.14a–e. The maximum transmittance of a coated glass substrate was as high as 96.1% at 530 nm, while that of a blank glass substrate was only 91% (Figure 3.14f). The coating could endure the water drop test. However, the surface structures of the coating are easy to destroy by external forces, meaning that the robustness and the adhesion-to-substrate of the coating need to be improved.

To enhance the surface robustness and adhesion-to-substrate of coatings, a tetraethyl orthosilicate (TEOS) CVD treatment was used to deposit

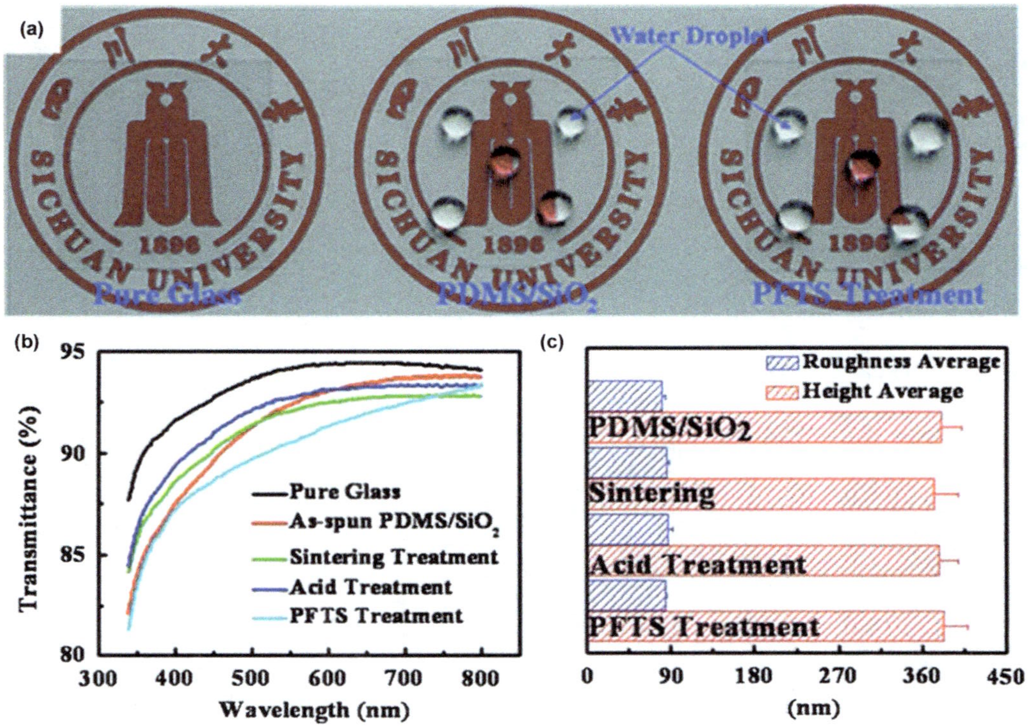

Figure 3.12 Optically transparent images of pure glass, as-spun PDMS/SiO_2 and PFTS treatment (a), transmittance of different samples measured with a UV-vis spectrometer (b) and statistical data of roughness average and height average were obtained by the software SPI4000 and WSxM4.0 with five 10 μm × 10 μm images (c).
Reproduced from ref. 13 with permission from the Royal Society of Chemistry.

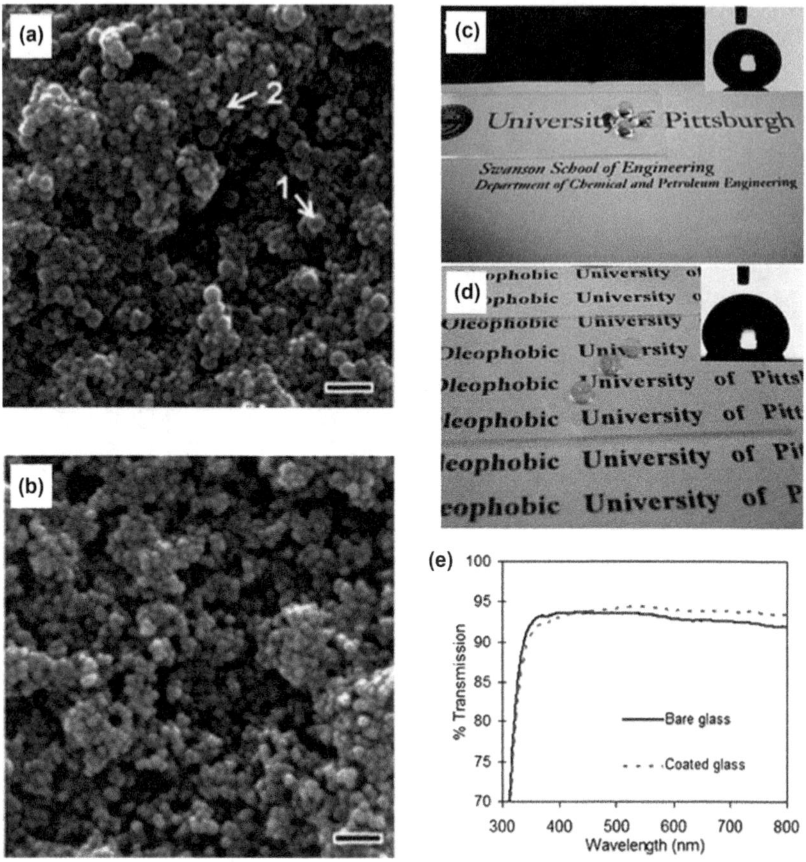

Figure 3.13 Figure 3.13 (a) and (b) are SEM images of the coatings with 25 bilayers before and after removing the sacrificial polystyrene particles, respectively. Labeled in (a) are a representative 60 nm polystyrene particle (labeled as "1") and a representative 20 nm silica particle (labeled as "2"). The polystyrene particles disappear in (b) after calcination. The scale bars are 200 nm. Photographs of a coated glass slide with droplets of (c) water and (d) hexadecane to demonstrate the superhydrophobicity, high oil-repellency, and transparency of the coating. Insets in (c) and (d) are images taken during the contact angle measurement. (e) Transmission spectrum of the glass slide before and after the coating process. The glass slide is coated with 25 bilayers, and is calcined and functionalized with FTS.
Reproduced from ref. 56 with permission from the Royal Society of Chemistry.

silica on the coatings, which was catalyzed by ammonia. The TEOS CVD treatment was carried out in an oven at 60 °C for 3 h, followed by hydrothermal treatment at 125 °C for 1 h. At the end, dynamic hardening at 700 °C for 2 min was applied to further reinforce the coatings.[58]

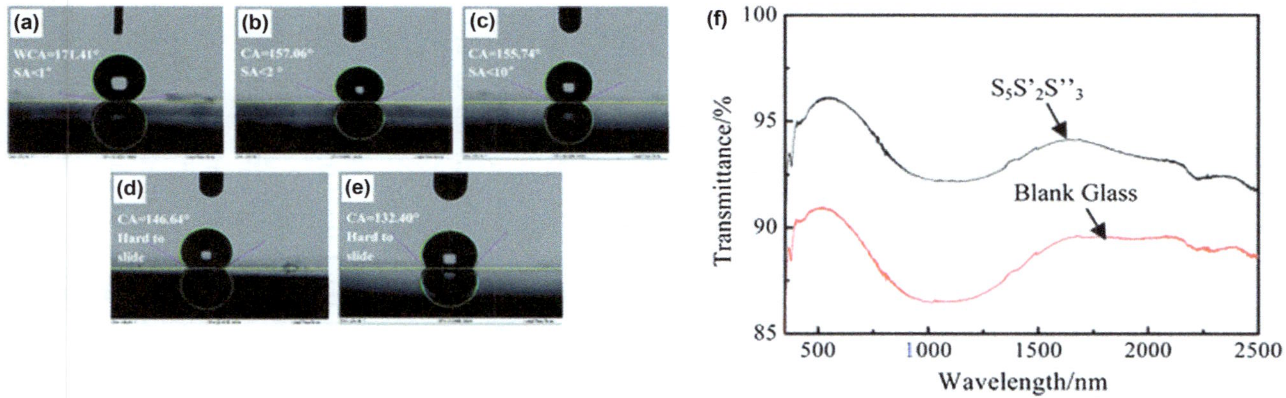

Figure 3.14 Digital images of the contact angles of water (a), diiodomethane (b), ethylene glycol (c), phenylcarbinol (d) and hexadecane (e), on the $S_5S'_2S''_3$ coating. (f) Transmission spectra of a blank glass substrate and a $S_5S'_2S''_3$-coated glass substrate recorded by an integrating sphere.
Reproduced from ref. 57 with permission from the Royal Society of Chemistry.

Figures 3.15a and b show the surface morphology of the coating before and after the TEOS CVD treatment. After the treatment, the size of the particles became larger, and the distance between the particles became closer than before. The nanoparticles that constructed the coating cross-linked with each other, and the whole surface turned into a three-dimensional network structure. As a result, the surface robustness and the adhesion-to-substrate were enhanced significantly compared to the original coating. The coating after the treatment could withstand a 4H pencil scratching test, a tape peeling test using 3M scotch tape 40 times, a water drop impact test and a sand abrasion test. The contact angles of the coating were 171° for water (Figure 3.15c) and 152° for ethylene glycol (Figure 3.15d). The maximum transmittance of the coating was 98.7% at 1580 nm. Although the TEOS CVD treatment significantly enhanced the surface robustness and adhesion-to-substrate, more efforts have yet to be made while maintaining the high transmittance and superamphiphobicity of the coating. For example, the coating hardness and its transmittance in the range of visible light should be improved, and especially, the fabrication process should be simplified.

The above coatings have significantly enhanced the transmittance of substrate, which would ignite the hope of applying superamphiphobic coatings to such fields as solar energy utilization and optical devices. These coatings possessed a disordered surface texture and porous film structure, and their surface roughness was less than 100 nm. According to recent research, when the sizes of particles in coatings are smaller than 100 nm, the Mie scattering in the visible region would be negligible,[40,41] and thus light scattering would not significantly influence light transmittance. Meanwhile, a porous film structure could decrease the refractive index of the coating and might also restrain the incident light from reflection. This strategy would become an effective approach to the fabrication of antireflective superamphiphobic coatings.

3.3.10 Slippery Liquid Infused Porous Surfaces

Slippery liquid infused porous surfaces (SLIPS) were inspired by the *Nepenthes* pitcher plant, which exhibits antiwetting behavior to almost all fluids and shows extreme temperature and pressure stability. Different from traditional superhydrophobic/superamphiphobic coatings, SLIPS possess a thin layer of a liquid lubricant trapped inside the porous structure of the coating.[23,59–63] Aizenberg's group prepared several SLIPS on different kinds of substrates.[23,59,60] Low-surface-tension perfluorinated liquids were chosen to lubricate coatings (for example 3 M Fluorinert FC-70, $\gamma_B = 17.1$ mN m^{-1}; or DuPont Krytox oils). Figure 3.16a–c shows the process of fabricating SLIPS and the surface morphologies of the coatings.

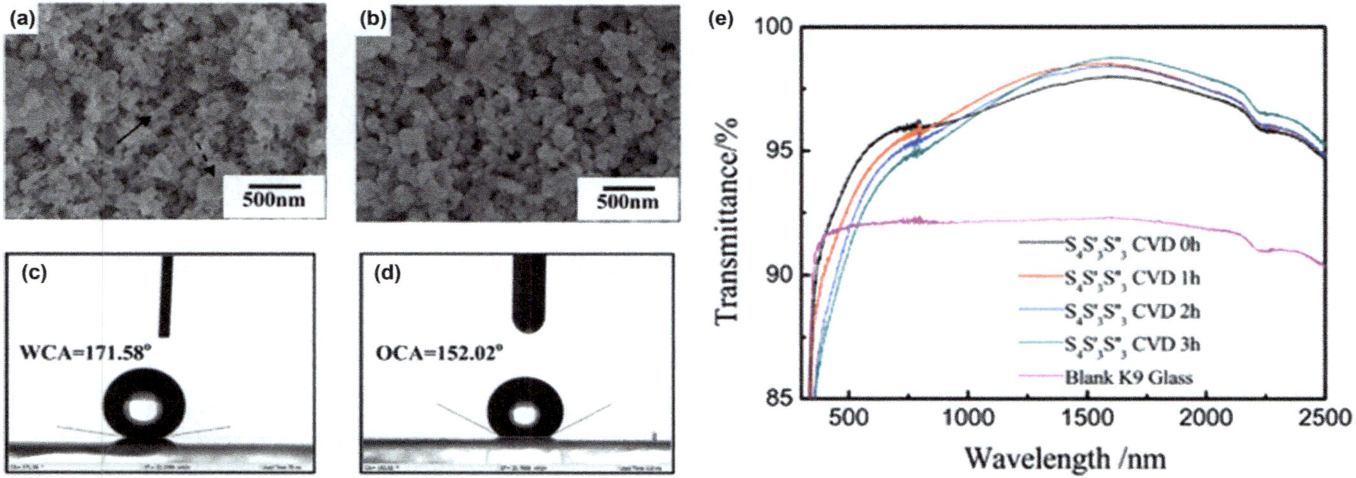

Figure 3.15 SEM images of the coating before (a) and after (b) 3 h TEOS CVD treatment. Contact angles of the coating of water (c) and ethylene glycol (d). (e) Transmission spectra of the coatings by 0 h, 1 h, 2 h and 3 h CVD TEOS treatments, respectively. Reproduced from ref. 58 with permission from the Royal Society of Chemistry.

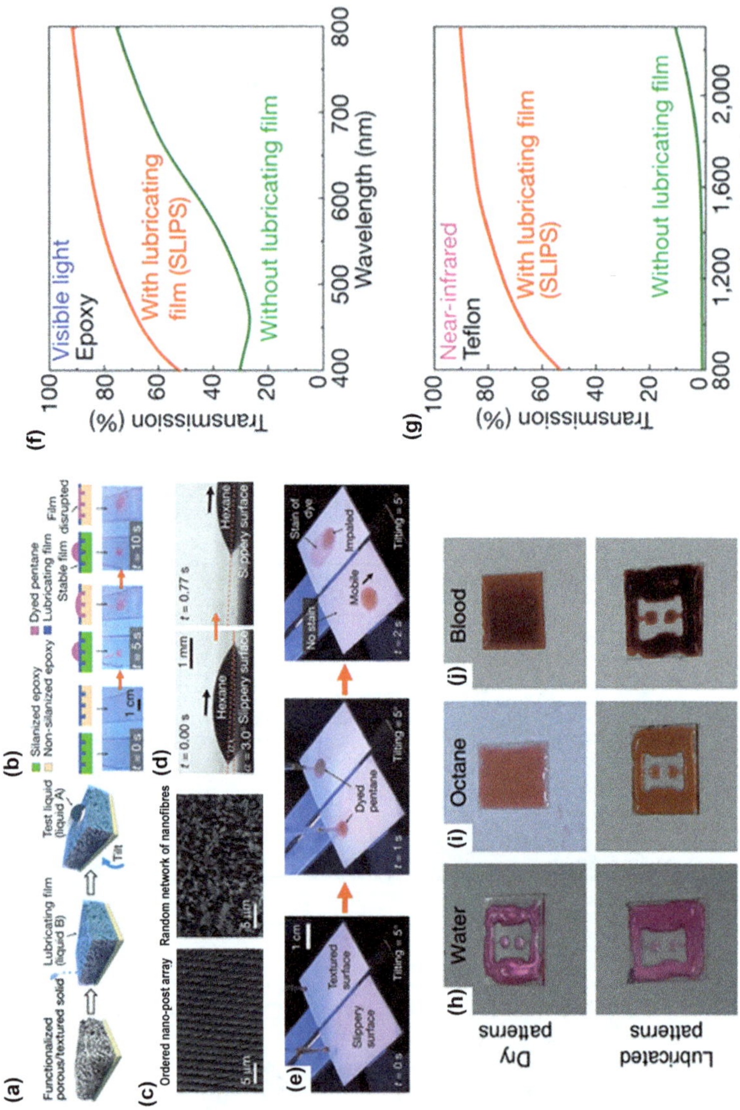
(a)
Functionalized porous/textured solid
Lubricating film (liquid B)
Test liquid (liquid A)
Tilt
(b)
Silanized epoxy
Non-silanized epoxy
Dyed pentane
Lubricating film
Stable film
Film disrupted
1 cm
t = 0 s
t = 5 s
t = 10 s
(c)
Ordered nano-post array
Random network of nanofibres
5 μm
(d)
t = 0.00 s
1 mm
t = 0.77 s
Hexane
Slippery surface
(e)
Textured surface
Slippery surface
1 cm
Tilting = 5°
Dyed pentane
No stain
Mobile
Stain of dye
Impaled
t = 0 s
t = 1 s
t = 2 s
(f)
Visible light
Epoxy
With lubricating film (SLIPS)
Without lubricating film
Transmission (%)
Wavelength (nm)
(g)
Near-infrared
Teflon
With lubricating film (SLIPS)
Without lubricating film
Transmission (%)
Wavelength (nm)
(h) Water
(i) Octane
(j) Blood
Dry patterns
Lubricated patterns

The SLIPS possessed low sliding angles to water and oil although they did not have large contact angles (Figure 3.16d–e). The lubricated surface was repellent to water, oil, blood, and ice (Figure 3.16h–j). The surface also possessed self-healing ability. Compared to the dry surface, a significant increase appeared in the transmittance of the lubricated coating (Figure 3.16f–g), because the lubricating liquid infused the surface texture and reduced the light scattering. SLIPS show an effective way to build high transmittance superamphiphobic coatings. However, the SLIPS would lose the omni-repellency if the lubricating liquid was dried up.

To give an easy and comprehensive comparison, we have summarized in Table 3.1 the above methods in terms of the number of process steps, adopted substrates, technical characteristics, transmittance and coating durability. Clearly, each method has its advantages and potential applications.

Figure 3.16 (a) Schematics showing the fabrication of SLIPS by infiltrating a functionalized porous/textured solid with a low-surface energy, chemically inert liquid to form a physically smooth and chemically homogeneous lubricating film on the surface of the substrate. (b) Comparison of the stability and displacement of lubricating films on silanized and non-silanized textured epoxy substrates. Top panels show schematic side views; bottom panels show time-lapse optical images of top views. Dyed pentane was used to enhance visibility. (c) Scanning electron micrographs showing the morphologies of porous/textured substrate materials: an epoxy-resin-based nanofabricated post array (left) and a Teflon-based porous nanofiber network (right). (d) Optical micrographs demonstrating the mobility of a low-surface-tension liquid hydrocarbon—hexane ($\gamma_A = 18.6 \pm 0.5$ mN m^{-1}, volume ~3.6 μl)—sliding on a SLIPS at a low angle ($\alpha = 3.0°$). (e) Time-sequence images comparing mobility of pentane droplets ($\gamma_A = 17.2 \pm 0.5$ mN m^{-1}, volume ~30 μl) on a SLIPS and a superhydrophobic, air-containing Teflon porous surface. Pentane is repelled on the SLIPS, but it wets and stains the traditional superhydrophobic surface. Optical transmission measurements for an epoxy resin-based SLIPS (f) and a Teflon-based SLIPS (g) in the visible light range (400–750 nm).[59] Reprinted with permission from Macmillan Publishers Ltd: *Nature*, Copyright 2011. (h)–(j) Patterning of various liquids based on surface chemistry contrasts. The upper row shows the addition of liquids to the dry, patterned substrates. The lower row shows the addition of liquids to the pre-lubricated substrates. (h) Rhodamine B-dyed water: the pattern is revealed both in a dry state and on the lubricated sample; (i) DFSB-K175-dyed octane and (j) sheep blood: the liquids wet and contaminate the entire surface in a dry state, while successful patterning occurs using the repellent properties of SLIPS surface areas.[60] Reprinted from Macmillan Publishers Ltd: *Nature*, Copyright 2013.

Table 3.1 Summary of fabrication methods of superamphiphobic coatings with high transmittance.

Fabrication method	Process steps	Application fields	Technical characteristics	Transmittance	Durability
Colloidal lithography and plasma nanotexturing[18]	Two	Polymeric microfluidics	Ordered surfaces on a large scale	Transparent	N. A.
3-D diffuser lithography[17]	Four	Large flexible substrates	Ordered surfaces on flexible substrates	77%	Pass water drop impact test
Electrospinning[21]	Four	Glass	Self-cleaning coatings on glass	85%	Hardness: 1.6 GPa Modulus: 24.7 GPa
Spray-coating[51]	One	Flat and curved substrates	Large area surfaces without pre- or post-treatment	91.5%	Pass water jetting and sand abrasion tests
Micromolding[20]	Two	Thermal or UV-curable materials	Simple and low-expertise	88–89%	Pass 6 months' durability test
Template-based method[19]	Four	Heat-resistant substrates	Easy-to-fabricate	<90%	Pass sand abrasion test
Spin-coating[13]	Four	Glass substrates	Low cost and easy manipulation	Nearly 93%	Pass weathering and ultrasonic damage tests
Layer-by-layer assembly[58]	Four	Glass substrates, *etc.*	Precise control over coating structure	98.7%	Pass pencil scratch, sand abrasion, tape peeling and water impact tests
Slippery liquid infused porous surfaces[60]	Two	Low energy surfaces	Low contact angles with low sliding angles	95%	Withstand strong shear forces and mechanical damage

3.4 Summary and Outlook

Superamphiphobic coatings with high transmittance have recently gained much attention due to both scientific interest and potential applications, especially in solar energy utilization and optical devices. Although many efforts have so far been made to fabricate superamphiphobic coatings with high transmittance, fundamental aspects are yet to be clarified for designing the optimal structure of superamphiphobic coatings with high transmittance and several significant challenges are yet to be met toward practical applications. First and foremost, most coated substrates mentioned above have transmittance lower than their substrates and it is difficult to meet the requirements of many practical applications, especially infrared detector protecting panels, building-integrated photovoltaics (BIPV), solar cell protecting panels, and self-cleaning greenhouses, *etc*. The design and fabrication of novel surface and film structures toward antireflective superamphiphobic coatings are thus emerging as an urgent future direction. How to design and fabricate structures that could harmonize the contradictory requirements of superamphiphobicity and antireflection would be the key point in future research. Only a few coated substrates have so far achieved enhanced transmittance compared to blank substrates,[52,57,58] and the coatings possessed a disordered surface texture and porous film structure. These successes at least shed partial light on the design and fabrication of antireflective superamphiphobic coatings with ultra-high transmittance. Secondly, the micro-nano sized features of superamphiphobic coatings can usually be destroyed by external forces easily and irreversibly, which would lead to the loss of superamphiphobicity. Therefore, most superamphiphobic coatings do not have sufficient mechanical strength and stability, which are, however, prerequisites for a thin film to be possibly useful in many field applications. The ammonia-catalyzed chemical vapor deposition of tetraethoxysilane (TEOS) proved to be an effective way to enhance the surface robustness of superamphiphobic coatings by reducing the gaps of nanoparticles in coatings and forming a three-dimensional cross-linked network structure.[19,58] Assembling nanoparticles with binder solution would be another method to enhance the robustness of superamphiphobic coatings. Last but not least, the complicated fabrication processes of the coatings mentioned above are not suited to mass and large-area production; novel and facile fabrication methods should be invented for future practical applications. Dip-coating, spray-coating, and roll-to-roll coating might be ideal methods toward the low-cost and large-scale production of coatings. It is anticipated that multi-functional coatings with high transmittance, superamphiphobicity and excellent robustness would be fabricated in future *via* simple and easy approaches, and would be widely used in a broad spectrum of applications (Figure 3.17).

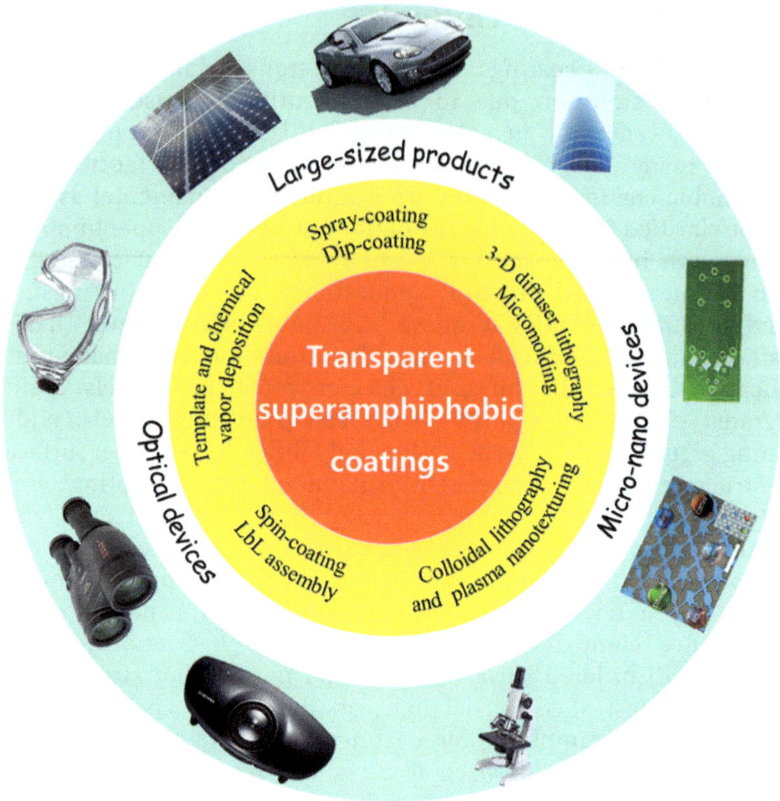

Figure 3.17 Summary of the review.

Acknowledgements

Financial support from the National Natural Science Foundation of China (Grant Nos. 21571182, 21271177), a Chinese Academy of Sciences Grant (CXJJ-14-M38), the National High Technology Research and Development Program ("863" Program) of China (Grant No. 2011AA050525), the Knowledge Innovation Program of the Chinese Academy of Sciences (CAS) (Grant Nos. KGCX2-YW-370, KGCX2-EW-304-2), and Key Laboratory of Space Energy Conversion Technology, TIPC, CAS is greatly appreciated.

References

1. R. N. Wenzel, *Ind. Eng. Chem.*, 1936, **28**, 988.
2. A. B. D. Cassie and S. Baxter, *Trans. Faraday Soc.*, 1944, **40**(0), 0546.
3. A. Nakajima, A. Fujishima, K. Hashimoto and T. Watanabe, *Adv. Mater.*, 1999, **11**(16), 1365.
4. L. Yao and J. He, *Prog. Mater. Sci.*, 2014, **61**, 94.

5. L. Jiang, Y. Zhao and J. Zhai, *Angew. Chem., Int. Ed.*, 2004, **43**(33), 4338.
6. X. Gao and L. Jiang, *Nature*, 2004, **432**(7013), 36.
7. L. Xu and J. He, *Langmuir*, 2012, **28**(19), 7512.
8. X. Du, X. Li and J. He, *ACS Appl. Mater. Interfaces*, 2010, **2**(8), 2365–2372.
9. D. Schaeffer, G. Polizos, D. Smith, D. Lee, S. Hunter and P. Datskos, *Nanotechnology*, 2015, **26**, 055602.
10. Y. Sun, L. Wang, Y. Gao and D. Guo, *Appl. Surf. Sci.*, 2015, **324**, 825.
11. K. Tsujii, T. Yamamoto, T. Onda and S. Shibuichi, *Angew. Chem., Int. Ed. Engl.*, 1997, **36**, 1012.
12. S. Shibuichi, T. Yamamoto, T. Onda and K. Tsujii, *J. Colloid Interface Sci.*, 1998, **208**, 287.
13. Z. He, M. Ma, X. Lan, F. Chen, K. Wang, H. Deng, Q. Zhang and Q. Fu, *Soft Matter*, 2011, **7**(14), 6435.
14. A. Tuteja, W. Choi, M. Ma, J. M. Mabry and S. A. Mazzella, *Science*, 2007, **318**(5856), 1618.
15. L. Cao, H. H. Hu and D. Gao, *Langmuir*, 2007, **23**, 4310.
16. K. Golovin, D. H. Lee, J. M. Mabry and A. Tuteja, *Angew. Chem.*, 2013, **52**, 13007.
17. M. Im, H. Im, J. H. Lee, J. B. Yoon and Y. K. Choi, *Soft Matter*, 2010, **6**(7), 1401–1404.
18. K. Ellinas, A. Tserepi and E. Gogolides, *Langmuir*, 2011, **27**(7), 3960.
19. X. Deng, L. Mammen, H. J. Butt and D. Vollmer, *Science*, 2012, **335**(6064), 67.
20. S. M. Kang, S. M. Kim, H. N. Kim, M. K. Kwak, D. H. Tahk and K. Y. Suh, *Soft Matter*, 2012, **8**(33), 8563.
21. V. A. Ganesh, S. S. Dinachali, H. K. Raut, T. M. Walsh, A. S. Nair and S. Ramakrishna, *Rsc Adv.*, 2013, **3**(12), 3819.
22. Z. Geng and J. He, *Adv. Mater. Interfaces*, 2015, **2**, 1500196.
23. P. Kim, M. J. Kreder, J. Alvarenga and J. Aizenberg, *Nano Lett.*, 2013, **13**(4), 1793.
24. S. G. Lee, D. S. Ham, D. Y. Lee, H. Bong and K. Cho, *Langmuir*, 2013, **29**(48), 15051.
25. K. Fukada, S. Nishizawa and S. Shiratori, *J. Appl. Phys.*, 2014, **115**, 103516.
26. X. Zhu, Z. Zhang, G. Ren, X. Men, B. Ge and X. Zhou, *J. Colloid Interface Sci.*, 2014, **421**, 141.
27. X. Deng, F. Schellenberger, P. Papadopoulos, D. Vollmer and H. Butt, *Langmuir*, 2013, **29**, 7847.
28. J. Zhang, A. Wanga and S. Seegerb, *Polym. Chem.*, 2014, **5**, 1132.
29. L. Yuan, T. Wu, W. Zhang, S. Ling, R. Xiang, X. Gui, Y. Zhua and Z. Tang, *J. Mater. Chem. A*, 2014, **2**, 6952.
30. P. Muthiah, B. Bhushan, K. Yun and H. Kondo, *J. Colloid Interface Sci.*, 2013, **409**, 227.
31. A. Tuteja, W. Choi, J. M. Mabry, G. H. McKinley and R. E. Cohen, *Proc. Natl. Acad. Sci. U. S. A.*, 2008, **105**(47), 18200.

32. Z. Xue, M. Liu and L. Jiang, *J. Polym. Sci., Part B: Polym. Phys.*, 2012, **50**(17), 1209.
33. L. Cao, T. P. Price, M. Weiss and D. Gao, *Langmuir*, 2008, **24**(5), 1640.
34. K. Rei, P. Laurent and M. Jonasz, *Appl. Opt.*, 2007, **46**, 8118.
35. S. Walheim, E. Sch€affer, J. Mlynek and U. Steiner, *Science*, 1999, **283**, 520.
36. H. K. Raut, V. A. Ganesh, A. S. Nair and S. Ramakrishna, *Energy Environ. Sci.*, 2011, **4**(10), 3779.
37. T. Li, J. He, L. Yao and Z. Geng, *J. Colloid Interface Sci.*, 2015, **444**, 67.
38. X. X. Zhang, S. Cai, D. You, L. H. Yan, H. B. Lv, X. D. Yuan and B. Jiang, *Adv. Funct. Mater.*, 2013, **23**(35), 4361.
39. Y. Du, L. E. Luna, W. S. Tan, M. F. Rubner and R. E. Cohen, *ACS Nano*, 2010, **4**(7), 4308.
40. Y. Rahmawan, L. Xu and S. Yang, *J. Mater. Chem. A*, 2013, **1**(9), 2955.
41. K. Manabe, S. Nishizawa, K. H. Kyung and S. Shiratori, *ACS Appl. Mater. Interfaces*, 2014, **6**(16), 13985.
42. J. C. Hulteen and R. P. Vanduyne, *J. Vac. Sci. Technol., A*, 1995, **13**(3), 1553.
43. S. M. Yang, S. G. Jang, D. G. Choi, S. Kim and H. K. Yu, *Small*, 2006, **2**, 458.
44. X. J. Huang, J. H. Lee, J. W. Lee, J. B. Yoon and Y. K. Choi, *Small*, 2008, **4**(2), 211.
45. M. Ma, R. M. Hill, J. L. Lowery, S. V. Fridrikh and G. C. Rutledge, *Langmuir*, 2005, **21**(12), 5549.
46. D. H. Reneker and A. L. Yarin, *Polymer*, 2008, **49**(10), 2387.
47. A. Steele, I. Bayer and E. Loth, *Nano Lett.*, 2009, **9**(1), 501.
48. A. Das, T. M. Schutzius, I. S. Bayer and C. M. Megaridis, *Carbon*, 2012, **50**(3), 1346.
49. X. L. Wang, H. Y. Hu, Q. Ye, T. T. Gao, F. Zhou and Q. J. Xue, *J. Mater. Chem.*, 2012, **22**(19), 9624.
50. S. Nishizawa and S. Shiratori, *J. Mater. Sci.*, 2013, **48**, 6613.
51. D. Ge, L. Yang, Y. Zhang, Y. Rahmawan and S. Yang, *Part. Part. Syst. Charact.*, 2014, **31**, 763.
52. Z. Geng, J. He and L. Yao, *RSC Adv.*, 2015, **5**, 89262.
53. F. Shi, Z. Liu, G. L. Wu, M. Zhang, H. Chen, Z. Q. Wang, X. Zhang and I. Willner, *Adv. Funct. Mater.*, 2007, **17**(11), 1821.
54. C. J. Lawrence, *Phys. Fluids*, 1988, **31**(10), 2786.
55. R. K. Iler, *J. Colloid Interface Sci.*, 1966, **21**(6), 569.
56. L. Cao and D. Gao, *Faraday Discuss.*, 2010, **146**, 57.
57. Z. Geng, J. He, L. Xu and L. Yao, *J. Mater. Chem. A*, 2013, **1**(31), 8721.
58. Z. Geng and J. He, *J. Mater. Chem. A*, 2014, **2**(39), 16601.
59. T. S. Wong, S. H. Kang, S. K. Tang, E. J. Smythe, B. D. Hatton, A. Grinthal and J. Aizenberg, *Nature*, 2011, **477**(7365), 443.
60. N. Vogel, R. A. Belisle, B. Hatton, T. S. Wong and J. Aizenberg, *Nat. Commun.*, 2013, **4**, 2167.

61. C. Urata, B. Masheder, D. Cheng and A. Hozumi, *Langmuir*, 2012, **28**, 17681.
62. C. Urata, B. Masheder, D. Cheng and A. Hozumi, *Chem. Commun.*, 2013, **49**, 3318.
63. K. Manabe, S. Nishizawa, K. Kyung and S. Shiratori, *ACS Appl. Mater. Interfaces*, 2014, **6**, 13985.

CHAPTER 4

Superhydrophilic and Superhydrophobic Thin Film Type of Photocatalysts with Self-cleaning Properties

TAKASHI KAMEGAWA[a,b] AND HIROMI YAMASHITA*[a]

[a] Division of Materials and Manufacturing Science, Graduate School of Engineering, Osaka University, 2-1 Yamadaoka, Suita, Osaka 565-0871, Japan; [b] Nanoscience and Nanotechnology Research Center, Osaka Prefecture University, 1-2 Gakuencho, Nakaku, Sakai, Osaka 599-8570, Japan
*Email: yamashita@mat.eng.osaka-u.ac.jp

4.1 Introduction

In our daily life, various types of functional materials with specific properties are used for realizing a comfortable environment with convenience. The flexible controlling of surface properties is a key issue for the design of fascinating materials with advanced functions. The surface property of materials is closely related to various phenomena such as surface wettability, adhesive joining, abrasion strength and friction behavior. For the evaluation of the surface wettability of various materials, the water contact angle measurement is often adopted as an easy and reliable method.[1,2] It is possible to split surface wettability into two categories—hydrophilic state and hydrophobic state—based on the behavior of a water droplet on the surface

RSC Smart Materials No. 21
Self-cleaning Coatings: Structure, Fabrication and Application
Edited by Junhui He

Published by the Royal Society of Chemistry, www.rsc.org

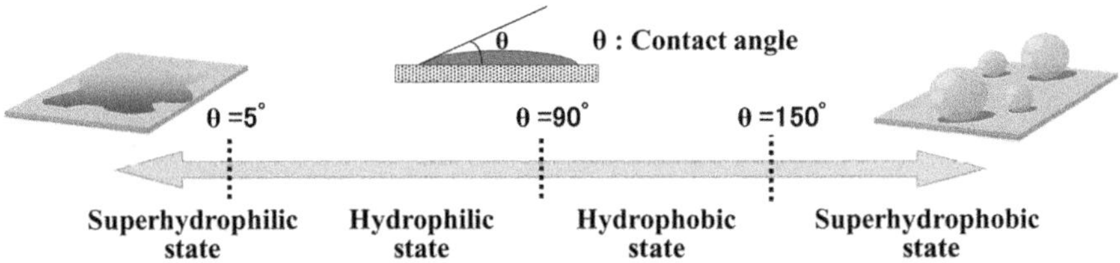

Figure 4.1 Schematic image of the classification of surface wettability on a solid surface.

of target materials (Figure 4.1). A water contact angle less than 90° is indicative of a hydrophilic state, while a water contact angle greater than 90° indicates a hydrophobic state. Even today, many researchers are continuously devoted to the design of a functional surface with a hydrophilic or hydrophobic property due to the existence of a potential demand in various areas of fundamental research and practical application.[3–8]

On a hydrophilic surface, especially a superhydrophilic surface, a water droplet gets wet and rapidly spreads to the entire surface with high uniformity. The superhydrophilic surface shows a water contact angle lower than 5° (Figure 4.1). Such a surface property, designed by the application of coating techniques, is already commercialized for antifogging mirrors, glasses and the outer walls of buildings.[3,4] Meanwhile, on a hydrophobic surface, especially a superhydrophobic surface, a water droplet has a spherical shape and shows a water contact angle larger than 150° (Figure 4.1). A superhydrophobic surface shows specific functions such as waterproofing, antifogging, the effect of decreasing surface friction and accretion of ice and snow.[5–8] The functionalization on a smooth solid surface has also a certain limitation in the improvement of surface wettability. Surface architectures observed in nature such as the leaves of plants, the legs of a water strider and wings of a butterfly are indispensable.[9–11] Diverse techniques for the design of a desired surface property and structure are indispensable in agreement with necessary requirements. The development of multifunctional thin films has attracted much attention in recent years. In this chapter, our recent advancements in coating technologies for the design of superhydrophilic and superhydrophobic surfaces are mainly featured. Special attention has been focused on the photocatalytic self-cleaning property with advanced functions.

4.2 TiO_2 Thin Films for Superhydrophilic Coatings

In environmental remediation and energy related issues, TiO_2 is well known as an efficient photocatalyst.[3,12–15] TiO_2 has the potential to promote the removal of most organic compounds and toxic ions from water, which is accomplished by way of the high reactivity of a photoformed electron and hole. Complete degradation of volatile organic compounds, dye molecules, endocrine disruptors and bacteria in air and water results in a better natural environment.[16–23] Unique photocatalytic reactions for the production of hydrogen and fine chemicals can also occur on TiO_2 under moderate conditions.[23–29] For the past several decades, thin film types of TiO_2 have also attracted much attention due to their technological importance. After the discovery of the photoinduced superhydrophilicity of a TiO_2 surface in 1997, commercially applicable photofunctional materials with antifogging and self-cleaning properties were designed using pure TiO_2 and nanocomposite coatings.[30,31] TiO_2 thin films show multifunctional properties such as photocatalytic activity and photoinduced surface hydrophilicity. Based on these functions, TiO_2 thin films are adopted in the coating process of target materials for utilization of self-cleaning performance. Oily dirt is easily removed by the photocatalytic property of TiO_2, leading to maintenance-free characteristics as an additional function of materials.

These fascinating properties strongly depend on their surface architectures. A bottom-up technology for the synthesis of nanostructured materials was also important in the preparation of TiO_2 photocatalysts.[32–38] In this regard, taking advantage of microspheres is one way to design unique architectures. Both powder and thin film materials are designed by applying template strategies. TiO_2 thin films are mainly prepared by both a dry-process and wet-chemical process such as sol–gel, chemical vapor deposition (CVD) and sputtering deposition. By combining conventional sol–gel techniques with a bottom-up approach using uniform poly(methyl methacrylate) (PMMA) microspheres (Figure 4.2), TiO_2 thin films with a unique macroporous structure were successfully prepared on a quartz substrate.[37,38]

Polymer microspheres as template materials are easily removed by heat treatment or solvent extraction. TiO_2 thin films with macropores are denoted as m_n-TiO_2/Q (PMMA concentration: $n = 10–20$ wt%). Each prepared TiO_2 thin film had an anatase crystalline structure, relatively high transparency, and was well fixed on quartz substrate. The formation of macropores was confirmed by SEM measurement (Figure 4.2). The quartz substrate was uniformly covered by TiO_2 thin films without formation of cracks. The number of macropores depended on the PMMA concentration of the pre-coated aqueous dispersions. Macropores were homogeneously formed on the entire surface of TiO_2 thin films in the case of m_{20}-TiO_2/Q. The diameters of the macropores (*ca.* 400 nm) were slightly small compared with original PMMA particles due to the shrinkage of the TiO_2 network during the calcination process. The film thickness was about 200 nm, which showed a good correspondence with the radius of the PMMA microspheres. The architectures of the TiO_2 thin film surface were important in relation to noticeable phenomena such as photoinduced hydrophilicity and photocatalytic self-cleaning properties. Figure 4.3(A) shows the water contact angle changes on these thin film surfaces under UV light irradiation. The water contact angle gradually decreased with increases in UV irradiation time. On each sample, the water contact angle became lower than 5°. The spreading of water droplets on m_{20}-TiO_2/Q was clearly observed after a short period of UV light irradiation, indicating the positive roles of macropores for the enhancement of surface hydrophilicity. The rougher surface of TiO_2 thin

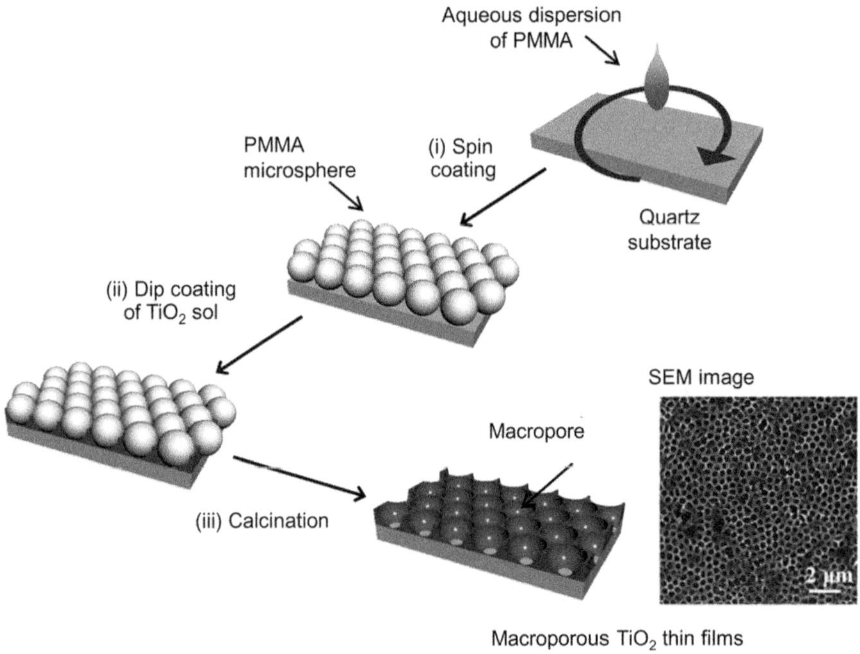

Figure 4.2 A schematic diagram of the procedures for the preparation of macroporous TiO_2 thin films.
Reproduced from ref. 38 with permission from the Royal Society of Chemistry.

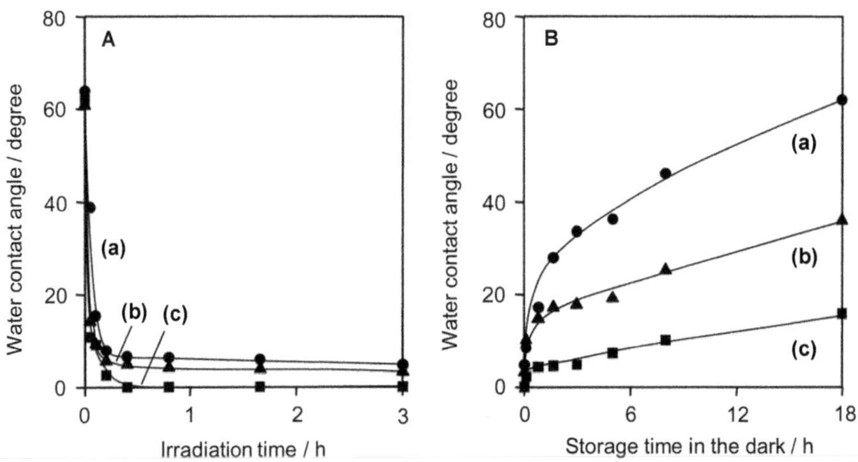

Figure 4.3 (A) Changes in the contact angle of water under UV light irradiation and (B) the recovery profiles in the dark after measurement of (A). Samples: (a) TiO_2/Q, (b) m_{10}-TiO_2/Q, and (c) m_{20}-TiO_2/Q.
Reproduced from ref. 38 with permission from the Royal Society of Chemistry.

films exhibits a faster change from the hydrophobic to hydrophilic state under irradiation of UV light.[39] During a period of being kept in the dark after the measurement of water contact angles (Figure 4.3(A)), the water contact angle on each sample gradually increased and recovered to almost the initial angle (Figure 4.3(B)). A slow recovery of water contact angles was observed in the following order: m_{20}-TiO_2/Q > m_{10}-TiO_2/Q > TiO_2/Q. TiO_2 thin films with macropores exhibited a good durability of the hydrophilic surface in the dark. The photocatalytic performance of these thin films was evaluated in the decomposition of 2-propanol in water and acetaldehyde in air under UV light irradiation. The reduction of waste in water and volatile organic compounds (VOCs) in air is important in relation to self-cleaning and keeping a clean environment. In both reaction systems, the reaction rate follows the order of m_{20}-TiO_2/Q > m_{10}-TiO_2/Q > TiO_2/Q. This tendency corresponds well with the increases of the surface area of samples. In relation to the self-cleaning, the construction of macropores was found to strongly enhance the photofunctional properties of TiO_2 thin films. The enhancement of these properties for efficient degradation of organic molecules and realizing a good surface hydrophilicity are advantages and possibilities in the design of self-cleaning materials.

4.3 Porous Silica Thin Films for Superhydrophilic Coatings

Isolated tetrahedrally coordinated transition metal species (M = Ti, V, Cr, Mo, W, *etc.*) can be designed within the matrixes of porous siliceous materials for applications in unique reaction systems.[40–44] These species are embedded within the silica matrix through the isomorphous substitution of Si atoms and tetrahedrally coordinated to oxygen atoms. Therefore, these species are considered to be highly dispersed at the atomic level within the frameworks and also to be well-defined and uniform active sites. The well-ordered porous structures and large surface areas originating from the silica frameworks provide condensation effects, high adsorptive capacities and high dispersivities of active species in both catalytic and photocatalytic reaction systems. In the photocatalytic reactions, unique photocatalytic performances were observed in the reduction of CO_2 with H_2O,[45–47] the decomposition of NO into N_2,[48,49] oxidation of CO into CO_2,[50–52] and the partial oxidation reaction of hydrocarbons.[53–56] The photocatalytic properties are attributed to the ligand-to-metal charge transfer (LMCT) processes of the isolated transition metal oxides with tetrahedral coordination, which could not be attained in bulk photocatalysts. The main stream of these works are carried out in the powder type of samples by many researchers.

Meanwhile, porous silica thin films have been prepared by spin-coating and dip-coating methods through a solvent evaporation process.[57,58] Based on these methods, isolated tetrahedrally coordinated transition metal species are incorporated within the porous silica framework in a thin film shape.

A tetrahedral Ti-containing porous silica thin film (TiMSF) was prepared on a quartz glass substrate *via* a spin-coating method. The precursor solution was synthesized by mixing water, a Si source, a Ti source and an organic surfactant as a structure-directing agent.[4,44,59–63] The sol solution obtained was spin-coated on the substrate, followed by calcination for surfactant removal. The film thickness was controllable by changing the amount of coated sol solution. TiMSF was completely colorless and transparent enough to see a background image, where transparent materials were used as substrates. The presence of a mesoporous structure in TiMSF was confirmed by XRD and TEM techniques. The typical and sharp XRD peak due to the (1 0 0) reflection of the mesoporous structure was observed at 2–3°. The TEM image of the TiMSF clearly demonstrated the presence of a uniform mesoporous structure. It was also confirmed that TiMSF showed high adhesion to substrate and was well-fixed on the quartz substrate after multiple peeling tests.

The local structure of the Ti-species was investigated by Ti K-edge X-ray absorption fine structure (XAFS) measurements. In the X-ray absorption near-edge structure (XANES) region, TiMSF showed a sharp pre-edge peak at 4967 eV. This peak is attributed to the 1s to 3d transition caused by a lack of an inversion center in the tetrahedron structure.[65–67] Depending on the environment, Ti-species show a pre-edge peak with a typical shape. Moreover, the Fourier transforms of the extended X-ray absorption fine structure (EXAFS) spectrum of TiMSF show an intense peak assigned to the existence of oxygen neighbors (Ti–O), indicating the successful incorporation of isolated tetrahedral Ti-species without the formation of aggregated Ti-oxide species. In a UV-vis absorption measurement, TiMSF only exhibited a typical absorption peak at around 210 nm with small absorption in the region from 240 to 280 nm, which is also clear evidence of the successful incorporation of tetrahedral Ti-species within silica matrixes. The absorption edge of embedded Ti-species within silica matrixes depends on the dispersion and their local structures. The absorption bands at 200–240 nm and 240–280 nm are assigned to the isolated tetrahedral Ti-species and dimers or small oligomers, respectively.[46,67,68] These absorption bands can be assigned to the LMCT transition of Ti-species with tetrahedral coordination.

In the thin film type of porous siliceous materials, it is possible to evaluate surface wettability from the water contact angle measurement. Figure 4.4 shows photographic images of a water droplet on a pure mesoporous silica thin film (MSF) and TiMSF. The water contact angle on MSF was 20–30°. The surface wettability changes to be more hydrophilic by incorporation of Ti-species within silica matrixes. The water contact angle on TiMSF was smaller than 10°. TiMSF also has a relatively hydrophilic surface as compared to TiO_2 thin films even before UV light irradiation. TiMSF responded to irradiation of UV light and induced a change of surface wettability to a highly hydrophilic state. The water contact angle on MSF was scarcely changed after UV light irradiation. TiMSF clearly showed photoinduced superhydrophilicity, which is similar to the property of bulk TiO_2 thin films.

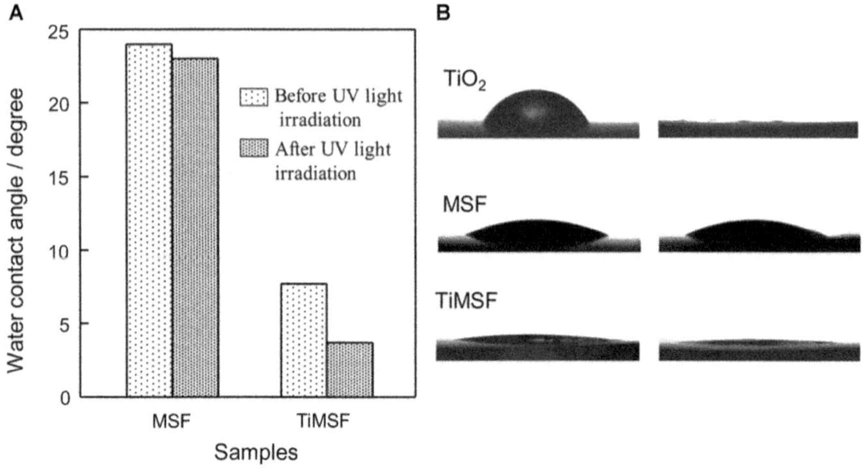

Figure 4.4 (A) Contact angle of a water droplet on MSF and TiMSF and (B) photographic images of a water droplet on each sample before and after UV light irradiation (irradiation time: 3 h).[4]

This phenomenon of TiMSF can be attributed to photoexcitation of Ti-species. Tetrahedral Ti-species were found to play a significant role in the improvement of the hydrophilic properties of MSF through the formation of a new adsorption site of water molecules by electron localization.

The photocatalytic property of TiMSF was also confirmed by the decomposition of oleic acid as a model of oily dirt.[4,63] Adhesion of oleic acid on the film surface leads to the disappearance of the highly hydrophilic property of TiMSF. Under UV light irradiation, the decreases in the water contact angles on TiMSF were observed by the photocatalytic removal of oleic acid (Figure 4.5). The water contact angle gradually decreased under UV light irradiation and finally reached a superhydrophilic state. The removal of oleic acid was also confirmed by the changes of FT-IR peaks ascribed to the C–H stretching vibration of alkyl groups in oleic acid. These results indicated that the TiMSF, having both hydrophilic property and photocatalytic performance, is applicable to hydrophilic coatings with a photocatalytic self-cleaning property.

The utilization of polymer microspheres as template materials was also investigated in the structural controlling of the TiMSF surface.[64] By combining a precursor solution for the synthesis of TiMSF with poly(methyl methacrylate) (PMMA) microspheres, tetrahedral Ti-containing porous silica thin films with a uniform mesoporous and macroporous structure (*m*-TiMSF) were successfully prepared on a quartz substrate with a high transparency (Figure 4.6). The macropores showed scraggly and distorted circular shapes. The combination of macroporous and mesoporous structures as well as the roles of tetrahedral Ti-species led to a significant improvement of surface hydrophilicity as compared to that on TiMSF. The water contact angle was smaller than 10° even before UV light irradiation. The water droplet was

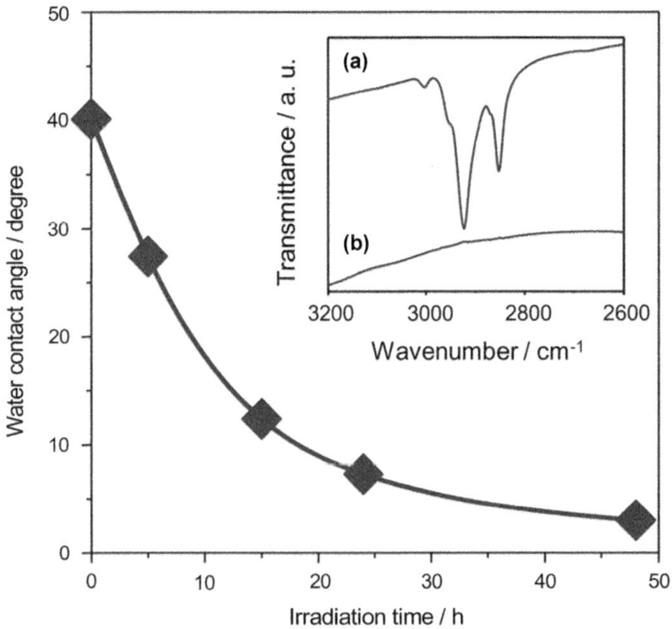

Figure 4.5 Time-dependent changes in the water contact angles of TiMSF after adhesion of oleic acid under UV light irradiation. Inset shows FT-IR spectra of TiMSF before (a) and after (b) UV light irradiation.[4]

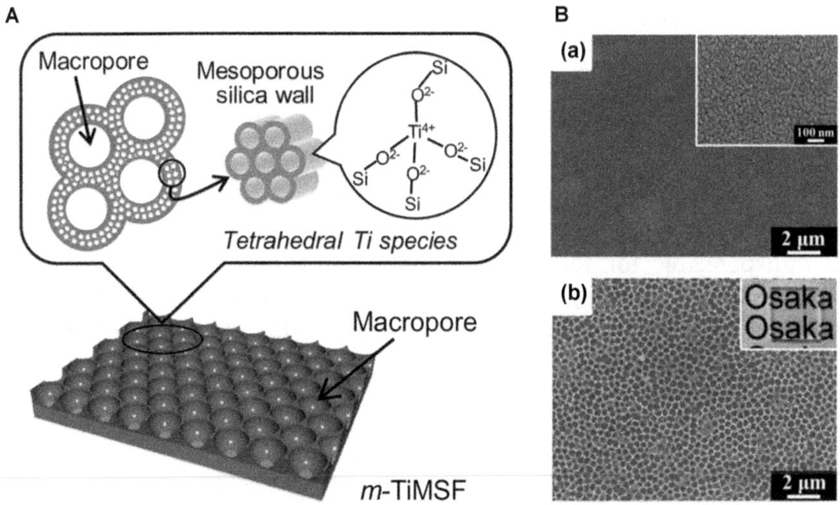

Figure 4.6 (A) Schematic image of Ti-containing porous silica thin films with a uniform mesoporous and macroporous structure (*m*-TiMSF) and (B) SEM image of (a) TiMSF and (b) *m*-TiMSF. Inset of (B,b) shows a photographic image of *m*-TiMSF.[64]

entirely spread out on *m*-TiMSF after UV light irradiation. Differences in the architectures directly influence the surface properties of thin films. In the case of *m*-TiMSF, the mesoporous structures are divided by macropores, leading to an increase in the fine roughness on the film surface. The strong hydrophilic nature of *m*-TiMSF attributes to the positive effects by combination of tetrahedral Ti-species as well as macroporous–mesoporous structures.

Moreover, the surface hydrophilicity of MSF was improved by the incorporation of transition metal species (*e.g.* V, Cr, Mo, W, *etc.*) instead of the tetrahedral Ti-species.[4,69–72] These samples were prepared through almost the same method by using a corresponding metal source (*i.e.* NH_4VO_3, $Cr(NO_3)_3 \cdot 9H_2O$, $(NH_4)_6Mo_7O_{24} \cdot 4H_2O$, or $(NH_4)_{10}W_{12}O_{41} \cdot 5H_2O$). Among them, the tetrahedral W-containing mesoporous silica thin film (WMSF) shows the highest hydrophilic properties.[4,69,70] The water contact angle on WMSF was *ca.* 3°, and changed to almost 0° after UV light irradiation. WMSF also exhibited a long durability superhydrophilic state. In the dark, the water contact angle on WMSF was slowly increased by the adhesion of oily dirt. However, the superhydrophilic state was easily recovered by using a self-cleaning property of WMSF under UV light irradiation. In the case of a tetrahedral Cr-containing mesoporous silica thin film (CrMSF), a surface wettability change was observed under UV and visible light irradiation. CrMSF has the possibility to be applied for indoor use as a self-cleaning material with surface hydrophilicity.[4,71,72]

4.4 Nanocomposite Materials for Superhydrophobic Coatings

4.4.1 Superhydrophobic Surface Designed by Catalytic Property of Coated Materials

Nanocomposite coatings are effective techniques for accumulating multiple functions, enabling maintenance-free characteristics. The selection and precise compositional control of key building blocks in composite materials are indispensable for realizing fascinating functions in various areas of fundamental research and practical application. Artificial hydrophobic surfaces are commonly designed through controlling chemical compositions for lowering the surface energy and geometric structures of solid surfaces on a micro- or nano-scale.[2,73–76] Considering both factors is indispensable for achieving sufficient superhydrophobic states. In numerous instances, carbon materials and fluorine-containing polymers are adopted and combined with other desired materials for the design of a superhydrophobic surface.[77–80]

The superhydrophobic surface has been designed by the synthesis of carbon nanotubes (CNTs) on Ti-containing mesoporous silica thin films (TiMSF) with a Co–Mo catalyst (Figure 4.7). In this case, Co–Mo catalysts

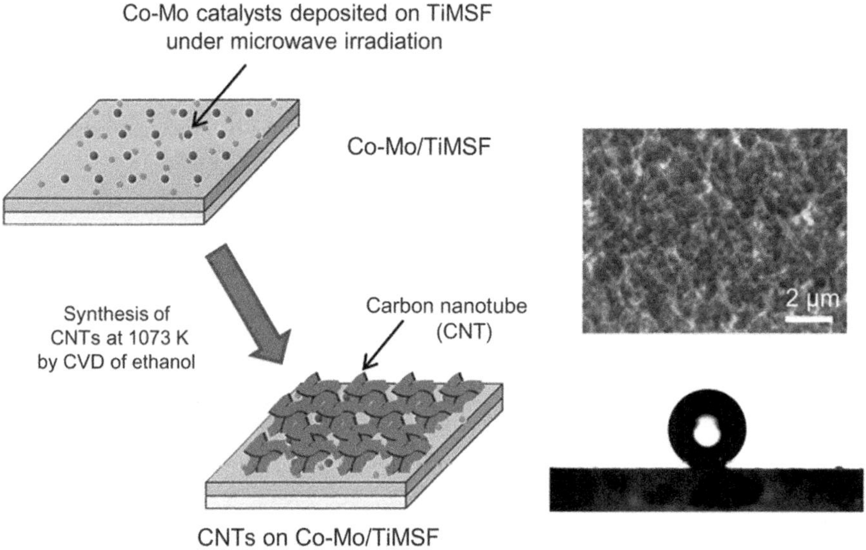

Figure 4.7 Preparation procedure of CNTs-modified TiMSF with a Co–Mo catalyst, SEM image of surface and photographic image of water droplet. Reproduced from ref. 86 with permission from the PCCP Owner Societies.

were deposited on TiMSF under microwave irradiation (500 W; 2450 MHz). It is well known that microwave irradiation enables rapid and uniform heating, which is also applicable to the synthesis of metallic colloidal nanoparticles with a narrow size distribution.[23,81–86] Highly dispersed metal nanoparticles are deposited on Ti-containing mesoporous silica by using the interaction between isolated Ti-species within the silica frameworks and microwaves.[84–86] For the design of a superhydrophobic surface, the Co–Mo deposited TiMSF surface was modified by CNTs through the catalytic reaction of Co–Mo and alcohol vapor at 1073 K.[86] The surface of the thin film was densely covered with CNTs (diameter: *ca.* 15 nm). A surface fine structure was constructed by the formed CNTs. On this film surface, the water contact angle reached 165°, indicating the formation of a superhydrophobic surface. The Co–Mo catalysts are not uniformly formed on the MSF without the incorporation of Ti-species or by conventional thermal heating. Increases in the water contact angle were observed after surface modification by CNTs. However, the superhydrophobic state was hardly realized on the thin film surface.

Tetrahedral Cr-containing colloidal silica nanoparticles (Cr/SNPs) were also used in the design of a superhydrophobic surface.[87] Cr/SNPs were fixed on a quartz substrate, and played roles in the formation of surface roughness and photocatalyst for modification. In this case, the surface was modified by polyethylene (PE) *via* photocatalytic ethylene polymerization. The tetrahedral Cr-species show photocatalytic activity for polymerization of ethylene under UV and visible light irradiation.[71,72,88] The surface of the substrate was firstly

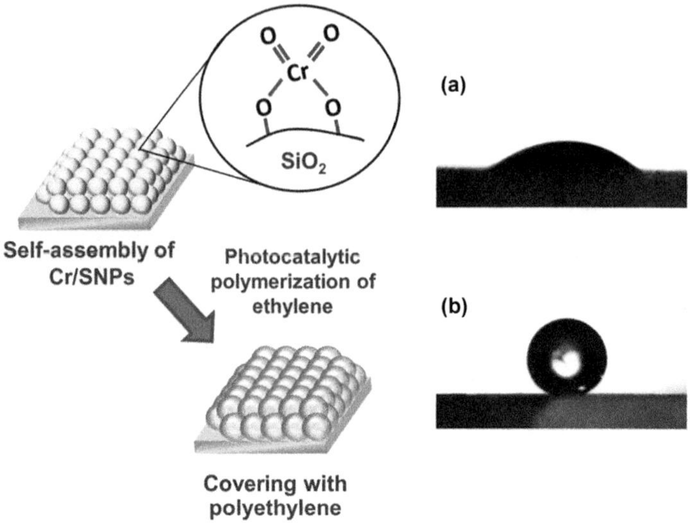

Figure 4.8 Schematic image of Cr/SNPs-coated glass and photographic images of a water droplet before (a) and after (b) photocatalytic polymerization of ethylene.
Reproduced from ref. 87 with permission from the Royal Society of Chemistry.

covered by Cr/SNPs through the spin coating process. Cr/SNPs have a uniform spherical shape *ca.* 80 nm in diameter (Figure 4.8). Photocatalytic ethylene polymerization on the surface was performed by UV light irradiation in the presence of ethylene gas at 298 K. The formation of PE was confirmed by FT-IR measurement. The spherical shape of Cr/SNPs was hardly changed before and after formation of PE. These thin films have a high transparency. After PE modification, the water contact angle reached *ca.* 162° (Figure 4.8). A sufficient superhydrophobic property was achieved by PE modification on the Cr/SNPs surface using the photocatalytic property of coated materials.

4.4.2 Superhydrophobic Surface with Photocatalytic Self-cleaning Property

The nanocomposite thin films mentioned above show sufficient superhydrophobicity, however, these functions are limited and are scarcely used with a self-cleaning property. The design of superhydrophobic surfaces with additional functions have been investigated in recent years. For instance, the reversible switching of surface wettability between superhydrophobic and superhydrophilic states was studied using external stimuli-sensitive materials.[7,89–95] External stimuli, *i.e.*, light, temperature, pH and others, induce the reversible switching of surface wettability.[89–95] These are nanocomposite materials designed by combining each component with desired functions. In practical uses, the performances of superhydrophobic coatings are slowly decreased by adhesion of oily dirt. TiO_2 is one of the promising materials for superhydrophilic coatings with a self-cleaning property and has enough potential for easy removal of oily dirt under UV light irradiation.[95–98] However, the design of durable superhydrophobic surfaces with photocatalytic self-cleaning properties *via* combination of hydrophobic materials and TiO_2 is not so easy due to the photoinduced wettability changes of the TiO_2 surface. The composite thin film photocatalyst of hydrophilic materials and TiO_2 is well-investigated by many researchers, whereas the combination of hydrophobic materials and TiO_2 has not yet received a full investigation. The simultaneous uses of the hydrophobic property and the functions of TiO_2 lead to durable superhydrophobic surfaces with a photocatalytic self-cleaning property.

A nanocomposite coating of TiO_2 and polytetrafluoroethylene (PTFE) by radio frequency-magnetron sputtering (RF-MS) deposition provides superior hydrophobicity with a photocatalytic self-cleaning effect on a fine-structured substrate.[97] Dry processes for thin film coatings are very attractive for mixing several components, which has some difficulties in wet-chemical processes. The TiO_2 and PTFE composite film was prepared on quartz and a structured Ti substrate (denoted as TiO_2-PTFE/Q and TiO_2-PTFE/Ti). The TiO_2 and PTFE were accumulated on the substrate at the same time (molar ratio: Ti–F = *ca.* 0.15). A FE-SEM image showed that TiO_2 and PTFE deposited separately on the quartz substrate (TiO_2/Q and PTFE/Q) was composed of accumulated small roundish particles without any large cracks (Figure 4.9(a)–(c)). The sea-island surface structure was observed in TiO_2-PTFE/Q. The sea and island parts consist of PTFE and TiO_2, respectively. The structured Ti substrate has a needle-like structure. In the deposition of PTFE and PTFE-TiO_2, each component was uniformly accumulated on the structured Ti substrate without large geometric changes (Figure 4.9(d)–(f)). Several Ti needles were covered with each component and thus formed a relatively thick rod-like structure. FT-IR investigations also indicated the state of PTFE in composite samples. Absorption peaks were observed at *ca.* 1725, 1250, and 740 cm^{-1} attributed to the existence of CF=CF and CF_n ($n = 1$–3) groups.[99,100] PTFE was not completely decomposed during the sputtering deposition process.

The measurement of the water contact angle shows the importance of the surface fine structure of coated materials. The water contact angles on PTFE/Q and TiO_2-PTFE/Q were only about 116° and 111°, respectively. The water contact angle is only about 120° on flat fluorine-containing polymers such as a PTFE surface. The water contact angle on PTFE/Ti (168°) and TiO_2-PTFE/Ti (157°) fulfilled the definition of a superhydrophobic surface. A water droplet with a spherical shape was clearly observed on these surface.[96] Under UV light irradiation, TiO_2/Ti was gradually hydrophilized and finally achieved a superhydrophilic state. The water contact angles on PTFE/Ti and TiO_2-PTFE/Ti were almost the same values even after UV light irradiation (Figure 4.10). The deposition of TiO_2 on PTFE/Ti induced surface

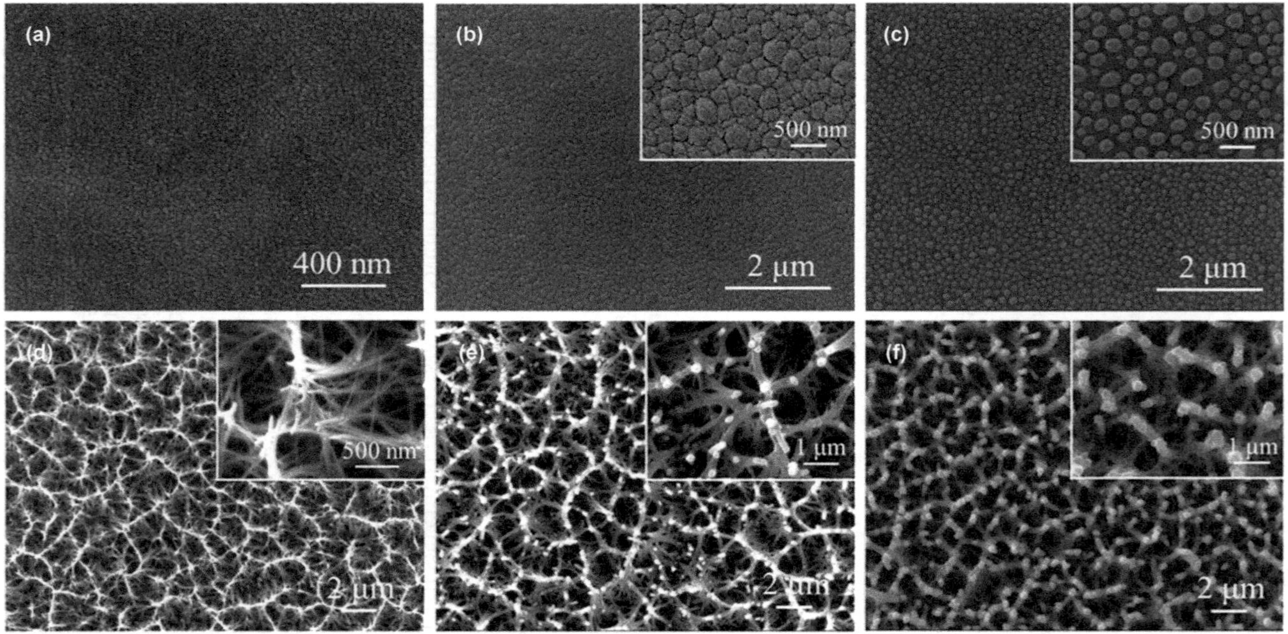

Figure 4.9 SEM images of (a) TiO_2/Q, (b) PTFE/Q, (c) PTFE-TiO_2/Q, (d) structured Ti substrate, (e) PTFE deposited on (d), and (f) PTFE-TiO_2 deposited on (d).[97]

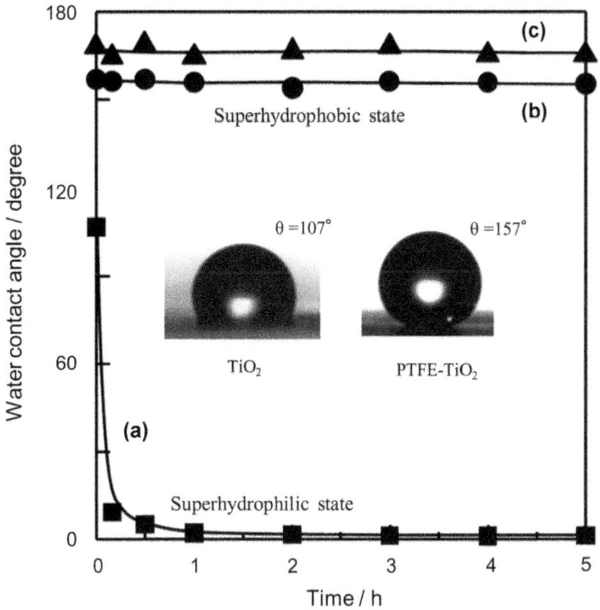

Figure 4.10 Changes in the water contact angle (θ) on each sample ((a) TiO_2/Ti, (b) PTFE-TiO_2/Ti, and (c) PTFE/Ti) under UV light irradiation. Inset: photographic images of the water droplet before UV light irradiation.[97]

wettability changes to the superhydrophilic state, which were similar to that on TiO_2/Ti. The nanocomposite coating of TiO_2-PTFE restricted the surface wettability changes. The two domains mixed on the nano-scale possess hydrophilic and hydrophobic natures, which maintained a good water-repellent property without spreading of the water droplet. The increases of TiO_2 content in the composite of TiO_2-PTFE led to surface wettability changes under the same conditions. Chemical composition plays an important role in the design of an artificial superhydrophobic surface. The self-cleaning property of each sample was confirmed by the decomposition of oleic acid as a model contaminant of oily dirt under UV light irradiation (Figure 4.11). After adhesion of oleic acid, the water contact angle on each film was immediately changed to around 65°. The water contact angle on PTFE/Ti was almost the same value after UV light irradiation. Once the surface was polluted, the superhydrophobicity was hardly recovered without washing by an organic solvent. In the case of TiO_2/Ti, the water droplet was completely spread out on the surface after UV light irradiation, which realized both effects of the photocatalytic degradation of oleic acid and photoinduced superhydrophilicity. In contrast, TiO_2-PTFE/Ti showed a unique response and a gradual increase in the water contact angle to its original superhydrophobic state. The removal of oleic acid was achieved by the photocatalytic self-cleaning property of the TiO_2 domain. TiO_2-PTFE/Ti also restored its superhydrophobic state even after five cycles of adhesions of

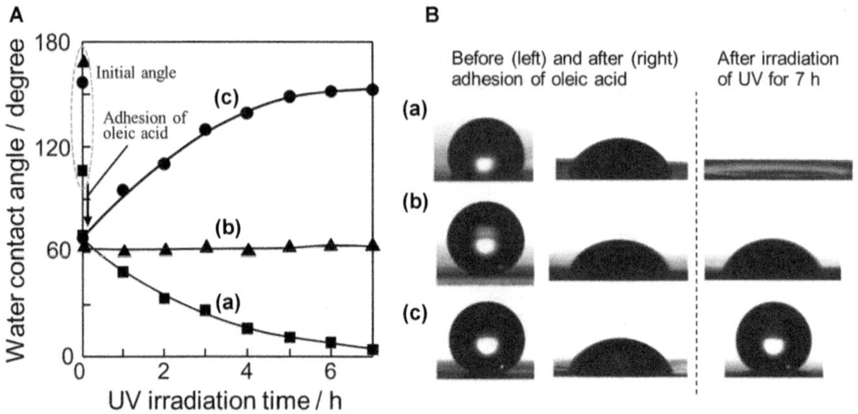

Figure 4.11 (A) Water contact angle changes and (B) photographic images of a water droplet on each sample by adhesion of oleic acid and following UV light irradiation ((a) TiO_2/Ti, (b) PTFE/Ti, and (c) PTFE-TiO_2/Ti).[97]

oleic acid and UV light irradiation. The compatibility between hydrophobic materials and the photocatalytic self-cleaning effect of TiO_2 was attained in TiO_2-PTFE/Ti through the mixing of each component.

4.5 Summary

Coating technologies for the design of superhydrophilic and superhydrophobic surfaces have attracted much attention in recent years. The control of chemical composition and architecture of the film surface is indispensable for realizing fascinating functions. TiO_2 and transition metal-containing porous silica thin films exhibit superhydrophilicity and a self-cleaning property. Superhydrophobic surfaces were also designed using the catalytic property of coated materials *via* formation of a carbon nanotube and polyethylene. A nanocomposite of TiO_2 and PTFE realized the use of a photocatalytic self-cleaning property on a superhydrophobic surface. The simultaneous use of the self-cleaning property as an additional function is useful for keeping a clean surface with superhydrophilicity and superhydrophobicity. The usage of light energy, including natural solar light, opens the way to keeping clean surfaces without a complicated system, leading to energy-saving and maintenance-free coatings.

References

1. G. Kumar and K. N. Prabhu, *Adv. Colloid Interface Sci.*, 2007, **133**, 61.
2. A. Lafuma and D. Quéré, *Nat. Mater.*, 2003, **2**, 457.
3. A. Fujishima, T. N. Rao and D. A. Tryk, *J. Photochem. Photobiol., C*, 2000, **1**, 1.
4. Y. Horiuchi and H. Yamashita, *Appl. Catal., A*, 2011, **400**, 1.
5. X. Feng and L. Jiang, *Adv. Mater.*, 2006, **18**, 3063.

6. P. Roach, N. J. Shirtcliffe and M. I. Newton, *Soft Matter*, 2008, **4**, 224.
7. B. Xin and J. Hao, *Chem. Soc. Rev.*, 2010, **39**, 769.
8. T. Darmanin and F. Guittard, *J. Mater. Chem. A*, 2014, **2**, 16319.
9. W. Barthlott and C. Neinhuis, *Planta*, 1997, **202**, 1.
10. L. Feng, S. H. Li, Y. S. Li, H. J. Li, L. J. Zhang, J. Zhai, Y. L. Song, B. Q. Liu, L. Jiang and D. B. Zhu, *Adv. Mater.*, 2002, **14**, 1857.
11. X. F. Gao and L. Jiang, *Nature*, 2004, **432**, 36.
12. X. Chen and S. S. Mao, *Chem. Rev.*, 2007, **107**, 2891.
13. A. Kubacka, M. Fernández-García and G. Colón, *Chem. Rev.*, 2011, **112**, 1555.
14. Y. Kuwahara, T. Kamegawa, K. Mori and H. Yamashita, *Curr. Org. Chem.*, 2010, **14**, 616.
15. X. Qian, K. Fuku, Y. Kuwahara, T. Kamegawa, K. Mori and H. Yamashita, *ChemSusChem*, 2014, **7**, 1528.
16. H. Yamashita, M. Harada, J. Misaka, M. Takeuchi, K. Ikeue and M. Anpo, *J. Photochem. Photobiol., A*, 2002, **148**, 257.
17. R. Asahi, T. Morikawa, T. Ohwaki, K. Aoki and Y. Taga, *Science*, 2001, **293**, 269.
18. T. Ohno, M. Akiyoshi, T. Umebayashi, K. Asai, T. Mitui and M. Matsumura, *Appl. Catal., A*, 2004, **265**, 115.
19. Y. Kuwahara, K. Maki, Y. Matsumura, T. Kamegawa, K. Mori and H. Yamashita, *J. Phys. Chem. C*, 2009, **113**, 1552.
20. T. Kamegawa, D. Yamahana and H. Yamashita, *J. Phys. Chem. C*, 2010, **114**, 15049.
21. X. F. Qian, T. Kamegawa, K. Mori, H. X. Li and H. Yamashita, *J. Phys. Chem. C*, 2013, **117**, 19544.
22. T. Kamegawa, Y. Ishiguro, H. Seto and H. Yamashita, *J. Mater. Chem. A*, 2015, **3**, 2323.
23. K. Fuku, T. Kamegawa, K. Mori and H. Yamashita, *Chem. – Asian J.*, 2012, **7**, 1366.
24. G. Palmisano, E. Garcia-Lopez, G. Marci, V. Loddo, S. Yurdakal, V. Augugliaro and L. Palmisano, *Chem. Commun.*, 2010, **46**, 7074.
25. X. Lang, W. Ma, C. Chen, H. Ji and J. Zhao, *Acc. Chem. Res.*, 2013, **47**, 355.
26. T. Kamegawa, H. Seto, S. Matsuura and H. Yamashita, *ACS Appl. Mater. Interfaces*, 2012, **4**, 6635.
27. T. Kamegawa, S. Matsuura, H. Seto and H. Yamashita, *Angew. Chem., Int. Ed.*, 2013, **52**, 916.
28. Y. Shiraishi, Y. Sugano, S. Tanaka and T. Hirai, *Angew. Chem., Int. Ed.*, 2010, **49**, 1656.
29. H. Kominami, S. Yamamoto, K. Imamura, A. Tanaka and K. Hashimoto, *Chem. Commun.*, 2014, **50**, 4558.
30. R. Wang, K. Hashimoto, A. Fujishima, M. Chikuni, E. Kojima, A. Kitamura, M. Shimohigoshi and T. Watanabe, *Nature*, 1997, **388**, 431.
31. R. Wang, K. Hashimoto, A. Fujishima, M. Chikuni, E. Kojima, A. Kitamura, M. Shimohigoshi and T. Watanabe, *Adv. Mater.*, 1998, **10**, 135.

32. J. Liu, M. Li, J. Wang, Y. Song, L. Jiang, T. Murakami and A. Fujishima, *Environ. Sci. Technol.*, 2009, **43**, 9425.
33. T. Kawahara, Y. Konishi, H. Tada, N. Tohge, J. Nishii and S. Ito, *Angew. Chem., Int. Ed.*, 2002, **41**, 2811.
34. X. Chen, Z. Li, J. Ye and Z. Zou, *Chem. Mater.*, 2010, **22**, 3583.
35. Y. Li, T. Sasaki, Y. Shimizu and N. Koshizaki, *J. Am. Chem. Soc.*, 2008, **130**, 14755.
36. G. Kaune, M. Memesa, R. Meier, M. A. Ruderer, A. Diethert, S. V. Roth, M. D'Acunzi, J. S. Gutmann and P. Mller-Buschbaum, *ACS Appl. Mater. Interfaces*, 2009, **1**, 2862.
37. T. Kamegawa, N. Suzuki and H. Yamashita, *Chem. Lett.*, 2009, **39**, 610.
38. T. Kamegawa, N. Suzuki and H. Yamashita, *Energy Environ. Sci.*, 2011, **4**, 1411.
39. H. Y. Lee, Y. H. Park and K. H. Ko, *Langmuir*, 2000, **16**, 7289.
40. J. M. Thomas, *Angew. Chem., Int. Ed.*, 1999, **38**, 3588.
41. A. Corma, *Chem. Rev.*, 1997, **97**, 2373.
42. H. Yamashita and M. Anpo, *Curr. Opin. Solid State Mater. Sci.*, 2003, **7**, 471.
43. H. Yamashita and K. Mori, *Chem. Lett.*, 2007, **36**, 348.
44. H. Yamashita, K. Mori, S. Shironita and Y. Horiuchi, *Catal. Surv. Asia*, 2008, **12**, 88.
45. H. Yamashita, Y. Fuji, Y. Ichihashi, S. G. Zhang, K. Ikeue, D. R. Park, K. Koyano, T. Tatsumi and M. Anpo, *Catal. Today*, 1998, **45**, 221.
46. K. Ikeue, H. Yamashita, T. Takewaki and M. Anpo, *J. Phys. Chem. B*, 2001, **105**, 8350.
47. H. Yamashita, K. Ikeue, T. Takewaki and M. Anpo, *Top. Catal.*, 2002, **18**, 95.
48. H. Yamashita, Y. Ichihashi, M. Anpo, M. Hashimoto, C. Louis and M. Che, *J. Phys. Chem.*, 1996, **100**, 16041.
49. H. Yamashita, S. G. Zhang, Y. Ichihashi, Y. Matsumura, Y. Souma, T. Tatsumi and M. Anpo, *Appl. Surf. Sci.*, 1997, **121**, 305.
50. I. R. Subbotina, B. N. Shelimov, V. B. Kazansky, A. A. Lisachenko, M. Che and S. Coluccia, *J. Catal.*, 1999, **184**, 390.
51. T. Kamegawa, R. Takeuchi, M. Matsuoka and M. Anpo, *Catal. Today*, 2006, **111**, 248.
52. T. Kamegawa, J. Morishima, M. Matsuoka, J. M. Thomas and M. Anpo, *J. Phys. Chem. C*, 2007, **111**, 1076.
53. H. Yamashita, K. Yoshizawa, M. Ariyuki, S. Higashimoto, M. Che and M. Anpo, *Chem. Commun.*, 2001, 435.
54. H. Yamashita, M. Ariyuki, K. Yoshizawa, K. Kida, S. Ohshiro and M. Anpo, *Res. Chem. Intermed.*, 2004, **30**, 235.
55. S. Ohshiro, O. Chiyoda, K. Maekawa, Y. Masui, M. Anpo and H. Yamashita, *C. R. Chim.*, 2006, **9**, 846.
56. H. Yamashita, S. Ohshiro, K. Kida, K. Yoshizawa and M. Anpo, *Res. Chem. Intermed.*, 2003, **29**, 881.
57. M. Ogawa, *Chem. Commun.*, 1996, 1149.

58. N. Nishiyama, S. Tanaka, Y. Egashira, Y. Oku and K. Ueyama, *Chem. Mater.*, 2002, **14**, 4229.
59. H. Yamashita, S. Nishio, I. Katayama, N. Nishiyama and H. Fuji, *Catal. Today*, 2006, **111**, 254.
60. Y. Horiuchi, H. Ura, T. Kamegawa, K. Mori and H. Yamashita, *Appl. Catal., A*, 2010, **387**, 95.
61. H. Yamashita, S. Nishio, S. Imaoka, M. Shimada, K. Mori, T. Tanaka and N. Nishiyama, *Top. Catal.*, 2008, **47**, 116.
62. Y. Horiuchi, H. Ura, T. Kamegawa, K. Mori and H. Yamashita, *J. Mater. Chem.*, 2011, **21**, 236.
63. Y. Horiuchi, H. Ura, T. Kamegawa, K. Mori and H. Yamashita, *J. Phys. Chem. C*, 2011, **115**, 15410.
64. T. Kamegawa, Y. Masuda, N. Suzuki, Y. Horiuchi and H. Yamashita, *ACS Appl. Mater. Interfaces*, 2011, **3**, 4561.
65. J. M. Thomas and G. Sankar, *Acc. Chem. Res.*, 2001, **34**, 571.
66. H. Yamashita, S. Kawasaki, Y. Ichihashi, M. Harada, M. Takeuchi, M. Anpo, G. Stewart, M. A. Fox, C. Louis and M. Che, *J. Phys. Chem. B*, 1998, **102**, 5870.
67. L. Marchese, T. Maschmeyer, E. Gianotti, S. Coluccia and J. M. Thomas, *J. Phys. Chem. B*, 1997, **101**, 8836.
68. L. Marchese, E. Gianotti, V. Dellarocca, T. Maschmeyer, F. Rey, S. Coluccia and J. M. Thomas, *Phys. Chem. Chem. Phys.*, 1999, **1**, 585.
69. Y. Horiuchi, K. Mori, N. Nishiyama and H. Yamashita, *Chem. Lett.*, 2008, **37**, 748.
70. Y. Horiuchi, T. Kamegawa, K. Mori, H. Yamashita and N. Nishiyama, *e-J. Surf. Sci. Nanotechnol.*, 2009, **7**, 141.
71. K. Mori, S. Imaoka, S. Nishio, Y. Nishiyama, N. Nishiyama and H. Yamashita, *Microporous Mesoporous Mater.*, 2007, **101**, 288.
72. H. Yamashita, Y. Horiuchi, S. Imaoka, S. Nishio, N. Nishiyama and K. Mori, *Catal. Today*, 2008, **132**, 146.
73. D. Öner and T. J. McCarthy, *Langmuir*, 2000, **16**, 7777.
74. P. Aussillous and D. Quere, *Nature*, 2001, **411**, 924.
75. A. Steele, I. Bayer and E. Loth, *Nano. Lett.*, 2009, **9**, 501.
76. T. Onda, S. Shibuichi, N. Satoh and K. Tsuji, *Langmuir*, 1996, **12**, 2125.
77. A. Nakajima, A. Fujishima, K. Hashimoto and T. Watanabe, *Adv. Mater.*, 1999, **11**, 1365.
78. J. Genzer and K. Efimenko, *Science*, 2000, **290**, 2130.
79. D. Schondelmaier, S. Cramm, R. Klingeler, J. Morenzin, Ch. Zilkens and W. Eberhardt, *Langmuir*, 2002, **18**, 6242.
80. K. Tadanaga, J. Morinaga, A. Matsuda and T. Minami, *Chem. Mater.*, 2000, **12**, 590.
81. J. A. Gerbec, D. Magana, A. Washington and G. F. Strouse, *J. Am. Chem. Soc.*, 2005, **127**, 15791.
82. K. Fuku, R. Hayashi, T. Takakura, T. Kamegawa, K. Mori and H. Yamashita, *Angew. Chem., Int. Ed.*, 2013, **52**, 7446.

83. P. Verma, Y. Kuwahara, K. Mori and H. Yamashita, *J. Mater. Chem. A*, 2015, **3**, 18889.
84. S. Shironita, T. Takasaki, T. Kamegawa, K. Mori and H. Yamashita, *Catal. Lett.*, 2009, **129**, 404.
85. S. Shironita, T. Takasaki, T. Kamegawa, K. Mori and H. Yamashita, *Top. Catal.*, 2010, **53**, 218.
86. Y. Horiuchi, Y. Shimizu, T. Kamegawa, K. Mori and H. Yamashita, *Phys. Chem. Chem. Phys.*, 2011, **13**, 6309.
87. Y. Horiuchi, K. Fujiwara, T. Kamegawa, K. Mori and H. Yamashita, *J. Mater. Chem.*, 2011, **21**, 8543.
88. H. Yamashita, K. Kida, K. Ikeue, Y. Kanazawa, K. Yoshizawa and M. Anpo, *Stud. Surf. Sci. Catal.*, 2002, **146**, 597.
89. X. Zhang, M. Jin, Z. Liu, D. A. Tryk, S. Nishimoto, T. Murakami and A. Fujishima, *J. Phys. Chem. C*, 2007, **111**, 14521.
90. X. Feng, J. Zhai and L. Jiang, *Angew. Chem., Int. Ed.*, 2005, **44**, 5115.
91. Q. Fu, G. V. R. Rao, S. B. Basame, D. J. Keller, K. Artyushkova, J. E. Fulghum and G. P. Lopez, *J. Am. Chem. Soc.*, 2004, **126**, 8904.
92. F. Zhou and W. T. S. Huck, *Chem. Commun.*, 2005, 5999.
93. S. Wang, H. Liu, D. Liu, X. Ma, X. Fang and L. Jiang, *Angew. Chem., Int. Ed.*, 2007, **46**, 3915.
94. G. K. Jennings and E. L. Brantley, *Adv. Mater.*, 2004, **16**, 1983.
95. J. Lahann, S. Mitragotri, T. N. Tran, H. Kaido, J. Sundaram, I. S. Choi, S. Hoffer, G. A. Somorjai and R. Langer, *Science*, 2003, **299**, 371.
96. A. Nakajima, K. Hashimoto, T. Watanabe, K. Takai, G. Yamauchi and A. Fujishima, *Langmuir*, 2000, **16**, 7044.
97. T. Kamegawa, Y. Shimizu and H. Yamashita, *Adv. Mater.*, 2012, **24**, 3697.
98. T. Kamegawa, K. Irikawa and H. Yamashita, *Chem. Lett.*, 2015, **44**, 509.
99. M. A. Golub, T. Wydeven and A. L. Johnson, *Langmuir*, 1998, **14**, 2217.
100. C. Becker, J. Petersen, G. Mertz, D. Ruch and A. Dinia, *J. Phys. Chem. C*, 2011, **115**, 10675.

CHAPTER 5

Producing Self-cleaning, Transparent and Hydrophobic SiO_2-crystalline TiO_2 Nanocomposites at Ambient Conditions for Stone Protection and Consolidation

C. KAPRIDAKI,[a] L. PINHO,[b] M. J. MOSQUERA*[c] AND PAGONA MARAVELAKI-KALAITZAKI*[a]

[a] School of Architectural Engineering, Technical University of Crete, Akrotiri University Campus, 73100 Chania, Crete, Greece; [b] Laboratory of Inorganic Materials Chemistry (CMI), University of Namur, 61 rue de Bruxelles, B-5000 Namur, Belgium; [c] TEP-243 Nanomaterials Group, Departamento de Química-Física, Facultad de Ciencias, Campus Universitario Río San Pedro, Universidad de Cádiz, 11510 Puerto Real, Cádiz, Spain,

*Email: mariajesus.mosquera@uca.es; pmaravelaki@isc.tuc.gr

5.1 Introduction

The deterioration process of monuments and modern buildings originates from both the dramatic increase of the atmospheric pollution and the high levels of humidity. As a direct consequence of these external actions, ugly

RSC Smart Materials No. 21
Self-cleaning Coatings: Structure, Fabrication and Application
Edited by Junhui He

Published by the Royal Society of Chemistry, www.rsc.org

black crusts have been extensively formed on the building façades since organic matter and nitrogen oxides are deposited on the deteriorated surfaces. Therefore, the development of innovative versatile coatings for the protection and self-cleaning of building substrates has attracted the attention of researchers.[1,2]

TiO_2 nanoparticles (titania) have been widely used as semiconductor photocatalysts in many fields, such as water purification, self-cleaning materials, sensors and solar energy conversion, due to their unique properties.[3,4] The use of TiO_2 nanoparticles together with construction building materials has contributed to the development of versatile materials with continuously growing applications.[1] Titania has been successfully incorporated within mortars, concrete, exterior tiles and paving blocks during their manufacturing process, thus producing not only self-cleaning surfaces but also substrates with improved mechanical properties.[1,5–7] Moreover, the addition of TiO_2 nanoparticles within lime mortars has enhanced the carbonation process and the formation of hydraulic phases.[5]

However, the application of TiO_2 in the conservation of monuments and specifically, on their external façades, imposes several restrictions regarding the synthesis and application of the self-cleaning products.[8,9] Firstly, the applied materials should be synthesized *via* simple synthesis routes (one-pot synthesis), not demanding additional treatment such as heating, during *in situ* applications on the monuments or building constructions. Secondly, the production of self-cleaning materials should minimize the energy demand required as much as possible, giving rise for further scaling-up procedures. Thirdly, transparent and stable coatings are preferred in monument protection, since they comply with the aesthetic requirements imposed by the curators of cultural heritage.

Different strategies concerning the application of TiO_2 on exterior building surfaces have been recently reported in the literature. These include the use of aqueous TiO_2 colloidal dispersions,[10–12] TiO_2 nanoparticles incorporated in polymeric matrices[13,14] or even titanium isopropoxide solutions.[15,16] However, some of these strategies might entail remarkable drawbacks related both to the aesthetic parameters and the adhesion between coatings and substrates. The typical white colour of TiO_2 nanoparticles might induce a remarkable change in the aesthetic properties of the building materials after the application of the products. Moreover, due to the limited chemical affinity between stone substrates and products, the latter are easily removed from the surfaces, thus decreasing their lifetime. An alternative approach to improve the limited adhesion between TiO_2 and stone substrates is the combination of TiO_2 nanoparticles with a SiO_2 matrix.

It is already known that the silica matrix originating from the hydrolysis and condensation of the alkoxysilanes, is one of the most widely used bases in protective and consolidant agents.[17] On the one hand, the alkoxysilanes provide low viscosity, which enhances their penetration depth into the pore structure of the stone substrates. On the other hand, the main drawback of these products is the brittleness of the xerogels obtained after the polymerization, which are susceptible to cracking due to the high capillary pressures developed during drying.[18] Several attempts to avoid the cracking of these xerogels have been reported in the literature. Specifically, crack-free materials have been prepared in the presence of drying control chemical additives (DCCA)[19] and surfactants, such as *n*-octylamine,[20] to produce xerogels with large and uniform pores. In addition, the incorporation of TiO_2, SiO_2, Al_2O_3 and calcium oxalate (CaOx) particles into silica xerogels has also been reported to reduce cracking.[21–23] Finally, the addition of organosilanes into the silica precursors reduces the surface energy of the sols, thus preventing cracking and providing toughness and flexibility to the final xerogels.[24,25]

Therefore, products based on the incorporation of TiO_2 nanoparticles within a silica matrix, in the presence of *n*-octylamine, have been synthesized, as an alternative approach to innovative self-cleaning agents.[8,9] These silica–titania mesoporous crack-free materials induce improved adhesion properties, thus ensuring the long-term performance of the products. In addition, mixing Ti and Si alkoxide precursors has also been used to produce TiO_2 dispersions in a silica matrix, thus providing sols with lower values of viscosity than the sols containing pre-synthesized TiO_2 nanoparticles.

However, these synthesis routes exhibit limitations relating not only to the experimental procedures, but also to *in situ* applications. Firstly, it is well known that the Ti alkoxides hydrolyze much faster than Si alkoxides,[26,27] thus providing, after application on building substrates, inhomogeneous and cracked coatings. Secondly, the formation of the photoactive anatase TiO_2 crystals usually demands high temperatures during the synthesis routes.[28]

The innovative aspect of the work presented here refers to the synthesis of photoactive, transparent and hydrophobic nanocomposites at ambient conditions for the protection of monuments and buildings. We investigated nanocomposites by mixing Ti and Si alkoxides in the presence of Ox and hydroxyl-terminated polydimethylsiloxane (PDMS). In order to evaluate the effect of Ox concentration on the nanostructure and on the properties of the nanocomposites, we designed three nanocomposites with increasing Ox concentrations and different sequences of the reagent agents.

Hydroxyl-terminated PDMS was selected as an additive to the nanocomposite structure due to its ability to reduce the surface energy of the colloidal solutions providing elasticity, flexibility, hydrophobic and crack-free properties to the final xerogels.[29,30] Regarding the Ox addition, we primarily thought that it could behave as an additional protective agent for the stone substrate through its reaction with the calcium carbonate and the production of CaOx. It is well known that CaOx, which is more stable and resistant to weathering than calcium carbonate, constitutes the main component found in well-preserved surfaces of monuments, the so-called "patinas" layers.[31,32] In our previous studies, we have developed innovative and effective consolidants based on a silica matrix modified with CaOx nanoparticles.[22,23] Furthermore, the Ox might also play a role as a sol–gel catalyst for the alkoxysilane and as a chelating agent for the delaying of Ti alkoxide hydrolysis, thus providing homogeneous xerogels. One more possible advantage of adding Ox is the fact that this acid has already been established as a DCCA additive that can act as an additional medium for crack-free xerogel production.[33] Finally, according to the literature, the Ox could also behave as a hole scavenger during the photocatalytic procedure.[34–37] More specifically, the action of the oxalic acid ions as a hole-scavenger agent could be summarized to the following: (1) these ions characterized as good bidentate ligands, can be absorbed on the positively charged TiO_2 surfaces; (2) the oxalic ions participate in the photocatalysis firstly binding the photogenerated holes and secondly producing radicals that further promote the photocatalytic procedure.

The present chapter is organized as follows. Firstly, we report the synthesis and characterization of the three nanocomposites, in order to clarify the role both of the Ox concentration and the addition sequence of the precursor materials. Furthermore, we elucidate the reasons that contribute to the photocatalytic activity of the materials synthesized. To gain insight into the structure of the nanomaterials, several analytical techniques were used, such as FTIR, TEM/STEM, AFM and nitrogen physisorption. Finally, the assessment of the nanocomposites as protective, consolidant and self-cleaning materials on marble and calcareous stones for both monuments and modern buildings is also demonstrated.

5.2 Experimental

5.2.1 Materials

SiO_2–TiO_2–PDMS nanocomposites were synthesized by using tetraethylorthosilicate (TEOS, Sigma Aldrich), titanium(IV) isopropoxide (TTIP, Sigma Aldrich) and hydroxyl-terminated polydimethylsiloxane (PDMS, Sigma Aldrich), as raw materials. Ethanol (EtOH, Sigma Aldrich), isopropanol (ISP, Merck) and de-ionized water were used as solvents. Oxalic acid dihydrate (Ox, Panreac) was used as the catalyst. Finally, methyl orange (MO, Fluka) and methylene blue (MB, Panreac) were used for the photocatalytic experiments. All chemicals were used as received, without further purification prior to use.

5.2.2 Synthesis of STP Nanocomposites

In the present work, three nanomaterials were synthesized for producing innovative materials for stone protection and consolidation. The nanomaterials denoted as STP-1, STP-2 and STP-4 were inspired by the initial letters of silica, titania and PDMS, respectively. In order to estimate the influence of Ox on the properties and the structure of the final xerogels, we have synthesized three formulations with increasing Ox concentrations. Therefore, the number followed the names previously presented are related with the Ox concentration of the nanomaterials, number 1 corresponds to the lower Ox concentration, while 2 and 4 to the medium and high Ox content, respectively. Moreover, the STP-1 formulation differs from the other two formulations in the sequence of addition of the reactants during the synthesis. The synthesis routes of the new nanocomposites are illustrated in Scheme 5.1. Specifically, the steps of the STP-1 synthesis are as follows: (a) TEOS, EtOH, H_2O and Ox were stirred at ambient conditions for 140 min; (b) PDMS reagent was added drop by drop to the previous solution and the magnetic stirring was continued for another 15 min; and (c) thereafter the TTIP alkoxide was also added to the TEOS/PDMS solution under stirring. The solution obtained was left under strong magnetic stirring (1200 rpm) for 24 h, thus providing a transparent colloidal solution (Scheme 5.1a).

Regarding the synthesis of STP-2 and STP-4, the main steps are as follows: (a) PDMS, EtOH, H_2O and Ox were stirred at ambient conditions for 90 min; (b) TTIP diluted with EtOH was added with a very slow rate to the previous solution in order to prevent the hydrolysis of TTIP and simultaneous particle agglomeration;[38] and (c) the transparent colloidal solution obtained was left under magnetic stirring for 24 h. Specifically, in the case of STP-2, TTIP should be added in no less than 4 h, in order to obtain a transparent sol. Finally, after the addition of TEOS, the sol remained transparent (Scheme 5.1b).

The molar ratio of TEOS–EtOH–H_2O–PDMS–TTIP used in the synthesis of STP-1 was 1 : 4 : 4 : 0.04 : 0.017, whereas for STP-2 and STP-4, we used 1 : 5.6 : 4 : 0.04 : 0.017 (see Table 5.1).

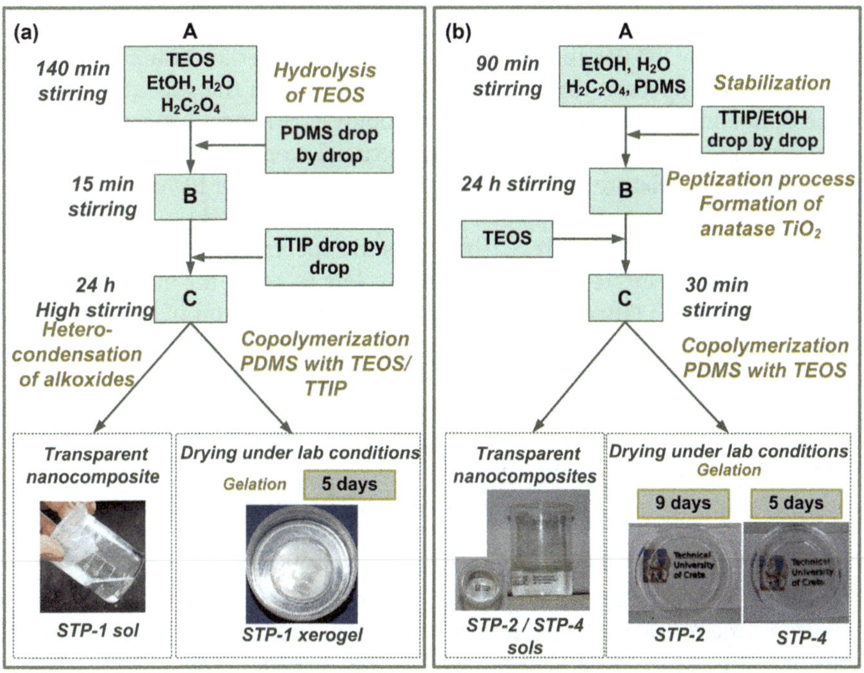

Scheme 5.1 Synthesis routes of: (a) STP-1 and (b) STP-2/STP-4 nanocomposites.

Table 5.1 Properties of the materials under study. Reprinted from C. Kapridaki, L. Pinho, M. Mosquera and P. Maravelaki-Kalaitzaki, Producing photoactive, transparent and hydrophobic SiO_2-crystalline TiO_2 nanocomposites at ambient conditions with application as self-cleaning coatings, *Appl. Catal., B*, **156–157**, 417–427, Copyright (2014), with permission from Elsevier.[69]

Sample	Ox moles[a]	Gelation time (days) Open vessel	Closed vessel	Vol. reduction%
STP-1	0.0001	5	35	82
STP-2	0.017	9	45	86
STP-4	0.036	5	7	84

[a]Referred to 1 mol of TEOS.

5.2.3 Characterization of the STP Nanocomposites

In order to observe the appearance of the xerogels obtained, the colloidal sols were cast into transparent molds and sealed by a moldable film on their tops (Parafilm). The film was perforated, thus increasing the solvent evaporation in order to simulate the phenomena taking place on the surface of building materials. These molds were kept under ambient conditions and monitored until a constant weight was reached and visually inspected in order to find any micro-cracking. In addition, the gelation times of the nanocomposites were also measured in both closed and open vessels.

The evolution of the sol–gel process along with the chemical bonds of the derived nanocomposites were investigated through Fourier transform infrared spectroscopy (FTIR) absorption spectra using a Perkin-Elmer 1000 spectrometer in the spectral range of 400–4000 cm^{-1}. Both the sols and powders of the nanocomposites were examined through FTIR. Specifically, the spectra shown in Figure 5.1(a) correspond to the crucial steps of STP-1 synthesis: (i) TEOS–EtOH–Ox mixture at 0 min; (ii) after the addition of PDMS to the latter mixture; (iii) after the addition of TTIP on the 4th day; and (iv) the xerogel of the STP-1 after 35 days. Figure 5.1(b) shows the spectra attributed to STP-2, as follows: (i) PDMS–EtOH–Ox sol after being left to react for 90 minutes; (ii) 24 h after adding TTIP; (iii) immediately after the addition of TEOS and (iv) STP-2 xerogel after 30 days. Since the FTIR spectra obtained for STP-4 and STP-2 were very similar; only the spectra for STP-2 are shown.

The thermal behaviour of the three nanomaterials was determined through differential thermal analysis (DTA) and thermogravimetric analysis (TGA) using a Setaram LabSysEvo 1600 °C at a heating rate 10 °C min^{-1} under a N_2 atmosphere from 27 to 1000 °C. The samples under study were previously dried at 60 °C for 24 h.

The specific surface area and pore volume of the nanocomposites were obtained from N_2 adsorption–desorption isotherms performed at 77 K in a Quantchrome Autosorb IQ device. Furthermore, the pore size distribution of materials was also estimated by analyzing the adsorption data by means of a hybrid NLDFT (non-local density functional theory) approach.[39]

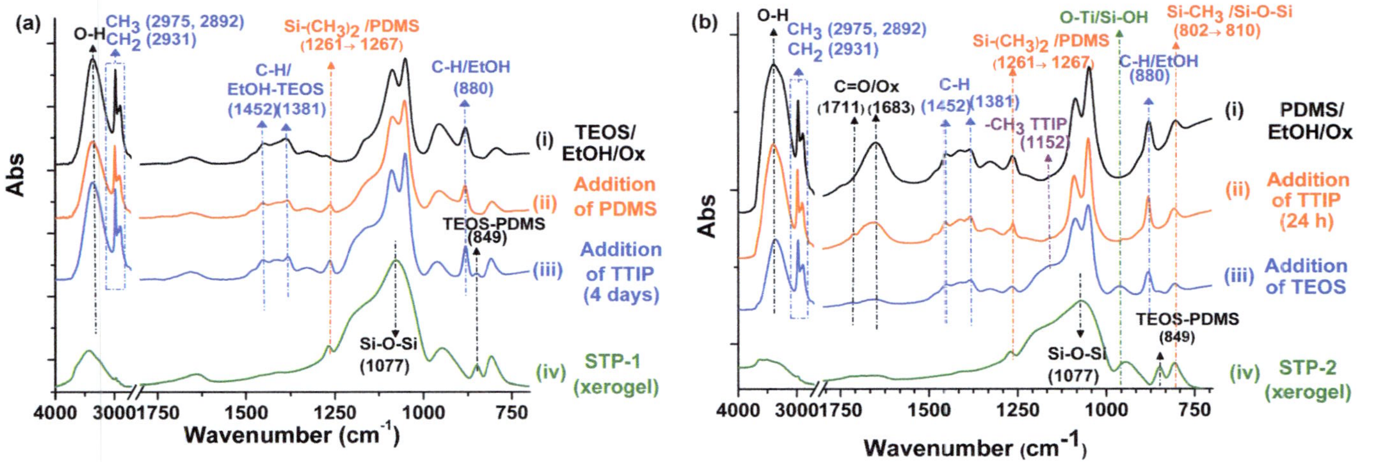

Figure 5.1 (a) FTIR spectra of STP-1 for increasing time intervals: (i) TEOS–EtOH–Ox mixture at 0 min; (ii) after the addition of PDMS to the latter mixture; (iii) after the addition of TTIP on the 4th day of curing time; (iv) STP-1 xerogel cured for 35 days. (b) FTIR spectra of STP-2 (i) PDMS–EtOH–Ox sol mixture after being left to react for 90 minutes; (ii) 24 h after the addition of TTIP to the latter sol; (iii) immediately after the addition of TEOS and (iv) STP-2 xerogel after 30 days.

Detailed microstructural features of the STP materials were examined by transmission electron microscopy (TEM) using a JEOL 2010F TEM/STEM microscope, operating at 200 kV with 0.19 nm spatial resolution, equipped with a high angle annular dark field (HAADF) detector and an Oxford X-Max silicon drift X-ray energy dispersive spectroscopy (XEDS) detector. The images were recorded in both TEM and STEM-HAADF modes.

X-Ray diffraction (XRD) analysis was performed in a Bruker D8 Advance diffractometer using Ni-filtered Cu Kα radiation (35 kV and 35 mA) with a Bruker Lynx Eye strip silicon detector.

The surface of the nanocomposites was also investigated by means of atomic force microscopy (AFM, Nanotec Electrónica S.L.) operated in tapping mode. Roughness measurements were performed on $1\times1\ \mu m^2$ scans.

Diffuse reflectance spectra (DRS) of the nanocomposites under study were recorded in the range of 200–800 nm by using a Varian Cart 3 spectrometer equipped with an integrating sphere. All the spectra were transformed to a magnitude proportional to the extinction coefficient through the Kubelka–Munk function F(R). For comparison purposes, we have also examined commercial TiO_2 nanoparticles (P25) from Evonik.

5.2.4 Photocatalytic Activity of the STP Nanocomposites

In order to evaluate the photocatalytic activity of the three nanocomposites, we assessed the photocatalytic degradation of methyl orange (MO) under UV irradiation. Specifically, 0.3 g of STP powders (previously dried at 60 °C for 24 h) were dispersed in MO aqueous solution with a concentration of 0.3 ppm.

Aiming at elucidating the role played by the TiO_2 prepared in the lab and the Ox added in the designed nanocomposites under the conditions of synthesis, we have also performed the photocatalytic experiments with the following TiO_2 formulations: (a) TiO_2 obtained from TTIP at ambient conditions; and (b) the previous TiO_2 formulation in presence of Ox. The synthesis of the as-prepared TiO_2 formulation used the following experimental procedure: (i) TTIP reagent was added, drop by drop, to a solution of H_2O and ISP, with a final molar ratio of TTIP–H_2O–ISP equal to 1 : 8.7 : 78; (ii) the solution obtained remained under magnetic stirring for 3 days, at constant laboratory temperature, in order for the chemical reaction to be accomplished; (iii) the powder produced was collected through a filtration procedure, washed with H_2O and dried at 60 °C for 24 h to remove the physically adsorbed solvents. This final TiO_2 powder, denoted as T_l formulation, was investigated for its ability to decompose the MO organic matter. For the photocatalytic procedure, T_l was dispersed in an aqueous solution of MO (2.5 ppm) in a T_l–MO molar ratio equal to 2.6×10^3, thus obtaining the dispersion named as $T_{l\text{-}d}$. Moreover, for the assessment of the T_l photocatalytic ability in the presence of oxalic acid, we prepared a new T_l dispersion, denoted with $T_{lOx\text{-}d}$, by adding 10 mg of Ox in the $T_{l\text{-}d}$ previous colloidal solution.

In addition, in both experimental procedures (STP and TiO_2 formulations), the photochemical degradation of MO was also assessed in order to determine the kinetic rate of the purely photochemical degradation mechanism. Prior to the photocatalytic assessment, the dispersed powders were magnetically stirred in the dark for 30 min, in order to achieve adsorption–desorption equilibrium. All the experiments were performed in a UV chamber equipped with four 8 W black light blue lamps emitting in the near-UV range (315–400 nm). The maximum absorbance of MO was monitored at 464–506 nm, depending on the pH values of the colloidal solution, by using a UV-VIS Varian (Cary 1E) spectrophotometer.

Considering that the photocatalytic degradation of MO follows a pseudo-first order kinetic model, the results obtained during the photocatalytic tests were fitted to a non-linear regression model (OriginLab). Therefore, the kinetic behaviour of the photocatalysts under study could be described according to eqn (5.1):

$$C = C_0 e^{-kt}, \tag{5.1}$$

where C corresponds to MO concentration at increasing times; k is the pseudo-first order kinetic constant; t is the reaction time; and C_0 is the initial MO concentration ($t = 0$).

The values of k as well as the values of the regression coefficients (r^2) were recorded and confirmed the good fitting between the experimental results and the model.[40]

5.2.5 Application and Characterization of the Nanocomposites on Stone Substrates

The performance of STP nanomaterials was evaluated on Thassos and Naxos marbles, along with Alfas calcareous limestone (composed of micritic calcite). Prior to the applications, polished marble and limestone samples were washed with deionized water and dried for 24 h at 60 °C. With the aim of evaluating the protective, self-cleaning and consolidation activities of the nanocomposites under study, the following applications were conducted: (a) STP-1 material was applied twice on Thassos marble as a coating; (b) STP-2 and STP-4 were applied as coatings and strengthening agents to Naxos marble and Alfas stone, respectively. All the substrates were first impregnated with EtOH, in order to facilitate the application of the sols. In the case of marble samples, the nanocomposites were applied by brush, whereas the Alfas with high porosity was treated by capillary absorption. The performance of the hybrid materials as protective and strengthening agents was evaluated, after a constant weight was reached, using the techniques described above. Untreated counterparts were also examined for comparison purposes.

The aesthetic changes produced on the stone samples were evaluated by means of the assessment of the total colour difference (ΔE^*) using a spectrophotometer (Konica Minolta, CM2600d).[41]

The hydrophobic behaviour of the treated substrates was investigated through water contact angle (CA) and water capillary absorption (WCA) measurements. CA measurements were determined using an optical tensiometer (Thetalite TL 101, KSV) under laboratory conditions, according to the sessile drop method.[24] Water droplets (~10 μL volume) were applied on three different points of the surface of each stone sample and the static contact angles (θ_s) were measured in the images produced. WCA was measured according to UNI-EN 15801:2010,[42] by using the gravimetric sorption technique. Moreover, the total water capillary absorption (TWCA), which expresses the total amount of water absorbed during the WCA test, was also recorded.

In order to test the ability of the treated specimens to allow water vapour evaporation, we have also carried out a water vapour permeability (WVP) test according to the procedure referred to in UNI-EN 15803:2010,[43] using special disc-shaped specimens.

Treated stone samples and their untreated counterparts were also visualized by scanning electron microscopy (SEM), using a FEI-Quanta Inspect (D8334) instrument. Moreover, in order to evaluate the chemical affinity between the STP nanocomposites and the stone substrates, we performed FTIR analyses of treated stone samples.

The mechanical properties of the treated Alfas stones were evaluated by measuring: (a) the indirect tensile strength (Brazilian test) and (b) the drilling resistance and penetration depth by using a drilling resistance measurement system (DRMS), a Sint Technology device.[44]

Finally, the effectiveness of the materials as self-cleaning agents on stone was assessed by monitoring the discolouration of methylene blue (MB) applied on the surfaces. Specifically, 1 mL of a solution of 1 mM MB dissolved in ethanol was deposited on the treated and untreated surfaces. The coloured specimens were later irradiated in the UV chamber (equipped with 8 W black light blue lamps, emitting in the near-UV range 315–400 nm; 3 mW cm^{-2}), and the values of the total colour differences (ΔE^*) at various time intervals were recorded using a spectrophotometer.[8] In particular, the photocatalytic activity of the treated specimens was demonstrated through the recorded ratios of $\Delta E^*/\Delta E_0^*$ at various irradiation times. ΔE^* corresponds to the total colour difference between the unstained surface and the corresponding stained surface at increasing time intervals; ΔE_0^* expresses the total colour difference between stained and unstained surface at time equal to 0 min.

5.3 Results and Discussion

5.3.1 Composition Characterization

The sol–gel transition of the synthesized nanocomposites took place spontaneously within 5–9 days under laboratory conditions (see Table 5.1). However, both the sequence of the reagents and their amounts have a clear effect on the gelation time of the sols at ambient conditions in open and closed vessels (Table 5.1). The product STP-1 cannot be directly compared with the other two nanocomposites, since both the Ox content and the sequence of addition of the materials is different. Therefore, we can compare only STP-2 and STP-4 nanocomposites, which revealed the shorter gelation time exhibited by the material with the higher Ox concentration (STP-4) (see Table 5.1). Taking into account the latter result, the role of Ox as a sol–gel catalyst has been demonstrated. Moreover, the shorter gelation time observed for STP-1 with respect to STP-2, seems to be attributed to the different sequence of addition of the reagents and not to the amount of Ox present in the formulation. Specifically, in the case of STP-1, TEOS was added in the first step of the synthesis, thus requiring a shorter time for achieving the sol–gel transition.

By comparing the shrinkage of the xerogels, we indicated that all the materials under study provided similar volume reductions (see Table 5.1). These findings are in good agreement with previous observations of Mosquera *et al.*,[45] thus confirming that the presence of PDMS affects the shrinkage of the xerogels. This shrinkage is attributed to the PDMS chains, which coil up during the drying process, thus providing high elasticity and rubbery behaviour to the derived nanocomposites. Therefore, the similar shrinkage behaviour for the three nanocomposites can be ascribed to the equal PDMS concentration included in the nanomaterials under study.

After the drying process, all the synthesized materials produced crack-free, colourless, homogeneous and highly transparent monolithic xerogels, as indicated in Scheme 5.1. Specifically, factors such as the low amount of Ti alkoxide added to the starting sols and the good dispersion of titanium domains by using strong stirring and slow addition of TTIP, provided transparency to the nanocomposites.

However, given the different hydrolysis and condensation rates of Si and Ti alkoxides, additional difficulties may arise if one wishes to obtain homogeneous gels employing these two precursors.[46] It is well known that the hydrolysis reactions of alkoxides proceed by nucleophilic substitution mechanisms and depend on the electronic partial charge of the metal atoms. However, Ti^{+4} presents a more electrophilic character (higher positive partial charge) than Si^{+4}.[46] Moreover, the chemical reactivity of the alkoxides is associated with the unsaturation degree of the metal or silicon which can be estimated from the difference between the usual coordination number (N) and the oxidation state (Z) of the metal coordination. The N–Z values obtained are equal to 0 and 2 for Si and Ti, respectively, justifying the higher reactivity of titanium comparing to silicon alkoxides.[27] Therefore, in any sol–gel synthesis including a Ti alkoxide, phase separation may easily take place instead of the formation of a homogeneous gel network.

We considered that the addition of Ox in synthesis could promote the formation of homogeneous gels. Specifically, the action of Ox could be summarized as follows: (1) the oxalate ions are strong ligands that may coordinate with Ti alkoxides, thus delaying the Ti alkoxide hydrolysis process;[46,47] (2) Ox acts as a catalyst in the hydrolysis and condensation of TEOS; and (3) Ox prevents the agglomeration of TiO_2 nanoparticles, generating stable colloidal dispersion solutions with pH values that are far from the isoelectric point.[38,48]

In this point, we may further discuss the decrease in gelation time observed for STP-4 (5 days) with respect to STP-2 (9 days). These results showed that the presence of higher amounts of Ox also shortened the gelation time of the nanocomposites. Therefore, Ox, apart from acting as a chelating agent, obviously acts as a sol–gel catalyst (Table 5.1).

Finally, crack-free nanomaterials were obtained due to the synergistic action of Ox and PDMS (Scheme 5.1). Particularly, Ox acts as DCCA,[33] while PDMS enhances the formation of a flexible silica network thus avoiding cracking during drying.[29,30]

FTIR provided further insights into the sol–gel process and the chemical structure of the nanocomposites under study. All the spectra obtained for the sols showed a typical wide band located at the spectral region of 3000–3700 cm^{-1}, attributed to O–H bonds. As expected, the intensity of this band decreased with time for all the nanocomposites, due to the condensation of the silica network.

In spectrum ii of Figure 5.1(a), we may clearly confirm the presence of PDMS which was added to the mixture TEOS–EtOH–Ox, due to the appearance of a peak at 1261 cm^{-1} attributed to the symmetrical bending of the –CH_3 groups in Si–(CH_3).[29] As expected, this peak is also present in the spectra of STP-2 (Figure 5.1(b)). Moreover, in the spectra of xerogels of Figure 5.1(a) and (b), we also observed a shift of this peak to higher wavenumber values (1267 cm^{-1}), together with an additional peak located at 849 cm^{-1}. This could probably be related to the copolymerization of PDMS and TEOS.[49] For STP-1, the peak at 849 cm^{-1} can be observed even after the 4th day of the sol formation (Figure 5.1(a), spectrum iii). Finally, in all of the xerogel materials, the peak at 1267 cm^{-1} gives us the opportunity to assume that the PDMS ensures hydrophobic properties to the final hybrid nanocomposites.

The addition of TTIP into the sols of STP-2 and STP-4 produced two low intensity peaks at 945 and 1152 cm^{-1}, which may be related to O–Ti and –CH_3 bonds, respectively (Figure 5.1(b), spectrum ii).[50] It is worth mentioning that in the cases of STP-2 and STP-4, the role of Ox as a chelating agent was established through the observation of the peak located at 1683 cm^{-1}. This peak corresponds to the C=O stretching of Ox, which presented a shoulder at 1711 cm^{-1} after the addition of TTIP (Figure 5.1(b), spectra ii). This observation could also justify a probable coordination between the nucleophilic oxalic ions with the Ti alkoxide added, thus deterring the instantaneous TTIP hydrolysis and polymerization.[51]

The completion of the condensation of the xerogels can be confirmed through the changes observed in several peaks (Figure 5.1(a) and (b), spectra iv).[49] The absence of the peaks located at 2931, 2975, 2892, 1452, 1381 and 880 cm^{-1} can be attributed to the elimination of C–H bonds of the groups $-CH_2/-CH_3$ during the hydrolysis and condensation of the alkoxides. The antisymmetric stretching of Si–O–Si bonds at 1077 cm^{-1}, a peak at 810 cm^{-1} attributed to the SiO_4 structure and the absorption of Si–OH groups at 954 cm^{-1} indicated the evolution of the formation of xerogels (Figure 5.1(a) and (b), spectra iv).

The results obtained through the FTIR analysis demonstrate the accomplishment of the hydrolysis and condensation of the alkoxides, the co-polymerization of PDMS within the silica network and the role of Ox as a chelating agent.

The TGA-DTA results for the nanocomposites under study revealed weight loss in two regions. First, weight losses together with an endothermic peak observed upon heating from 40 to 205 °C were attributed to the release of absorbed water. At higher temperatures and up to 950 °C, another reduction in weight was ascribed to the removal of organic solvents and the decomposition of the organic matter.[52] By taking into account the above results, we can conclude that PDMS was successfully incorporated within the silica matrix, thus producing organic–inorganic hybrid nanocomposites.

5.3.2 Textural and Microstructural Characterization

The textural properties of the nanocomposites under study were determined through N_2 sorption analysis. N_2 adsorption–desorption isotherms are given in Figure 5.2 together with the values obtained for surface area (S_{area}) and pore volume (V_{pore}) (Table 5.2).

As seen in Table 5.2, STP-2 presented a significantly smaller pore volume than the other materials under study. Therefore, the pore size distribution could not be calculated. The different textural behaviour of STP-2 is explained by taking into consideration the information provided by TEM analysis, which is presented in the following paragraphs.

As indicated in Figure 5.2(a), STP-1 and STP-4 exhibit type IVa isotherms, which can be associated with mesoporous materials.[39] Regarding the hysteresis of the isotherms, both materials present a type H4 loop with parallel and almost horizontal branches.[39] According to the literature,[53] a H4

Table 5.2 Textural parameters obtained from the isotherms and the NLDFT method for the materials under study. Reprinted from C. Kapridaki, L. Pinho, M. Mosquera and P. Maravelaki-Kalaitzaki, Producing photoactive, transparent and hydrophobic SiO_2-crystalline TiO_2 nanocomposites at ambient conditions with application as self-cleaning coatings, *Appl. Catal., B*, **156–157**, 417–427, Copyright (2014), with permission from Elsevier.[69]

	S_{area} ($m^2\ g^{-1}$)	V_{pore} ($cm^3\ g^{-1}$)
STP-1	153	0.136
STP-2	3	0.006
STP-4	440	0.260

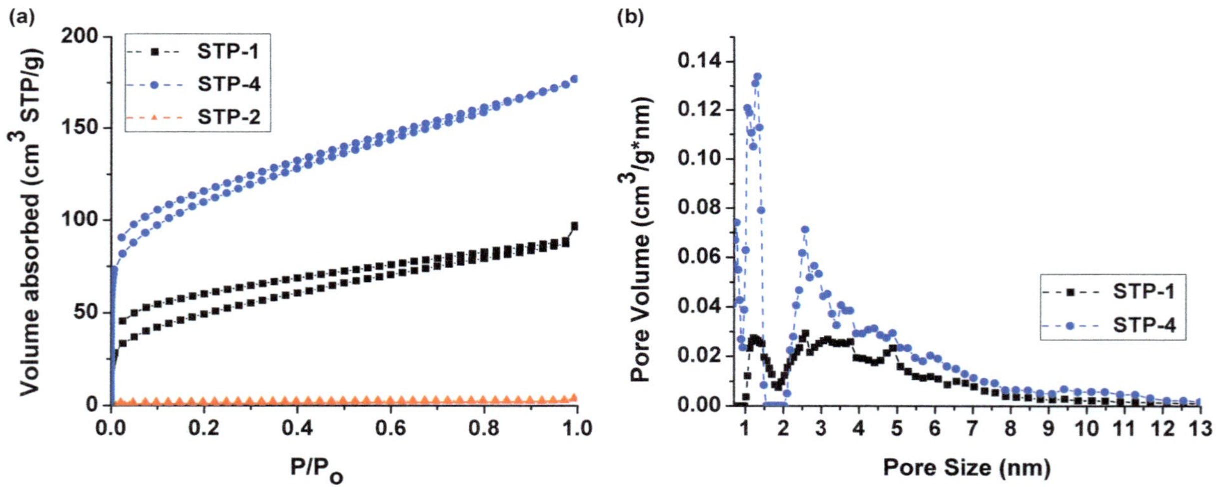

Figure 5.2 (a) Nitrogen adsorption–desorption isotherms and (b) NLDFT pore size distributions for the nanocomposites under study. Reprinted from C. Kapridaki, L. Pinho, M. Mosquera and P. Maravelaki-Kalaitzaki, Producing photoactive, transparent and hydrophobic SiO_2-crystalline TiO_2 nanocomposites at ambient conditions with application as self-cleaning coatings, *Appl. Catal., B*, **156–157**, 417–427, Copyright (2014), with permission from Elsevier.[69]

hysteresis loop is observed for materials that consist of particles with internal voids of irregular shape and broad size distribution.

Furthermore, the pore volume and the surface area observed for STP-1 nanocomposite showed lower values than those obtained for STP-4. The isotherms obtained for STP-1 and STP-4 (Figure 5.2(a)) also indicated adsorption of nitrogen at low pressure, which is associated with the presence of a microporous system.[28] Therefore the pore size distributions of these materials were also estimated using the NLDFT method which approaches both micro and mesopores.[39] The NLDFT pore size distribution obtained for the nanocomposites were calculated from the adsorption branches and are shown in Figure 5.2(b). STP-1 and STP-4 exhibited a bimodal distribution consisting of: (1) a microporous region with pores of sizes close to the mesopore range (from 1 to 2 nm); and (2) mesopores ranging from 2 to 8 nm.

For a deeper investigation into the microstructure of the nanomaterials, we also carried out TEM, HAAD-STEM and HRTEM, coupled with X-EDS analysis. In Figure 5.3, we present TEM images of the three synthesized nanomaterials. STP-1, exhibited a uniform SiO_2–TiO_2–PDMS network with aggregates ranging from 40–100 nm, probably due to the high connectivity of the Ti–O–Si bonds at atomic level (Figure 5.3(a)).[54] However, STP-2 and STP-4 with high concentrations of Ox exhibited aggregations, similar in size to STP-1, throughout their matrix (Figure 5.3(b)–(c). The TEM micrographs obtained for STP-2 and STP-4 allow us to observe the presence of nanoparticles, presumably TiO_2, incorporated within the SiO_2 network. This may probably be due to the formation of independent domains during the hydrolysis and condensation of Si and Ti alkoxides, which seem to be absent in STP-1. Gopala *et al.*[55] have obtained similar nanocomposites containing separate domains of silica and titania nanoparticles. In particular, STP-2 presented nanoparticles with sizes ranging from 5 to 10 nm in contrast to STP-4, which

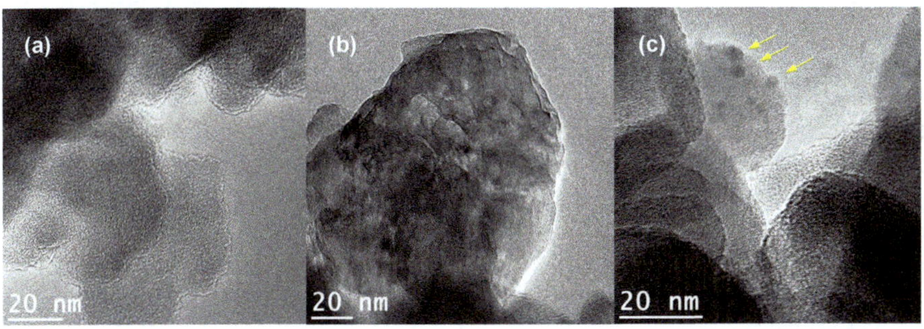

Figure 5.3 TEM images of: (a) STP-1, (b) STP-2 and (c) STP-4 nanocomposites. Reprinted from C. Kapridaki, L. Pinho, M. Mosquera and P. Maravelaki-Kalaitzaki, Producing photoactive, transparent and hydrophobic SiO_2-crystalline TiO_2 nanocomposites at ambient conditions with application as self-cleaning coatings, *Appl. Catal., B*, **156–157**, 417–427, Copyright (2014), with permission from Elsevier.[69]

contained particles with smaller sizes (2–5 nm, indicated by yellow arrows in Figure 5.3(c)).

In addition, we have observed that STP-2 presented a significantly higher amount of individual TiO_2 nanoparticles embedded within the silica matrix, with respect to STP-4. We should point out here that the excess reduction of pore volume along with the low surface area values observed by N_2 physisorption for STP-2 can be justified by this significantly large amount of aggregates and the separate domains of TiO_2 nanoparticles. We consider that the embedded non-porous TiO_2 aggregates into the STP-2 silica matrix have clearly negatively affected the porosity and the surface area of this nanocomposite. Facio *et al.*[24] have previously observed pore volume reductions of nanocomposites as the concentration of the embedded non-porous SiO_2 particles increased. Moreover, our conclusions correlated with the incorporation of the non-porous components into the silica matrix are in accordance with the TEM micrographs (Figure 5.3(a)–(c)). Specifically, by observing the TEM images of STP-1 and STP-4 (Figure 5.3(a) and (c)), the formation of porous materials can be confirmed, since a light texture of the matrix has been shown. However, STP-2 resulted in a dense texture without the presence of pores (Figure 5.3b).

In order to investigate the silica–titania distribution, we also carried out HAADF-STEM analysis for all the materials. The HAADF-STEM images for the nanocomposites under study are shown in Figure 5.4(a)–(c). In this image mode, the electrons that are subject to high-angle scattering are captured and the signal is approximately proportional to Z^2, where Z is the atomic number. Therefore, the obtained Z-contrasts of the HAADF-STEM images, attributed to different components of the nanocomposite, confirmed the previous discussion.[56] The STP-1 image (Figure 5.4(a)) does not clearly show any Z-contrast, thus indicating the connectivity at the atomic level.[54] The nanocomposites with higher Ox concentration (STP-2 and STP-4), as opposed to STP-1, presented a noticeable differentiation in visible contrast due to the discrimination between the nanoparticles and silica matrix (Figure 5.4(b) and (c)). A comparison between these nanomaterials clearly showed differences in Z-contrasts throughout their matrices. In the case of STP-2, the higher contrast of the observed zones may be associated with the TiO_2 nanoparticles, due to the higher atomic number of Ti compared to Si. In addition, the HAADF-STEM image of STP-2 indicated the presence of separate domains of TiO_2, which are hardly distinguishable by TEM analysis. In the STP-4 HAADF-STEM image, we confirmed the small and spherical TiO_2 particles observed by TEM study indicated with yellow arrows (Figure 5.4(c)).

X-EDS with elemental distributions of Si and Ti over the spotted area of the HAAD-STEM images are illustrated in Figure 5.4(a)–(c). The X-EDS maps obtained for STP-1 confirmed only the presence of Si referred to three specific individual spots. The absence of Ti peaks from these spectra could be associated with the homogeneous distribution of the Ti and Si alkoxides, at the atomic level. Therefore, the very low concentration of Ti produced throughout the matrix was hardly detected by the analyzer (Figure 5.4(a)).

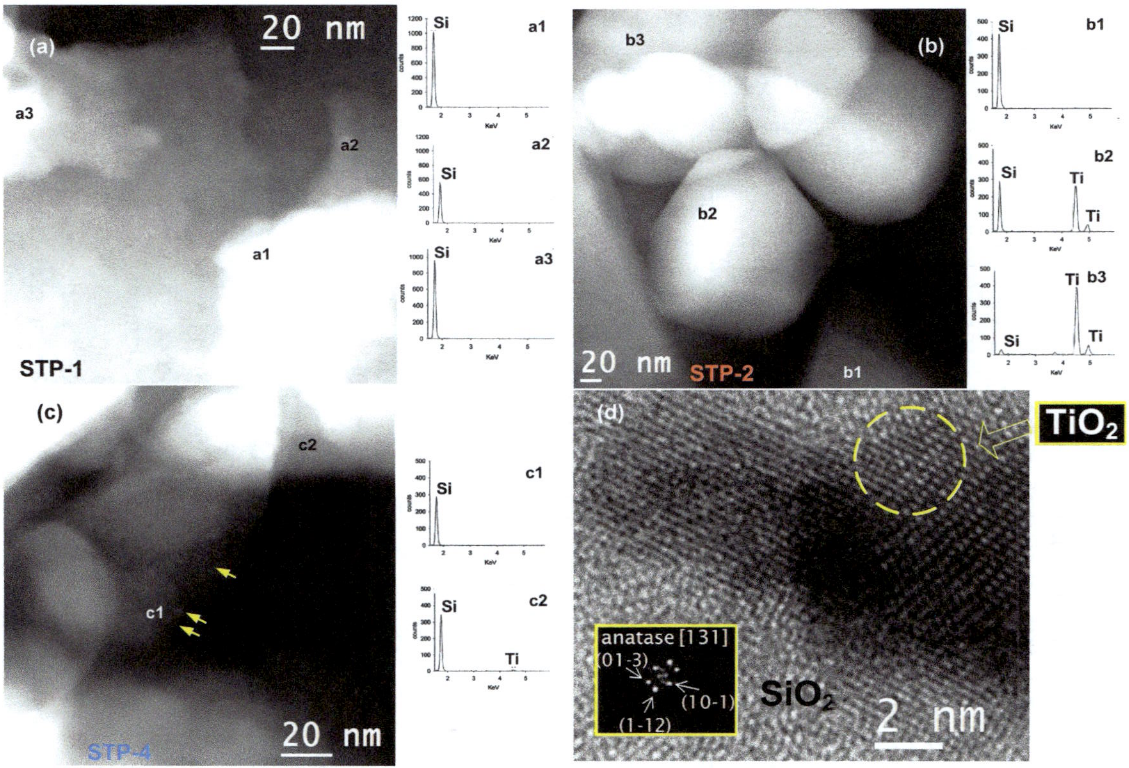

Figure 5.4 (a)–(c) STEM-HAADF images and X-EDS spectra for STP-1, STP-2 and STP-4, respectively and (d) HREM image corresponding to STP-2 nanocomposite. The latter image shows the TiO_2 anatase nanoparticles incorporated within the silica matrix. Reprinted from C. Kapridaki, L. Pinho, M. Mosquera and P. Maravelaki-Kalaitzaki, Producing photoactive, transparent and hydrophobic SiO_2-crystalline TiO_2 nanocomposites at ambient conditions with application as self-cleaning coatings, *Appl. Catal., B*, **156–157**, 417–427, Copyright (2014), with permission from Elsevier.[69]

In the case of STP-2, three X-EDS spectra were detected, due to the high visible contrasts obtained. As expected, Ti was not identified in the analyzed spot (Figure 5.4(b), spot b1) with lower intensity, contrary to the other two spots originating from zones with high contrast (b2 and b3). These findings along with the highest Ti peaks in the case of the highest contrast zone (b3 spot) further support the statement that the higher contrast zones can be attributed to separate domains of TiO_2. Moreover, in Figure 5.4(c), the TiO_2 nanoparticles located in independent domains inside the bulk of the STP-4, provided higher visible contrast than the SiO_2 matrix. The HAAD-STEM illuminated area in Figure 5.4(c), spot c2 further confirmed the presence of Ti.

The microstructure of the nanocomposites with high concentrations of Ox (STP-2 and STP-4) was further investigated through a HREM (high resolution electron microscopy) study. The representative HREM image of STP-2 illustrated in Figure 5.4(d) indicated the presence of TiO_2 nanoparticles with sizes of approximately 5 nm which were incorporated within the silica matrix. This HREM image provided dark zones associated with the overlapping of at least two crystalline nanoparticles. Furthermore, we should point out that the TiO_2 nanoparticles are more difficult to distinguish due to the fact that these domains are embedded within the silica network. The HREM image of the STP-4 material showed similar morphology features to those of STP-2, therefore it was not included.

At this scale, by analyzing the digital diffraction pattern of the HREM image (shown as an inset in Figure 5.4(d)), we demonstrated the crystallographic nature of the components contained in the nanocomposites. The presence of (01-3), (10-1) and (1-12) crystallographic planes associated with the anatase crystalline phase of TiO_2 was confirmed with the application of Fourier transformation to the visible lattice spacing present in the crystalline structure of our materials. These findings revealed the formation and incorporation of anatase TiO_2 nanoparticles, within the silica network of STP-2 and STP-4. Both the production and the embedment of the photoactive TiO_2 nanoparticles was achieved by the hydrolysis and condensation of Ti alkoxide, in the presence of Ox, at ambient temperature. Finally, the amorphous structure of the silica network was also confirmed.

The X-ray diffractogram patterns of the powder form of the nanomaterials prepared in our laboratory are also given in Figure 5.5. All the materials presented a typical amorphous structure, without the peaks associated with the crystalline phases (anatase, rutile or brookite) of TiO_2 particles. As previously reported, the crystalline structure of anatase TiO_2 detected by TEM analysis was hardly observed by an XRD study.[57] In the case of our materials, we assume that the absence of XRD characteristic peaks could be attributed either to the small size of the particles or to the lack of the orientation of the crystalline structures.

At this point, taking into account all the findings reported by TEM and physisorption analysis, we can draw a conclusion on the texture and microstructure of the nanocomposites under study. The sequence of the addition of the precursor materials is responsible for the difference of the Si

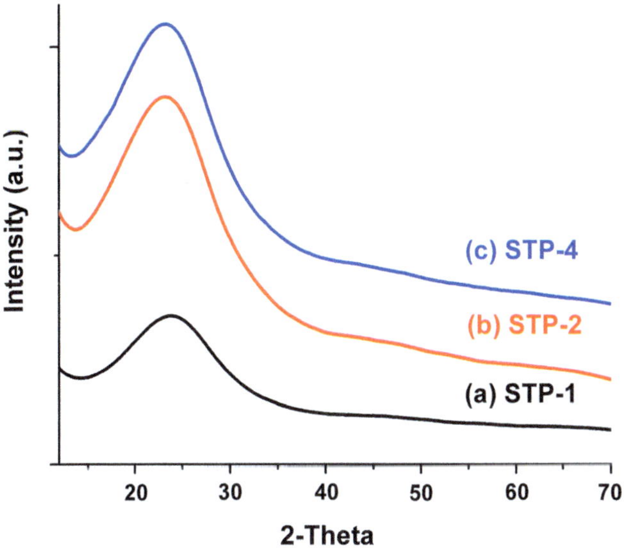

Figure 5.5 XRD patterns of the three nanocomposites under study.
Reprinted from C. Kapridaki, L. Pinho, M. Mosquera and P. Maravelaki-Kalaitzaki, Producing photoactive, transparent and hydrophobic SiO_2-crystalline TiO_2 nanocomposites at ambient conditions with application as self-cleaning coatings, *Appl. Catal., B*, **156–157**, 417–427, Copyright (2014), with permission from Elsevier.[69] Reprinted from C. Kapridaki and P. Maravelaki-Kalaitzaki, TiO_2–SiO_2–PDMS nanocomposite hydrophobic coating with self-cleaning properties for marble protection, *Prog. Org. Coat.*, **76**, 400–410, Copyright (2012), with permission from Elsevier.[70]

and Ti distribution between STP-1 and the other two materials. Specifically, in the syntheses of STP-2 and STP-4, the hydrolysis and homo-condensation of the Ti alkoxide took place prior to the addition of TEOS, thus promoting the formation of independent TiO_2 nanoparticles within the silica network. In contrast, in the STP-1 synthesis, the completion of the TEOS hydrolysis prior to the addition of TTIP is associated with the hetero-condensation of the two alkoxides (TTIP and TEOS), and therefore the production of Ti–O–Si connectivity at the atomic level.

Moreover, the production of anatase TiO_2 nanoparticles at ambient temperature is correlated to a peptization process occurring during the synthesis of STP-2 and STP-4. According to previously reported results, the presence of an acid medium enhances the re-dispersing process of the amorphous TiO_2 phase, thus facilitating the growth of crystalline particles (peptization).[58,59] In STP-2 and STP-4, the production of the amorphous phase of TiO_2 preceded the peptization process, which took place in the presence of Ox during the magnetic stirring for 24 h (Scheme 5.1). Consequently, in the materials containing higher Ox concentrations (STP-2 and STP-4), the anatase TiO_2 nanoparticles observed by TEM/STEM (Figures 5.3 and 5.4) were produced prior to the addition of TEOS. As opposed to these findings, in the STP-1 synthesis, the peptization process is restricted due to both the fast hetero-condensation of the alkoxides (Si and Ti) and the lower concentration of the acid medium (Ox).

By comparing the size of TiO_2 separate domains incorporated within the materials STP-2 and STP-4, we can conclude that these differences may correspond to the different Ox concentrations contained in these two nanocomposites. The results obtained by TEM showed that the nanomaterial with the highest Ox concentration (STP-4) consists of TiO_2 nanoparticles with the smallest size and the most spherical shape. Similar findings were also reported by Mahshid *et al.*,[48] confirming the formation of small and spherical TiO_2 nanoparticles at low pH values.

By taking into account all the results described above, we propose in Scheme 5.2 a schematic representation of the nanocomposites under study.

Figure 5.6(a)–(f) shows the 2D and 3D images of the surface topography of the nanocomposites under study observed by AFM. Specifically, in Figure 5.6(a)–(c), the 1 μm × 1 μm 2D top views of the three materials are illustrated, while the corresponding 3D images are depicted in Figure 5.6(d)–(f). The AFM images demonstrated that all the materials consist of aggregates with morphology similar to that observed in the micrographs of TEM analysis. Moreover, the AFM images clearly showed that the PDMS content modifies the surface morphology, producing roughness on top of the coatings. The presence of PDMS in our materials promotes the formation of tough and flexible xerogels, as a consequence of the creation of bridges linking silica particles.[29,30,60] During the drying process these PDMS chains coil up and therefore roughness on the surfaces is developed.

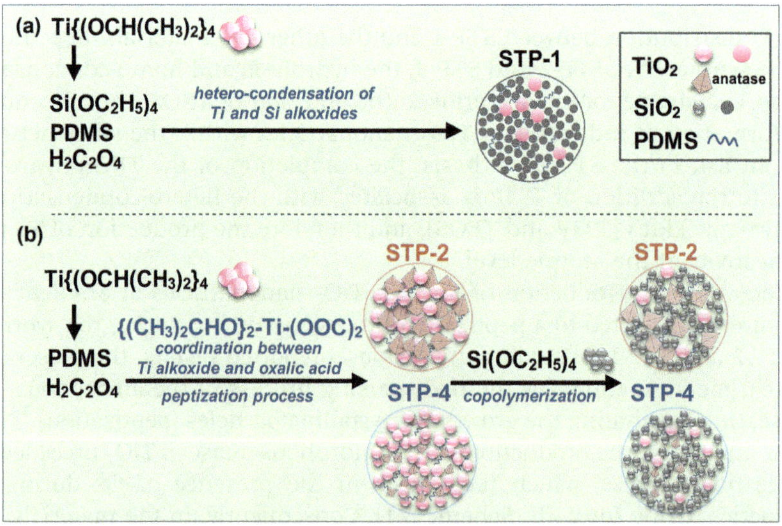

Scheme 5.2 Schematic representation of the synthesized nanocomposites: (a) STP-1 and (b) STP-2 and STP-4.

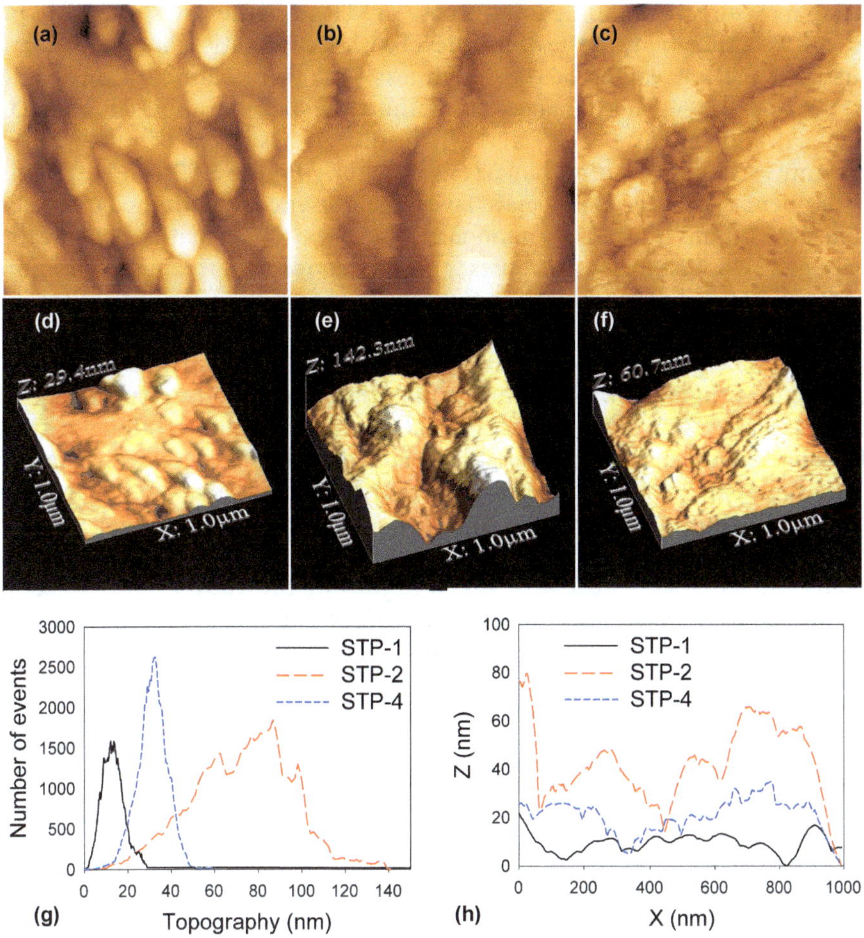

Figure 5.6 (a)–(c) 2D and (d)–(f) 3D AFM images of STP-1, STP-2 and STP-4, respectively. The (g) and (h) images show the corresponding roughness profiles.
Reprinted from C. Kapridaki, L. Pinho, M. Mosquera and P. Maravelaki-Kalaitzaki, Producing photoactive, transparent and hydrophobic SiO_2-crystalline TiO_2 nanocomposites at ambient conditions with application as self-cleaning coatings, *Appl. Catal., B*, **156–157**, 417–427, Copyright (2014), with permission from Elsevier.[69]

Despite the fact that PDMS concentration is the same in all of the materials under study, differences related to the roughness can be observed. The surface morphology of the STP-2 and STP-4 exhibited higher values of roughness compared to STP-1. We attribute this difference to the presence of the separate domains of TiO_2 particles obtained by TEM within the STP-2 and STP-4 silica networks. These findings are in accordance with the results of previous papers wherein nanoclusters (separate domains) of titania–silica materials were also observed by AFM analysis.[61,62] However, as we have already discussed in the case of STP-1 (Figure 5.6(a) and (d)), the Si and Ti connectivity at the atomic level without formation of nanoclusters justified the low values of the roughness (5.1 nm) obtained. Comparing STP-2 and STP-4, AFM images show that STP-4 containing a higher Ox concentration (Figure 5.6(c) and (f)) results in a surface of much lower roughness values (7.2 nm) than STP-2 (23.0 nm) (Figure 5.6(b) and (e)). This prominent difference is associated with the presence of both the larger size and higher amount of TiO_2 nanoparticles in the STP-2 matrix than in the case of STP-4. The lower pH values, as a consequence of the higher amount of Ox in STP-4, contribute to the formation of particles with a smaller size distribution.[48]

Figures 5.6(g) and (h) illustrate the particle height distribution (Z) and the typical roughness profiles for the materials under study, respectively. In the case of STP-2, the increase of the average roughness along with a broader Gaussian distribution is associated with the presence of larger TiO_2 nanoparticles, contrary to the other two materials. Comparing the height distribution of STP-1 and STP-4, we observed that STP-4 showed higher average roughness attributed to the presence of the separate TiO_2 domains, which were absent in STP-1. Regarding the findings illustrated in Figure 5.6(h), these typical roughness profiles are in accordance with previously published results on the SiO_2 composites with embedded TiO_2 nanoparticles.[62]

The UV-Vis diffuse reflectance spectra obtained for all the materials under study are shown in Figure 5.7. The calculated values of the energy gap (E_g) were obtained according to the method proposed by Tandon and Gupta.[63] Specifically, this method requires plotting of $(F(R)E)^{1/2}$ *versus* E and the E_g values as indicated in the inset of Figure 5.7. We should note that the absorption in the region 200–800 nm of the nanocomposites is associated with the presence of TiO_2, as the major SiO_2 component does not absorb in this range. However, the dependence of E_g values on the preparation method, impurities, crystalline effects, average crystal size and the method of calculation is well established.[64,65]

As expected, the pure TiO_2 of P25 presented the lowest value of E_g (3.33 eV). Our nanocomposites exhibited higher E_g values in the range of 3.49 up to 3.58 eV, influenced by the presence of the silica matrix from TEOS, the organosiloxane part from PDMS and the low quantity of TiO_2 compared to P25. By comparing the E_g values observed for the three nanomaterials, the STP-2 showed the lowest band gap. According to the quantum size effect, the band gap of a semiconductor decreases as the crystal size increases. Therefore, the presence of the largest TiO_2 crystals in STP-2, as obtained by TEM, induced the highest decrease in the band gap. Despite the absence of separate TiO_2 domains in the STP-1 material, we obtained a similar E_g value to STP-4. According to the results reported, amorphous and dispersed TiO_2 within SiO_2 matrixes can also absorb UV irradiation.[66]

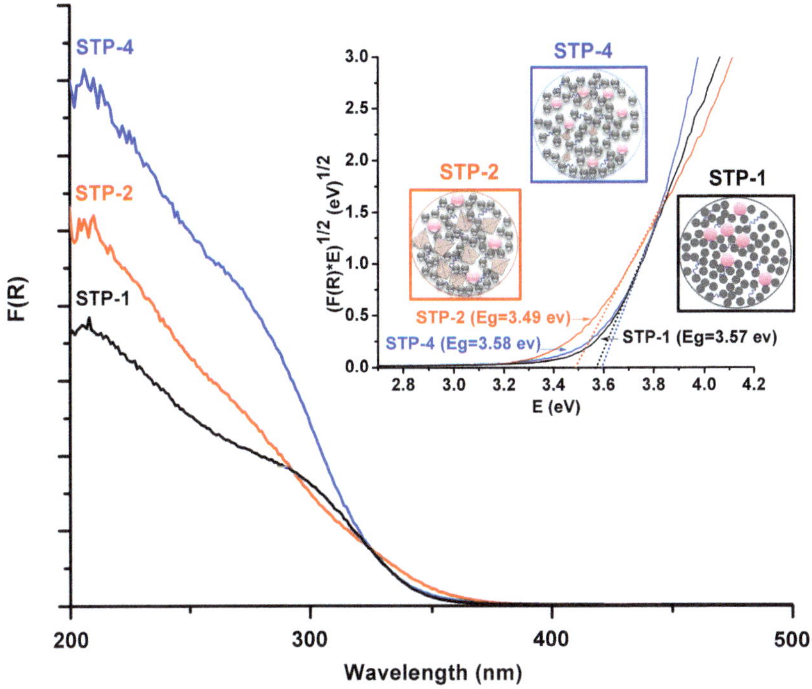

Figure 5.7 DR-UV-Vis spectra of the nanocomposites along with the plots of modified Kubelka–Munk as inset.
Reprinted from C. Kapridaki, L. Pinho, M. Mosquera and P. Maravelaki-Kalaitzaki, Producing photoactive, transparent and hydrophobic SiO_2-crystalline TiO_2 nanocomposites at ambient conditions with application as self-cleaning coatings, *Appl. Catal., B*, **156–157**, 417–427, Copyright (2014), with permission from Elsevier.[69]

5.3.3 Assessment of the STP Photocatalytic Activity

We also investigated the photocatalytic activity of the nanocomposites under study by monitoring the degradation of MO under UV-vis (see Figure 5.8(a)). Table 5.3 presents the pseudo-first order rate constants obtained. As expected, the photochemical degradation of MO is practically negligible during the evolution of this photocatalytic study. The results indicated that STP-1 presented a lower rate constant ($k = 6.2 \times 10^{-3}$ min^{-1}) than STP-2 and STP-4, which contain high concentrations of Ox. More specifically, STP-2 produced a kinetic rate constant equal to 113×10^{-3} min^{-1} while the rate constant estimated for STP-4 was 60×10^{-3} min^{-1}. The higher photocatalytic activity of STP-2 and STP-4 is associated with the independent TiO_2 domains and the role played by Ox as a hole scavenger.[34–37] According to the literature, Ox is characterized as a good bidentate ligand, showing the ability at low pH values to be strongly absorbed on the TiO_2 surfaces. In addition, the

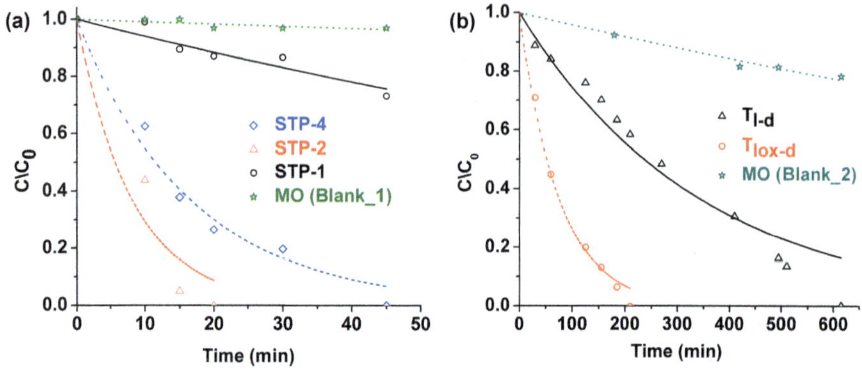

Figure 5.8 Photocatalytic degradation of methyl orange (MO) by: (a) STP nanocomposites and (b) TiO_2 formulations. The curves represent the fitting of the pseudo-first order equation to the experimental data.
Reprinted from C. Kapridaki, L. Pinho, M. Mosquera and P. Maravelaki-Kalaitzaki, Producing photoactive, transparent and hydrophobic SiO_2-crystalline TiO_2 nanocomposites at ambient conditions with application as self-cleaning coatings, *Appl. Catal., B*, **156–157**, 417–427, Copyright (2014), with permission from Elsevier.[69]

great contribution of Ox to the photocatalytic procedure can be attributed to $CO_2^{\bullet-}$ binded the already generated h^+ according to eqn (5.2).

$$C_2O_4^{2-} + h^+ \rightarrow CO_2 + CO_2^{\bullet-} \tag{5.2}$$

The $CO_2^{\bullet-}$ radicals produced probably have the ability to further enhance the photocatalytic activity. They might act either as a direct means of the decomposition of organic compounds or react with O_2 to produce photocatalytic active agents, such as $O_2^{\bullet-}/HO_2^{\bullet}$.[34–37]

In our photocatalytic study, the action of Ox as a hole scavenger produced oxidized radicals, which can reductively decompose the MO. In support of the above, the photocatalytic decomposition of MO was enhanced with the presence of EDTA, which is another good bidentate ligand.[67]

In order to estimate the Ox contribution to the photoactivity of the STP materials, we assessed a photodegradation study of MO by TiO_2 formulations, prepared at room temperature, under UV both in the presence and absence of Ox (Figure 5.8(b) and Table 5.3). The results clearly indicated that Ox improved the photoactivity of these TiO_2 formulations. More specifically, the formulation $T_{I\text{-}d}$ without Ox exhibited a lower rate constant ($k = 3 \times 10^{-3}$ min^{-1}) than the photocatalyst $T_{IOx\text{-}d}$ ($k = 13 \times 10^{-3}$ min^{-1}), which contained Ox. Therefore, these findings suggest that the presence of Ox in the STP-2 and STP-4 nanocomposites may improve their photocatalytic activity.

5.3.4 Treatment Assessment of the Nanocomposites

An overview of the treated samples (marbles and Alfas limestone) and the techniques performed for the assessment of the nanocomposites'

Table 5.3 Pseudo-first order kinetic rate constant (k) and regression coefficient (r^2) of methyl orange degradation, under UV irradiation, by the TiO_2 formulations and nanocomposites under study. Reprinted from C. Kapridaki, L. Pinho, M. Mosquera and P. Maravelaki-Kalaitzaki, Producing photoactive, transparent and hydrophobic SiO_2-crystalline TiO_2 nanocomposites at ambient conditions with application as self-cleaning coatings, *Appl. Catal., B*, **156–157**, 417–427, Copyright (2014), with permission from Elsevier.[69]

Nanocomposite/TiO_2 formulation	k (10^{-3} min^{-1})	r^2
MO (blank 1)	0.8	0.91
STP-1	6	0.90
STP-2	113	0.92
STP-4	60	0.98
MO (blank_2)	0.4	0.98
$T_{1Ox\text{-}d}$	13	0.99
$T_{1\text{-}d}$	3	0.95

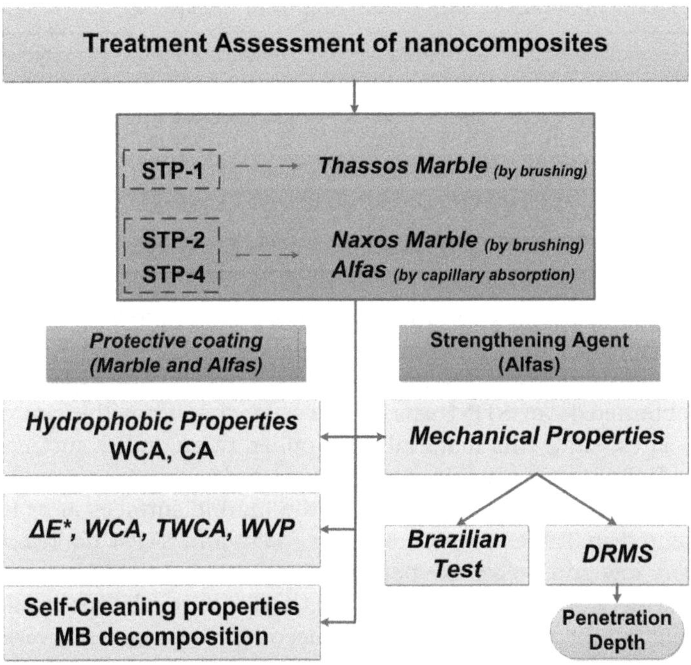

Scheme 5.3 Schematic representation of the assessment of STP treatments.

effectiveness as protective coatings, consolidants and self-cleaning agents are presented in Scheme 5.3.

5.3.4.1 STP as Hydrophobic and Self-cleaning Coatings

The effectiveness of STP-1, STP-2 and STP-4 nanocomposites as protective and self-cleaning coatings was investigated in Greek marble originating

Table 5.4 Physical properties of Thassos marble treated with STP-1. Reprinted from C. Kapridaki and P. Maravelaki-Kalaitzaki, TiO_2–SiO_2–PDMS nano-composite hydrophobic coating with self-cleaning properties for marble protection, *Prog. Org. Coat.*, **76**, 400–410, Copyright (2012), with permission from Elsevier.[70]

		Untreated	Treated with STP-1
Dry matter ($mg\,cm^{-2}$)		—	0.658 (±0.138)
ΔE^*		—	3.14 (±0.36)
Contact angle (°)	0 s	72.6 (±3.0)	114.8 (±0.4)
	20 s	59.9 (±2.7)	111.3 (±0.8)
WCA ($mg\,cm^{-2}\,s^{-1/2}$)		0.1067 (±0.0088)	0.0132 (±0.0067)
TWCA (%)		0.22 (±0.05)	0.13 (±0.03)
WVP ($g\,cm^{-2}\,h^{-1}$)		0.140 (±0.044)	0.093 (±0.017)

Table 5.5 Physical properties of Naxos marble treated with STP-2 and STP-4. Reprinted from C. Kapridaki, L. Pinho, M. Mosquera and P. Maravelaki-Kalaitzaki, Producing photoactive, transparent and hydrophobic SiO_2-crystalline TiO_2 nanocomposites at ambient conditions with application as self-cleaning coatings, *Appl. Catal., B*, **156–157**, 417–427, Copyright (2014), with permission from Elsevier.[69]

		Untreated	Treated with STP-2	Treated with STP-4
Dry matter ($mg\,cm^{-2}$)		—	0.4122 (±0.0115)	0.4237 (±0.0418)
ΔE^*		—	0.61 (±0.05)	1.88 (±0.05)
Contact angle (°)	0 s	56.2 (±3.0)	104.3 (±2.0)	95.4 (± 2.2)
	20 s	50.0 (±2.8)	103.2 (±1.2)	93.6 (±1.6)
WCA ($mg\,cm^{-2}\,s^{-1/2}$)		0.0485 (±0.0021)	0.0059 (±0.0003)	0.0096 (±0.0002)
TWCA (%)		0.57 (±0.04)	0.30 (±0.04)	0.36 (±0.02)
WVP ($g\,cm^{-2}\,h^{-1}$)		0.0040 (±0.0003)	0.0033 (±0.0004)	0.0031 (±0.0002)

from Thassos and Naxos and the results are provided in Tables 5.4 and 5.5. Table 5.4 shows the dry matter of STP-1 applied to Thassos marble; Table 5.5 provides the dry matter of STP-2 and STP-4 products. As expected, the dry matter obtained from STP-1 was higher than that from the two other nanocomposites, since this material was applied twice on the surfaces, due to its limited photoactivity as previously discussed.

The total colour differences (ΔE^*) of the marble surfaces after treatment are presented in Tables 5.4 and 5.5. For STP-2 and STP-4 nanocomposites, ΔE^* is very low and below the perceptibility threshold ($\Delta E^* \leq 5$).[21] Only the STP-1 treatment produced a higher ΔE^* (3.1) due to the double application of the product on the surface. Taking into account that chromatic variations in the range of 2 to 3 cannot be detected by the naked eye, our findings indicate that the application of our transparent nanocomposites as coatings does not induce adverse chromatic modifications on the surfaces.

The effectiveness of the STP nanocomposites as hydrophobic coatings was assessed by means of contact angle (CA) measurements and water capillary absorption (WCA).

Regarding the CA, the values of the static CA measured twice in succession are provided in Tables 5.4 and 5.5. These findings demonstrated that all the nanocomposites produced hydrophobic surfaces after treatment. We associate this with the presence of PDMS into the sols, allowing the formation of organic–inorganic hybrid nanocomposites. The organosilane PDMS reduces the surface energy of the nanocomposites, thus producing hydrophobic properties to the materials under study.[29,30] Although the content of PDMS is the same in all the composites, differences in CA values were observed. However, reliable comparisons can only be made between STP-2 and STP-4 treatments, as these materials were applied only once. The surfaces treated with STP-2 produced higher CA values than those treated with STP-4; these results correlated with the higher roughness value obtained by AFM analysis.[29]

The coefficients for water absorption along with the total amount of capillary absorbed water (TWCA) are shown in Table 5.4 for treatment with STP-1 and in Table 5.5 for treatment with STP-2 and STP-4. More specifically, reductions of the coefficients of 88, 88 and 80% for STP-1, STP-2 and STP-4, respectively, indicated the water repellency of each treated surface. In addition, these results were also confirmed by the TWCA values after treatment, where a remarkable decrease of the total amount of the absorbed water was recorded (41% for STP-1, 47% for STP-2 and 37% for STP-4).

Finally, we assessed the changes occurring in the water vapour permeability (WVP) of the marbles after treatment with the nanocomposites under study. The WVP coefficients are presented in Tables 5.4 and 5.5. According to the results obtained, the coefficients were reduced by approximately 34, 18 and 23% for STP-1, STP-2 and STP-4, respectively. These findings indicated that the changes of the WVP comply with the acceptable ranges of materials applied on building surfaces.[30]

SEM micrographs of marble surfaces treated with the materials under study are shown in Figure 5.9(a)–(c). For comparative purposes, we also took images of the xerogels of the nanocomposites by SEM (Figure 5.9(d)–(f)). As seen, all the applied materials on the surfaces exhibit complete absence of cracks, thus ensuring the preservation of the cohesive structure of the nanocomposites even after treatment. As already discussed, the formation of crack-free coatings is attributed to the synergistic effect of the PDMS and Ox.

In order to assess the chemical compatibility between the nanocomposites and marble surfaces, we carried out an FTIR study as depicted in Figure 5.10. From the spectra obtained, we confirmed that condensation and copolymerization reactions took place on the marble surface. Moreover, in the spectra corresponding to the treated surfaces with STP-4 (Figure 5.10, spectrum d), we can observe the presence of peaks (1618 and 1318 cm^{-1}) associated with the formation of calcium oxalate (CaOx).[23] The noticeable neo-formed CaOx on the treated surfaces produced by the reaction of excess Ox with calcium carbonate could also play a role as a preserving agent against weathering. We should note again that CaOx is the main component found on well-preserved surface layers of monuments.[31,32]

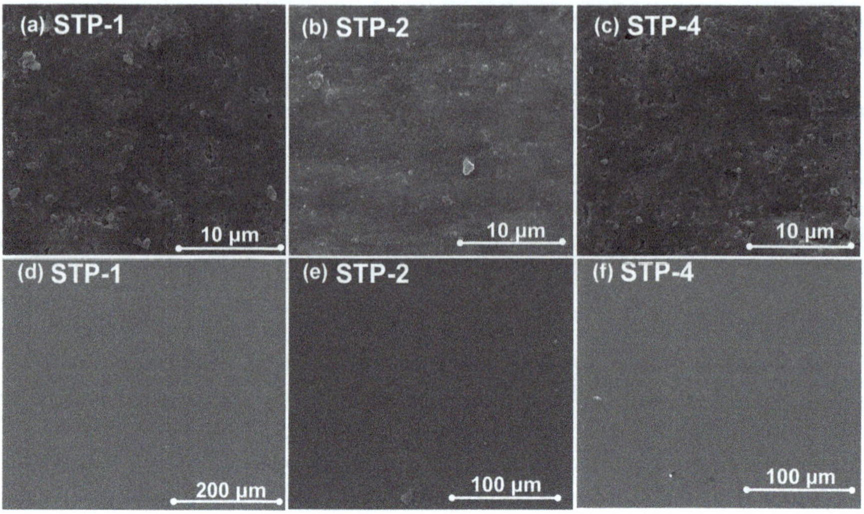

Figure 5.9 SEM micrographs of: (a)–(c) treated marble with the three STP nanomaterials and (d)–(f) STP xerogels.
Reprinted from C. Kapridaki and P. Maravelaki-Kalaitzaki, TiO_2–SiO_2–PDMS nano-composite hydrophobic coating with self-cleaning properties for marble protection, *Prog. Org. Coat.*, **76**, 400–410, Copyright (2012), with permission from Elsevier.[70]

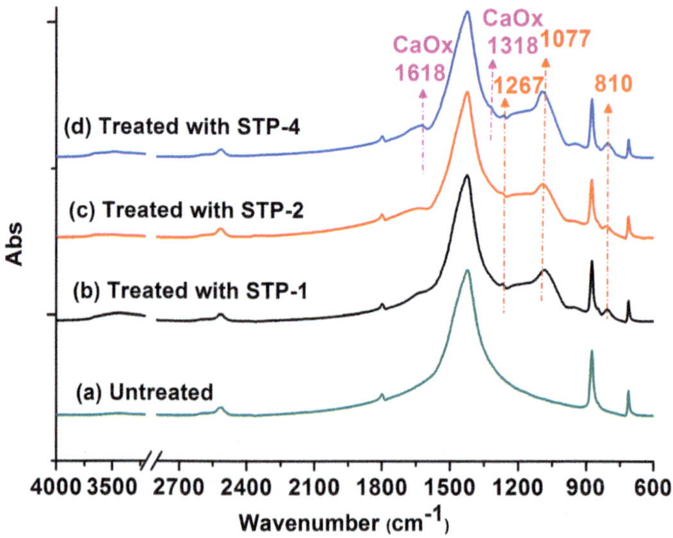

Figure 5.10 FTIR spectra of treated surfaces showing chemical compatibility between nanomaterials and marble surfaces.

Lastly and most importantly, the self-cleaning properties of the nanocomposites under study were assessed and the results obtained are illustrated in Figure 5.11. Specifically, the results of the photocatalytic

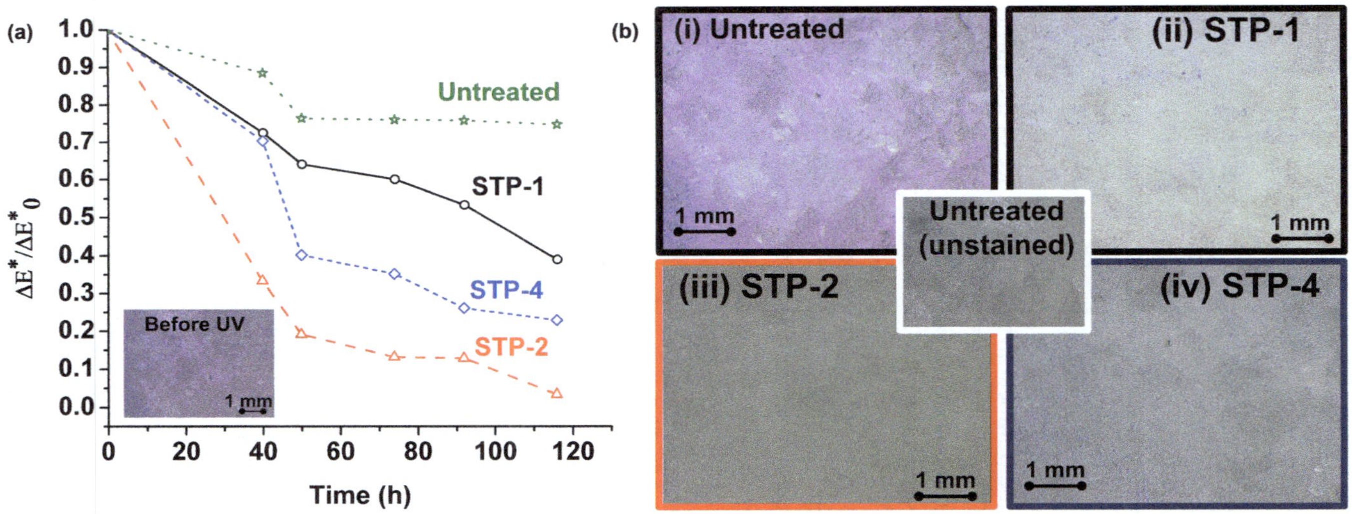

Figure 5.11 (a) Evolution of the total colour difference during the degradation of MB on the treated and untreated surfaces and (b) optical microscopy images of the stained surfaces after UV irradiation for 116 h.
Reprinted from C. Kapridaki, L. Pinho, M. Mosquera and P. Maravelaki-Kalaitzaki, Producing photoactive, transparent and hydrophobic SiO_2-crystalline TiO_2 nanocomposites at ambient conditions with application as self-cleaning coatings, *Appl. Catal., B*, **156–157**, 417–427, Copyright (2014), with permission from Elsevier.[69]

evolution on the treated surfaces are depicted in Figure 5.11(a), where the ratio $\Delta E^{*}/\Delta E_{0}^{*}$ under UV irradiation as a function of time is recorded. Figure 5.11(b) illustrates the optical microscopy images of the marble surfaces obtained at the end of the photocatalytic experiment.

The findings in Figure 5.11(a) are in accordance with the results obtained by the photocatalytic study of the pure nanocomposites (Figure 5.8). As observed, a slight MB degradation was produced on the untreated marble surface due to a photochemical mechanism.[68] The STP-1 treated surface revealed a slower rate of MB degradation, which was attributed to the absence of independent TiO_2 nanoparticles inside the material. However, very rapid degradation profiles were recorded on the treated marbles with STP-2 and STP-4, due to the presence of separate TiO_2 domains inside the silica matrix. By comparing the photocatalytic performance of the surfaces treated with STP-2 and STP-4, we found that STP-2 showed greater self-cleaning activity owing to the presence of a higher amount of independent anatase TiO_2 nanoparticles, as observed by TEM. Finally, the optical microscopy images (Figure 5.11(b)) related to treated and untreated surfaces established the effectiveness of our nanocomposites as self-cleaning agents. By observing these images, the highest self-cleaning activity of STP-2 is confirmed, since the most effective colour removal was obtained by this nanocomposite (Figure 5.11(b), image (iii)).

5.3.4.2 STP as Consolidants and Self-cleaning Agents

The effectiveness of STP-2 and STP-4 as consolidants and hydrophobic agents was also investigated. The dry matters of the materials are shown in Table 5.6, indicating similar behaviour for both.

The total chromatic variations (ΔE^{*}) induced after the consolidation procedure are presented in Table 5.6. These values confirmed that treatment with STP-2 and STP-4 resulted in minor colour modifications, ranging within limits acceptable for materials applied on buildings ($\Delta E^{*} \leq 5$). As already mentioned, chromatic alterations with $\Delta E^{*} = 2$–3 are not visible to the naked eye.

Moreover, both the CA and WCA measurements confirmed the hydrophobic effectiveness of the two STP nanocomposites. Specifically, the values of static CA corresponding to increasing time intervals (0, 20, 300, 600 and 1200 s) for both nanocomposites are shown in Table 5.6. According to these CA values, we concluded that the surfaces treated with STP-2 and STP-4 exhibited hydrophobic properties, as the water CA was increased compared to the untreated sample. As previously discussed, the water repellency is attributed to PDMS.

The remarkable decrease of both the water capillary coefficients (up to 95% for STP-2 and 97% for STP-4) and the TWCA demonstrate not only the water repellency of the treated surfaces, but also a probable occlusion of the pore size of the substrates (Table 5.6). The Alfas limestone treated with STP showed limited pore occlusion, due to deeper penetration of the nanocomposites.

Table 5.6 Physical properties and tensile strength of STP-2 and STP-4 as consolidant agents applied to Alfas limestones. Reproduced from C. Kapridaki and N. Maravelaki, TiO_2–SiO_2–PDMS nanocomposites with self-cleaning properties for stone protection and consolidation, in *Sustainable Use of Traditional Geomaterials in Construction Practice*, ed. R. Prikryl, A. Torok, M. Gomez-Heras, K. Miskovsky and M. Theodoridou, Special Publications, vol. 416, Copyright (2015), with permission from The Geological Society of London.[71]

Sample		Untreated	Treated with STP-2	Treated with STP-4
Dry matter ($mg\,cm^{-2}$)			0.112 (± 0.007)	0.097 (± 0.009)
ΔE^*		—	0.89 (±0.05)	0.65 (±0.03)
Contact angle (°)	0 s	21.80 (±1.18)	68.25 (±2.43)	77.00 (±1.23)
	20 s	—	62.55 (±1.59)	55.55 (±1.62)
	300 s	—	46.82 (±1.20)	42.30 (±1.34)
	600 s	—	35.18 (±1.83)	28.03 (±1.22)
	1200 s		14.88 (±1.94)	20.26 (±1.67)
WCA ($g\,cm^{-2}\,s^{-1/2}$)		0.0134 (±0.0009)	0.00063 (±0.00007)	0.00038 (±0.00003)
TWCA (%)		13.43 (±0.23)	2.99 (±0.28)	2.18 (±0.56)
WVP ($g\,cm^{-2}\,h^{-1}$)		0.0011 (±0.0001)	0.0010 (±0.0001)	0.0010 (±0.0001)
Tensile strength (MPa)		2.81 (±0.12)	3.43 (±0.58)	3.44 (±0.38)

Regarding the WVP values (see Table 5.6), no significant changes were observed after treatment of Alfas with STP-2 and STP-4. The reductions of the WVP undoubtedly conform to the already established ranges.

The effectiveness of STP-2 and STP-4 as strengthening agents was investigated through the Brazilian indirect tensile strength. According to the results presented in Table 5.6, the resistance to the tensile stress was increased by 22% in both nanocomposites. These findings proved that the consolidation treatment considerably strengthened the limestone.

In addition, by taking into account the results of the microdrilling resistance (DRMS), we estimated not only the robustness of the stone but also the penetration depth of the nanocomposite. In Figure 5.12(a), the DRMS profiles of the treated samples and the untreated counterparts are illustrated. These plots revealed that the Alfas treated with STP-2 and STP-4 showed increased mechanical resistance up to a depth of 15 mm from the surface. These results are in accordance with the strengthening improvement exhibited by the Brazilian tensile test.

Figure 5.12(b) depicts the FTIR spectra of the treated and untreated Alfas. In both treatments we identified CaOx, as observed on the treated marbles with STP-4 too (Figure 5.10). As mentioned above, the presence of the peaks located at 1618 and 1318 cm^{-1} confirms CaOx formation, which is more stable than the calcium carbonate and can also protect the stones against further weathering deterioration.[31,32]

Figure 5.13 shows the results obtained after the self-cleaning experiments. As indicated by the plots corresponding to the ratios $\Delta E^*/\Delta E_0^*$ under UV irradiation *versus* time (Figure 5.13(a)), STP-2 induced higher photocatalytic

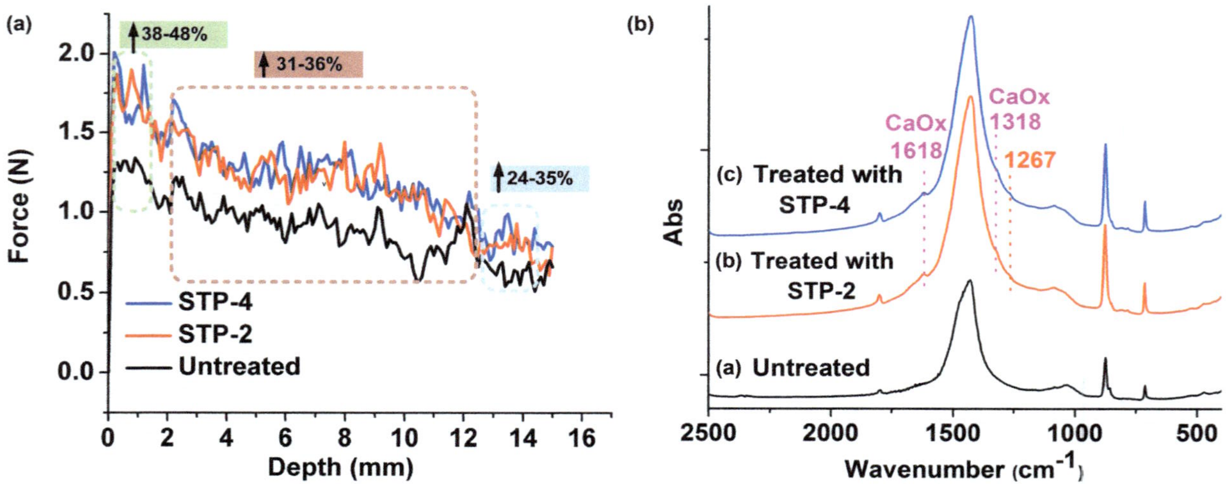

Figure 5.12 (a) Microdrilling profiles (DRMS) of Alfas limestone untreated and treated with STP-2 and STP-4 and (b) FTIR spectra of the untreated and treated Alfas surfaces.
Reproduced from C. Kapridaki and N. Maravelaki, TiO_2–SiO_2–PDMS nanocomposites with self-cleaning properties for stone protection and consolidation, in *Sustainable Use of Traditional Geomaterials in Construction Practice*, ed. R. Přikryl, Á. Török, M. Gómez-Heras, K. Miskovsky and M. Theodoridou, Special Publications, vol. 416, Copyright (2015), with permission from The Geological Society of London.[71]

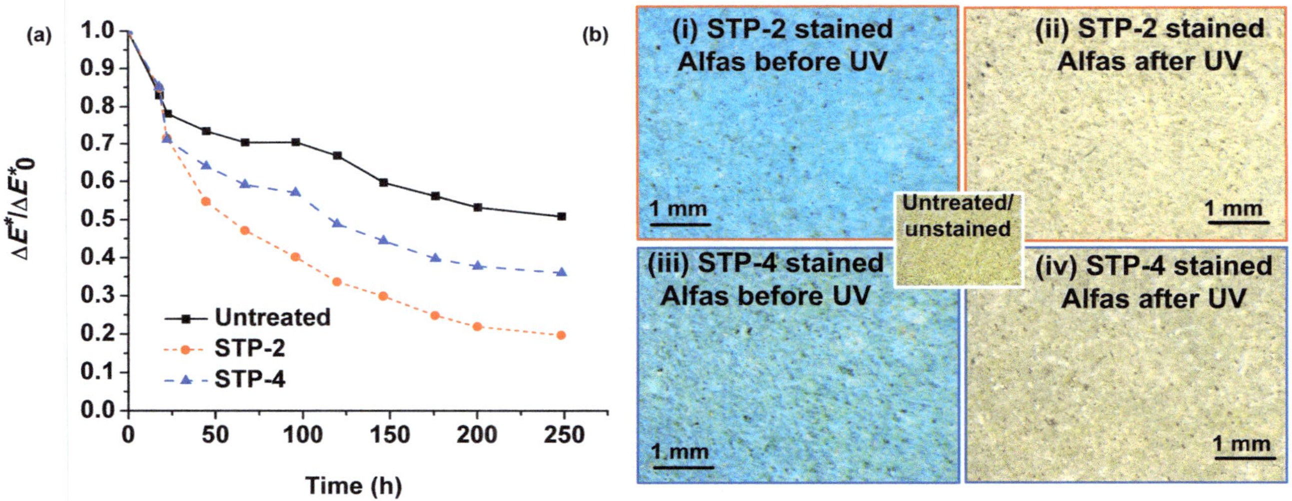

Figure 5.13 (a) Evolution of the total colour difference during the degradation of MB on the treated and untreated Alfas surfaces and (b) optical microscopy images of the stained surfaces before and after UV irradiation for 250 h.
Reproduced from C. Kapridaki and N. Maravelaki, TiO_2–SiO_2–PDMS nanocomposites with self-cleaning properties for stone protection and consolidation, in *Sustainable Use of Traditional Geomaterials in Construction Practice*, ed. R. Přikryl, Á. Török, M. Gómez-Heras, K. Miskovsky and M. Theodoridou, Special Publications, vol. 416, Copyright (2015), with permission from The Geological Society of London.[71]

effectiveness than STP-4, due to the greater amount of separate anatase TiO_2 nanoparticles. Figure 5.13(b) illustrates the optical microscopy images of the Alfas surfaces before and after photocatalytic experiments, indicating self-cleaning capability. These images established the effectiveness of STP-2 and STP-4 on removing the MB organic compound from the treated surfaces (Figure 5.13(b) (ii) and (iv)).

5.4 Conclusion

In the present chapter, we have presented the development of a simple, innovative and low-cost synthesis to produce transparent, photoactive and hydrophobic nanomaterials for stone conservation. The novel nanocomposites were synthesized mixing titanium and silicon alkoxides in the presence of oxalic acid. An organosilane additive, PDMS, was also incorporated within the SiO_2 matrix, thus reducing the surface energy and promoting hydrophobic properties.

We have demonstrated that the sequence of the reagents along with the concentration of the oxalic acid has a clear effect on the structural formation of the final xerogels. In the case of the STP-1 nanocomposite, connectivity at atomic level was achieved. In contrast, in the STP-2 and STP-4 nanocomposites, independent anatase crystalline TiO_2 particles were incorporated into the silica network. Moreover, the presence of oxalic acid plays a crucial role in both the microstructure and the photocatalytic behaviour of the nanocomposites acting as: (1) a sol–gel catalyst of the alkoxides; (2) a chelating agent delaying the hydrolysis of titanium alkoxide and facilitating the formation of homogeneous gels; (3) an acid medium to promote the production of anatase TiO_2 nanoparticles at ambient temperature through the peptization process; (4) a hole scavenger during the photocatalytic procedure, thus increasing the photoreactivity of the nanocomposites; and (5) a DCCA component, which enhances the cohesion of the microstructure and prevents the final xerogels from cracking. In addition, the formation of independent anatase TiO_2 nanoparticles with suitable crystalline size and their incorporation within the SiO_2 network is related to the high self-cleaning activity of the nanocomposites.

Finally, we have demonstrated that these new nanocomposites are capable of: (1) promoting hydrophobic properties of the lithic substrates without altering the colour parameters; (2) imparting self-cleaning properties to the surfaces; (3) allowing the permeability of the substrates; and (4) improving the mechanical properties of the calcareous stones.

Acknowledgements

We wish to express our gratitude to the European Union-European Social Fund (ESF) and National Sources, for providing a grant for this study in the framework of the "HERACLEITUS II" program. We wish also to thank the Spanish Government/FEDER-EU (MAT2013-42934-R) and Project Geopetra (IPT-2012-0959-310000, Innpacto subprogramme), and the Regional Government of Andalusia (project TEP-6386 and Group TEP-243) for financial support.

References

1. J. Chen and C. Poon, *Build. Environ.*, 2009, **44**, 1899.
2. P. Munafò, G. B. Goffredo and E. Quagliarini, *Constr. Build. Mater.*, 2015, **84**, 201.
3. K. Nakata and A. Fujishima, *J. Photochem. Photobiol., C*, 2012, **13**, 169.
4. Y. Lan, Y. Lu and Z. Ren, *Nano Energy*, 2013, **2**, 1031.
5. P. Maravelaki-Kalaitzaki, Z. Agioutantis, E. Lionakis, M. Stavroulaki and V. Perdikatsis, *Cem. Concr. Compos.*, 2013, **36**, 33.
6. I. Karatasios, M. S. Katsiotis, V. Likodimos, A. I. Kontos, G. Papavassiliou, P. Falaras and V. Kilikoglou, *Appl. Catal., B*, 2010, **95**, 78.
7. A. Nazari and S. Riahi, *Mater. Sci. Eng., A*, 2011, **528**, 2085.
8. L. Pinho and M. J. Mosquera, *J. Phys. Chem. C*, 2011, **115**, 22851.
9. L. Pinho, F. Elhaddad, D. S. Facio and M. J. Mosquera, *Appl. Surf. Sci.*, 2013, **275**, 389.
10. E. Quagliarini, F. Bondioli, G. B. Goffredo, C. Cordoni and P. Munafò, *Constr. Build. Mater.*, 2012, **37**, 51.
11. M. Lettieri, A. Calia, A. Licciulli, A. E. Marquardt and R. J. Phaneuf, *Bull. Eng. Geol. Environ.*, 2015, 1.
12. F. Gherardi, A. Colombo, M. D'Arienzo, B. Di Credico, S. Goidanich, F. Morazzoni, R. Simonutti and L. Toniolo, *Microchem. J.*, 2015, **126**, 54.
13. M. F. La Russa, S. A. Ruffolo, N. Rovella, C. M. Belfiore, A. M. Palermo, M. T. Guzzi and G. M. Crisci, *Prog. Org. Coat.*, 2012, **74**(1), 186.
14. A. Colombo, F. Gherardi, S. Goidanich, J. K. Delaney, E. R. de la Rie, M. C. Ubaldi, L. Toniolo and R. Simonutti, *RSC Adv.*, 2015, **5**, 84879.
15. L. Bergamonti, I. Alfieri, A. Lorenzi, A. Montenero, G. Predieri, G. Barone, P. Mazzoleni, S. Pasquale and P. P. Lottici, *Appl. Surf. Sci.*, 2013, **282**, 165.
16. L. Bergamonti, I. Alfieri, A. Lorenzi, G. Predieri, G. Barone, G. Gemelli, P. Mazzoleni, S. Raneri, D. Bersani and P. P. Lottici, *J. Sol–Gel Sci. Technol.*, 2015, **75**, 141.
17. G. Wheeler and E. S. Goins, *Alkoxysilanes and the Consolidation of Stone*, Getty Publications, Los Angeles, 2005, 160.
18. G. W. Scherer, *J. Am. Ceram. Soc.*, 1990, **73**(1), 3.
19. L. Nikolić and L. Radonjić, *Ceram. Int.*, 1994, **20**, 309.
20. M. J. Mosquera, D. M. de los Santos, L. Valdez-Castro and L. Esquivias, *J. Non-Cryst. Solids*, 2008, **354**, 645.
21. C. Miliani, M. L. Velo-Simpson and G. W. Scherer, *J. Cult. Herit.*, 2007, **8**, 1.
22. A. Verganelaki, V. Kilikoglou, I. Karatasios and P. Maravelaki-Kalaitzaki, *Constr. Build. Mater.*, 2014, **62**, 8.
23. A. Verganelaki, C. Kapridaki and P. Maravelaki-Kalaitzaki, *Ind. Eng. Chem. Res.*, 2015, **54**(29), 7195.
24. D. S. Facio and M. J. Mosquera, *ACS Appl. Mater. Interfaces*, 2013, **5**, 7517.
25. Y. Luo, L. Xiao and X. Zhang, *J. Cult. Herit.*, 2014, **16**(4), 470.
26. R. J. Davis and Z. Liu, *Chem. Mater.*, 1997, **9**(11), 2311.

27. P. Judeinstein and C. Sanchez, *J. Mater. Chem.*, 1996, **6**, 511.
28. A. L. Pénard, T. Gacoin and J. P. Boilot, *Acc. Chem. Res.*, 2007, **40**(9), 895.
29. J. F. Illescas and M. J. Mosquera, *J. Phys. Chem. C*, 2011, **115**, 14624.
30. J. F. Illescas and M. J. Mosquera, *ACS Appl. Mater. Interfaces*, 2012, **4**, 4259.
31. L. Rampazzi, A. Andreotti, I. Bonaduce, M. P. Colombini, C. Colombo and L. Toniolo, *Talanta*, 2004, **63**(8), 967.
32. P. Maravelaki-Kalaitzaki, *Anal. Chim. Acta*, 2005, **532**, 187.
33. L. Hench and G. Orcel, U.S. Pat., 4851150 A, 1989.
34. Y. Wang and P. Zhang, *J. Hazard. Mater.*, 2011, **192**(3), 1869.
35. Y. Li and F. Wasgestian, *J. Photochem. Photobiol., A*, 1998, **112**(2–3), 255.
36. S. Bassaid, D. Robert and M. Chaib, *Appl. Catal., B*, 2009, **86**(1–2), 93.
37. W. Gao, R. Jin, J. Chen, X. Guan, H. Zeng, F. Zhang and N. Guan, *Catal. Today*, 2004, **90**(3–4), 331.
38. G. Fu, P. S. Vary and C. Lin, *J. Phys. Chem. B*, 2005, **109**(18), 8889.
39. M. Thommes, K. Kaneko, A. V. Neimark, J. P. Olivier, F. Rodriguez-Reinoso, J. Rouquerol and K. S. W. Sing, *Pure Appl. Chem.*, 2015, **87**(9–10), 1051.
40. L. M. Pastrana-Martínez, S. Morales-Torres, S. K. Papageorgiou, F. K. Katsaros, G. E. Romanos, J. L. Figueiredo, J. L. Faria, P. Falaras and A. M. T. Silva, *Appl. Catal., B*, 2013, **142–143**, 101.
41. R. S. Berns, *Billmeyer and Saltzman's Principles of Color Technology*, John Wiley, New York, 2000, 272.
42. UNI-EN 15801:2010, *Conservation of Cultural Property—Test Methods—Determination of water absorption by capillarity*, Official Italian version EN 15801:2009, European Standard issued by CEN, technical body CEN/TC346–Conservation of cultural property, Date of availability (DAV), 2009.
43. UNI-EN 15803:2010, *Conservation of Cultural Property—Test Methods—Determination of water vapor permeability (dp)*, Official Italian version of EN 15803:2009 European Standard issued by CEN, technical body CEN/TC 346–Conservation of cultural property, Date of availability(DAV), 2009.
44. P. Tiano, *The Use of Microdrilling Techniques for the Characterization of Stone Materials*, ed. L. Binda and R. C. de Vekey, RILEM Publications, Paris, 2003, PRO 26: International RILEM Workshop On Site Control and Evaluation of Masonry Structures and Materials, p. 369.
45. M. J. Mosquera, D. M. de los Santos and T. Rivas, *Langmuir*, 2010, **26**(9), 6737.
46. H. S. Kibombo, R. Peng, S. Rasalingam and R. T. Koodali, *Catal. Sci. Technol.*, 2012, **2**, 1737.
47. S. Doeuff, M. Henry, C. Sanchez and J. Livage, *J. Non-Cryst. Solids*, 1987, **89**, 206.
48. S. Mahshid, M. Askari and M. S. Ghamsari, *J. Mater. Process. Technol.*, 2007, **189**, 296.

49. L. Téllez, J. Rubio, F. Rubio, E. Morales and J. L. Oteo, *Spectrosc. Lett.*, 2004, **37**, 11.
50. P. D. Moran, G. A. Bowmaker, R. P. Cooney, K. S. Finnie, J. R. Bartlett and J. L. Woolfrey, *Inorg. Chem.*, 1998, **37**(11), 2741.
51. T. Lopez, E. Sanchez, P. Bosch, Y. Meas and R. Gomez, *Mater. Chem. Phys.*, 1992, **32**, 141.
52. P. Novotná, J. Zita, J. Krýsa, V. Kalousek and J. Rathouský, *Appl. Catal., B*, 2008, **79**, 179.
53. M. Kruk and M. Jaroniec, *Chem. Mater.*, 2001, **13**(10), 3169.
54. X. Gao and I. E. Wachs, *Catal. Today*, 1999, **51**, 233.
55. D. S. Gopala, R. R. Bhattacharjee, R. Haerr, B. Yeginoglu, O. D. Pavel, B. Cojocaru, V. I. Parvulescu and R. M. Richards, *ChemCatChem*, 2011, **3**, 408.
56. J. M. Thomas, P. A. Midgley, T. J. V. Yates, J. S. Barnard, R. Raja, I. Arslan and M. Weyland, *Angew. Chem., Int. Ed.*, 2004, **43**, 6745.
57. M. Bonne, S. Pronier, Y. Batonneau, F. Can, X. Courtois, S. Royer, P. Marécot and D. Duprez, *J. Mater. Chem.*, 2010, **20**, 9205.
58. H. Shin, H. S. Jung, K. S. Hong and J.-K. Lee, *Chem. Lett.*, 2004, **33**, 1382.
59. B. Fei, Z. Deng, J. H. Xin, Y. Zhang and G. Pang, *Nanotechnology*, 2006, **17**, 1927.
60. X. Zhang, H. Ye, B. Xiao, L. Yan, H. Lv and B. Jiang, *J. Phys. Chem. C*, 2010, **114**(47), 19979.
61. Y. Von Lim, H. Fan, Z. Shen, C. H. Kang, Y. Feng and S. Wang, *Appl. Phys. A: Mater. Sci. Process.*, 2008, **95**(2), 555.
62. L. Pinho, J. C. Hernández-Garrido, J. J. Calvino and M. J. Mosquera, *Phys. Chem. Chem. Phys.*, 2013, **15**, 2800.
63. S. P. Tandon and J. P. Gupta, *Phys. Status Solidi B*, 1970, **38**, 363.
64. B. Llano, M. C. Hidalgo, L. A. Rios and J. A. Navío, *Appl. Catal., B*, 2014, **150–151**, 389.
65. H. Dimitroula, V. M. Daskalaki, Z. Frontistis, D. I. Kondarides, P. Panagiotopoulou, N. P. Xekoukoulotakis and D. Mantzavinos, *Appl. Catal., B*, 2012, **117–118**, 283.
66. H. Yamashita, S. Kawasaki, Y. Ichihashi, M. Harada, M. Takeuchi, M. Anpo, G. Stewart, M. A. Fox, C. Louis and M. Che, *J. Phys. Chem. B*, 1998, **102**(30), 5870.
67. L. M. Pastrana-Martínez, S. Morales-Torres, A. G. Kontos, N. G. Moustakas, J. L. Faria, J. M. Doña-Rodríguez, P. Falaras and A. M. T. Silva, *Chem. Eng. J.*, 2013, **224**, 17.
68. A. Mills and J. Wang, *J. Photochem. Photobiol., A*, 1999, **127**, 123.
69. C. Kapridaki, L. Pinho, M. Mosquera and P. Maravelaki-Kalaitzaki, *Appl. Catal., B*, 2014, **156**, 417.
70. C. Kapridaki and P. Maraveleki-Kalaitzaki, *Prog. Org. Coat.*, 2012, **72**, 400.
71. C. Kapridaki and N. Maravelaki, *Sustainable Use of Traditional Geomaterials in Construction Practice*, ed. R. Přikryl, Á. Török, M. Gómez-Heras, K. Miskovsky and M. Theodoridou, Special Publications, 2015, vol. 416.

CHAPTER 6

Self-cleaning Coatings on Polymeric Substrates

R. FATEH,[a] R. DILLERT[a] AND D. BAHNEMANN*[a,b]

[a] Institute of Technical Chemistry, Gottfried Wilhelm Leibniz University Hannover, Callinstr. 3, 30167 Hannover, Germany; [b] Laboratory for Nanocomposite Materials, Department of Photonics, Faculty of Physics, Saint-Petersburg State University, Ulianovskaia str. 3, Peterhof, Saint Petersburg 198504, Russia
*Email: bahnemann@iftc.uni-hannover.de

6.1 Introduction

For over 30 years, it has been known that titanium dioxide has the ability to oxidize a wide range of organic compounds upon UV light irradiation. This phenomenon, which is called the photocatalytic effect of the titanium dioxide, is due to the reactions of the photogenerated electron–hole pairs created upon excitation by UV light with an energy larger or equal to its band gap energy ($E_g = 3.2$ eV in the case of anatase TiO_2, corresponding to $\lambda \leq 380$ nm). In addition to the photocatalytic effect, changes of the wettability of the surface of titanium dioxide by water or organic liquids are usually observed. Under irradiation by UV light, the contact angle of water on the titanium dioxide surface decreases and water begins to spread over the surface. This phenomenon, which is called photo-induced superhydrophilicity, is important from a technical point of view for the production of so-called self-cleaning surfaces. Recently, thin transparent layers containing TiO_2 have been intensively studied based upon their interesting

RSC Smart Materials No. 21
Self-cleaning Coatings: Structure, Fabrication and Application
Edited by Junhui He

Published by the Royal Society of Chemistry, www.rsc.org

application potential including the formation of photocatalytically active and/or self-cleaning surfaces.[1–11]

The first self-cleaning window glass was produced in 2001 by the major glass manufacturing company Pilkington Glass (Pilkington Activ™). The transparent coating on Pilkington Activ™ comprising a film of nanocrystalline titanium dioxide as the active photocatalyst layer is just 40 nm thick, and is applied to the glass by chemical vapour deposition during the manufacturing process at temperatures of 600 °C.[12] This layer is applied to clear float glass to produce a hard, thin, transparent, mechanically robust, photoactive coating of titanium dioxide. This material appears to be the ideal, readily available titanium dioxide photocatalyst film. Therefore, Pilkington Activ™ has been used as a benchmark for all other photocatalyst films currently being produced by research groups in academia and industry alike for light-driven air and water purification and for self-cleaning purposes.[13]

In the near future, rigid plastic optical components will replace parts made from glass whenever improved properties and/or lower costs can be achieved. The unique properties of a transparent polymer such as polycarbonate, including break resistance, weather resistance, non-flammability, and safety, make it an amazing candidate for a lot of applications. The applications of polymeric materials range from plastic vessels and machine parts to optical grades for compact discs and optical fibres. However, the properties of the polymeric substrates have to be improved by functional coatings in order to match the enhancement in the materials' quality to meet the future demand of industries especially the automotive industry. This can be achieved by providing the material sheets with a self-cleaning superhydrophilic and photocatalytically active surface layer. Demands such as high transparency, low reflectivity, and high mechanical stability of the applied layers have to be taken into consideration.

Ideally, a self-cleaning surface must comply with two essential criteria: (1) high photo-induced superhydrophilic properties, and (2) uninterrupted performance in the absence of UV light.[14] The superhydrophilic property of the surface allows water to spread completely across the surface rather than to remain in droplet form, which provides the surface with an additional important property besides the self-cleaning properties.[15,16]

This chapter summarizes the concepts of photocatalysis and of the light-induced changes in wettability. It demonstrates methods of preparing self-cleaning coatings on polymer surfaces and how to deposit the coatings onto polymeric substrates by sol–gel techniques. The progress that has been achieved in the production of transparent, self-cleaning coatings on polycarbonate is presented.

6.2 Self-cleaning Coatings

Self-cleaning technology is as old as nature since numerous surfaces in nature show self-cleaning properties such as the wings of butterflies and the leaves of plants, *e.g.*, cabbage or lotus.[10] Providing polymers with

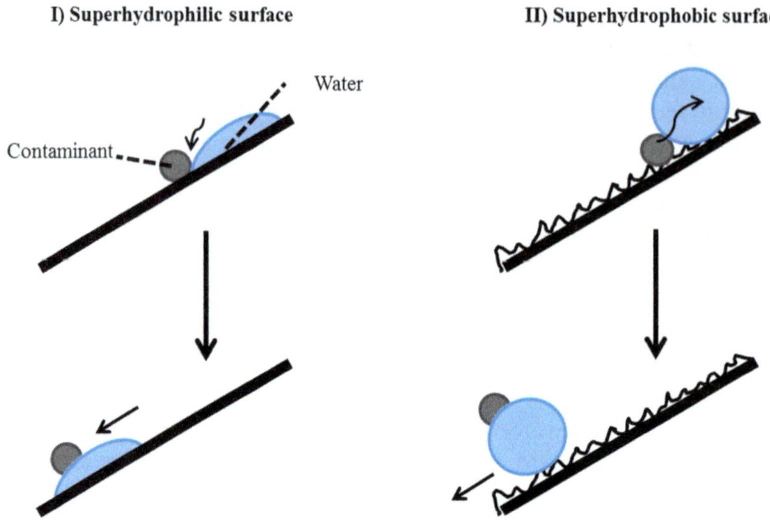

Figure 6.1 Schematic illustrations of self-cleaning processes on (I) a superhydrophilic surface, and (II) a superhydrophobic surface.
Reproduced from ref. 10 with permission from the Royal Society of Chemistry.

self-cleaning coatings is one of the research fields which have attracted the attention of a lot of scientists all over the world due to its industrial importance. Superhydrophobic surfaces with a water contact angle higher than 150° and superhydrophilic surfaces with a water contact angle smaller than 5° have prompted extensive interests for both fundamental research and practical applications.

In general, self-cleaning coatings are divided into two types: hydrophobic and hydrophilic. The cleaning of the surface of these two categories is through the action of water; *i.e.*, in the hydrophobic type by rolling droplets whereas the hydrophilic one functions by sheeting water that carries away dirt (Figure 6.1).[10]

Moreover, the hydrophilic coatings may also have an additional property, *i.e.*, they can chemically break down adsorbed dirt upon illumination with suitable light sources such as sunlight (Figure 6.2).[2,3]

6.2.1 Wettability and Photoinduced Hydrophilicity

The wettability, that is how liquids behave on a surface, is one of the fundamental properties of every solid and, thus, is important for a wide range of natural systems as well as in many technical applications. It can be assessed using contact angle measurements.[17,18]

The contact angle θ is defined as the angle between the tangent to the liquid–air interface and the tangent to the solid–air interface (Figure 6.3). In the special case of water being the wetting liquid, when the contact angle is

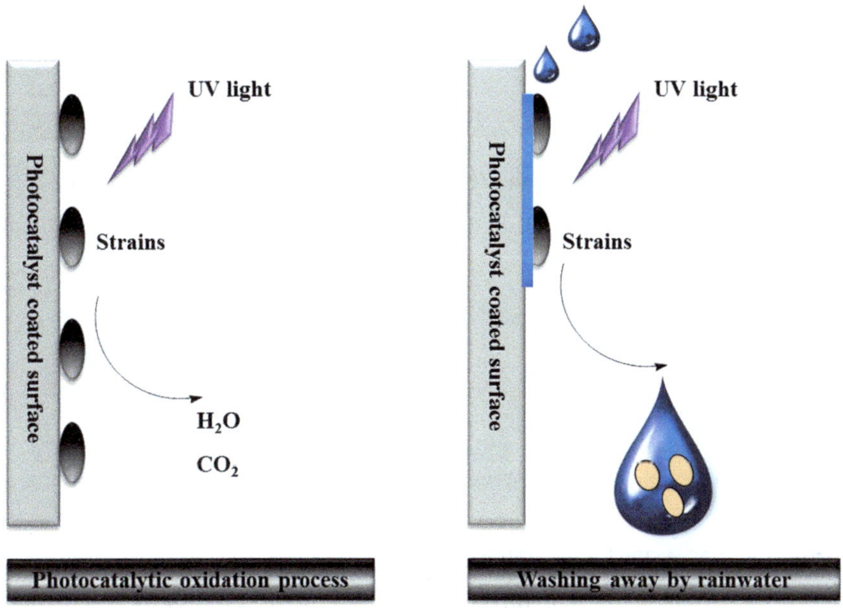

Figure 6.2 The self-cleaning process on a photocatalytically active surface: conventional photocatalytic oxidation process (left) and spreading of water droplets on the superhydrophilic surface (right).
Reproduced from ref. 10 with permission from the Royal Society of Chemistry.

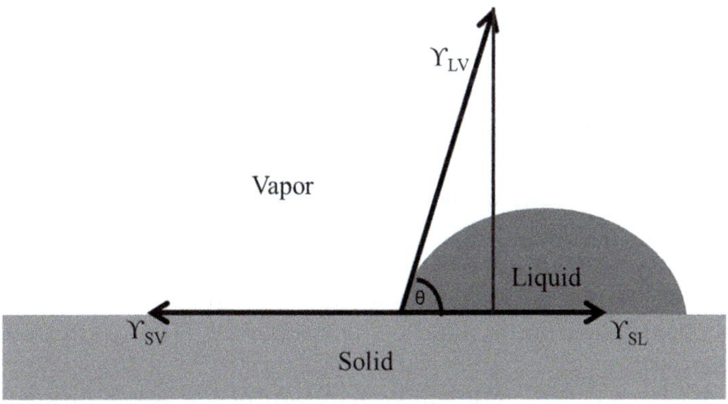

Figure 6.3 Relationship between the interfacial energies per unit area at the point of the three-phase equilibrium.
Reproduced from ref. 9 with permission from the Royal Society of Chemistry.

less than 90°, the water droplet spreads over the entire surface and the surface is called a hydrophilic surface (when $\theta \approx 0°$, the surface is superhydrophilic). If the contact angle exceeds 90° and the water forms droplets,

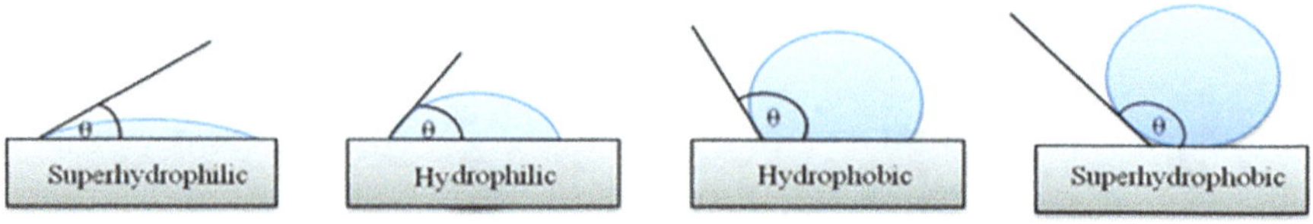

Figure 6.4 Schematic of the contact angle of a droplet on different surfaces.

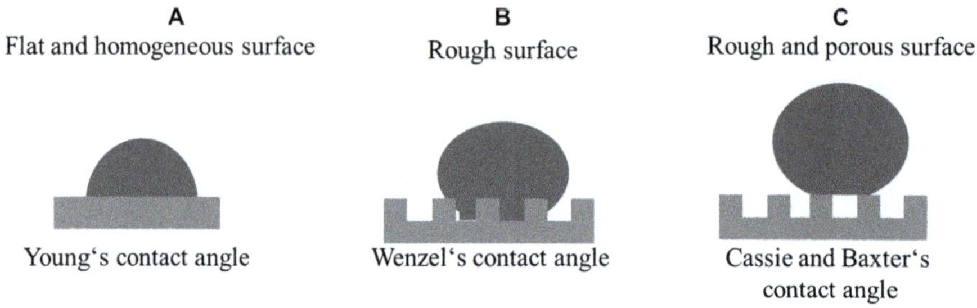

Figure 6.5 The contact angle models.
Reproduced from ref. 19 with permission from John Wiley and Sons. Copyright © 2006 WILEY-VCH Verlag GmbH & Co. KGaA, Weinheim.

the surface is called a hydrophobic surface; when $\theta \approx 180°$, the surface is said to be superhydrophobic (Figure 6.4).[17,18]

The contact angle θ_Y on an ideal (flat and homogeneous) surface was first defined by Young's equation in 1805.[18]

$$\cos\theta_Y = (\Upsilon_{SV} - \Upsilon_{SL})/\Upsilon_{LV} \tag{6.1}$$

Here, Υ_{SV}, Υ_{SL}, and Υ_{LV} are the interfacial energy of the solid–vapour interface, the solid–liquid interface, and the liquid–vapour interface, respectively (*cf.* Figure 6.5A).

However, Young's equation neglects the effect of gravity and of the roughness of the surface. In reality, the contact angles are influenced by the specific kinds of atoms and surface terminations present at the liquid–solid–vapour interfaces. Nevertheless, surface roughness plays an equally important role for the wettability.[18] This means that Young's equation is not accurate.

Wenzel assumed the complete penetration of the water droplet into the grooves of the rough surface and suggested the following equation:

$$\cos\theta_W = r\cos\theta_Y \tag{6.2}$$

where θ_W is the predicted Wenzel contact angle and r defines the roughness of a solid surface. According to Wenzel's equation, increasing the surface roughness of a hydrophilic surface enhances the wettability thus enhancing the hydrophilicity of a hydrophobic surface (*cf.* Figure 6.5B).[18]

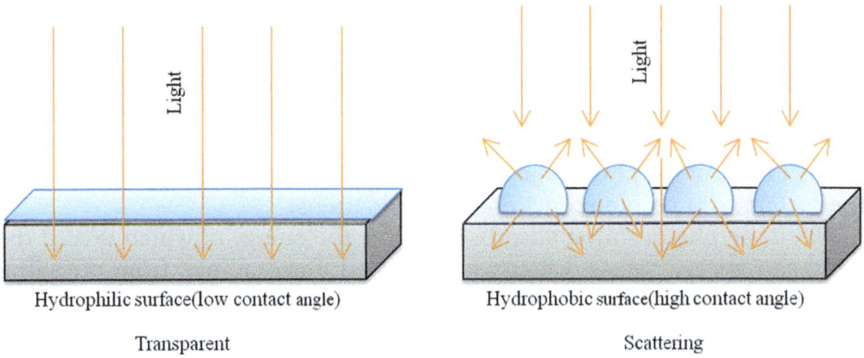

Figure 6.6 The difference between a hydrophilic and a hydrophobic surface.

On the other hand, Cassie and Baxter assumed that when a liquid spreads over a rough porous surface, the solid–vapour interface is destroyed and two new interfaces, *i.e.*, the solid–liquid and the liquid–vapour interface, are formed. Thus, the contact angle on this surface is given by:

$$\cos\theta_{CB} = f\cos\theta_W + (1-f)\cos\theta_V \tag{6.3}$$

where θ_{CB} is the predicted Cassie–Baxter contact angle, and f is defined to be the fractional projected area of a material with a smooth surface contact angle, θ, hence, the $(1-f)$ term reflects the contribution of air remaining under the drop.[18] In this case, the vapor pockets are trapped underneath the liquid; this means the droplet does not wet entirely as a result of the trapped superhydrophobic vapour pocket (*cf.* Figure 6.5C).[19]

Moreover, the individual water droplets will form a film of water on the hydrophilic surfaces. The formation of this film provides the surfaces with new features, *e.g.*, the surfaces will have an anti-fogging effect. Here, the water forms a film on the surface and no small droplets, thus, it is not clouded in the field of view (Figure 6.6).

6.2.2 Photocatalysts and Photocatalysis Mechanism

The photocatalysis mechanism has been extensively studied over the past 30 years. Generally, when a semiconductor is irradiated by light of sufficient energy, electrons are excited from its valence band to the conduction band. The photogenerated charge carriers (conduction band electron and valence band hole) may migrate to the surface of the crystallite where they are able to reduce and oxidize, respectively, adsorbed electron acceptors and donors by interfacial charge transfer. However, this is not the only fate of these photogenerated charge carriers since they may also recombine and transform their energy to heat.

In order to be reduced, the relevant potential level of the acceptor species is thermodynamically required to lie below (*i.e.*, less negatively on the electrochemical scale) the conduction band potential of the semiconductor.

On the other hand, the potential level of the donor needs to be located above the valence band position of the semiconductor to be able to donate an electron to the vacant hole. Actually, the energy level at the bottom of the conduction band is the reduction potential of the photogenerated electrons, whereas the energy level at the top of the valence band determines the oxidizing ability of the photogenerated holes, each value reflecting the ability of the system to promote reductions and oxidations.

When the photocatalyst adsorbs photons, the energy of which exceed the band gap energy and which can be provided, for example, from the ultraviolet radiation of sunlight or from a suitable illuminating light source, it will produce pairs of electrons and holes. This occurs since an electron from the valance band of the photocatalyst becomes excited upon illumination by light. The excess light energy thus promotes the electron to the conduction band of the photocatalyst consequently creating the pair consisting of the negatively charged electron (e^-) and the positively charged hole (h^+). The energy difference between the valance and the conduction band is the band gap energy. For example, the band gap energy of TiO_2 is 3.2 eV. Excited-state electrons and holes can recombine and dissipate the input energy as heat or light, get trapped in metastable surface states, or react with electron donors and acceptors adsorbed at the semiconductor surface. For example, after their reaction with water, the holes can produce hydroxyl radicals with high oxidation potential. Depending upon the exact conditions, the holes themselves, $^\bullet OH$ radicals, superoxide radicals $O_2^{\bullet -}$, H_2O_2 and O_2 itself can play important roles in the photocatalytic reaction mechanism (Figure 6.7).[1] The photocatalytic activity of a given system is usually tested by measuring the degradation of model pollutants. Due to the ease of the determination of the concentration of compounds such as stearic acid, methylene blue, and chlorophenol quantitatively by chromatography and spectroscopy, they are popular choices as model pollutants.[2]

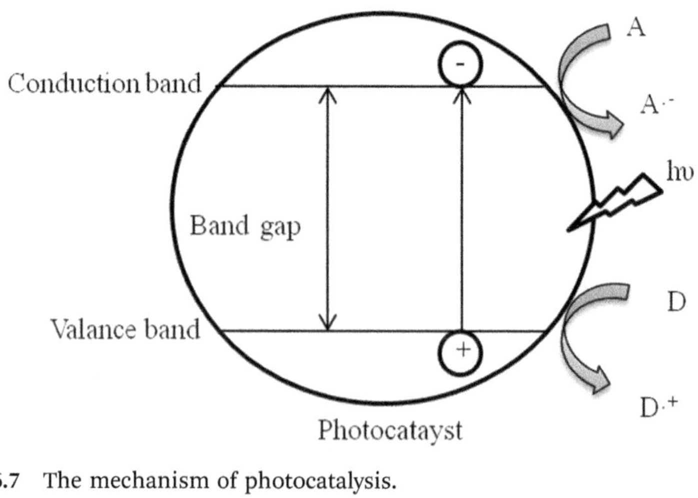

Figure 6.7 The mechanism of photocatalysis.

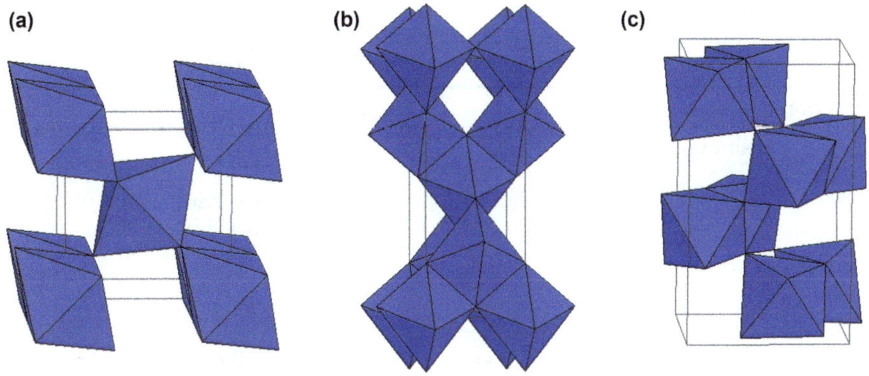

Figure 6.8 Crystal structures of TiO_2 (a) anatase, (b) rutile, and (c) brookite. Reprinted from O. Carp *et al.*, *Prog. Solid State Chem.*, Photoinduced reactivity of titanium dioxide, **32**, 33–177, Copyright (2004) with permission from Elsevier.

Titanium dioxide is the most famous photocatalyst and it has been used frequently for self-cleaning windows and hydrophilic self-cleaning surfaces due to its favourable physical and chemical properties. In addition to its high photocatalytic activity under UV(A) illumination and its photoinduced superhydrophilicity, it is also non-toxic, chemically inert in the absence of light, inexpensive, relatively easy to handle and it can be conveniently deposited into thin films.[2]

TiO_2 is an n-type semiconductor because of its oxygen vacancies. This material exists in nature in three crystal phases: anatase (tetragonal), rutile (orthorhombic), and brookite (tetragonal). Each crystal phase exhibits slightly different physical and chemical properties.[1] Figure 6.8 shows the crystal structure of the three types of TiO_2.

The differences in lattice structure of anatase and rutile TiO_2 cause different charge densities and electronic band structures, leading to different band gaps (for bulk materials: anatase 3.20 eV and rutile 3.02 eV).[1] Therefore, the absorption thresholds correspond to wavelengths of 384 and 410 nm, respectively, for the two titania forms. The values mentioned have been determined for single crystals or at least well-crystallized samples. However, higher values of band gaps are usually obtained for weakly crystallized thin films or nanosized materials.

6.3 Photocatalytically Active Films

Recently, thin transparent layers containing TiO_2 have been intensively studied based upon their interesting application potential including photocatalytic water and air purification.[1] The self-cleaning properties of TiO_2 films derive from two unique light-induced properties: photocatalysis[1] and superhydrophilicity.[20] That is, organic pollutants adsorbed on the surface of a TiO_2 film can be decomposed under illumination with ultraviolet light

Table 6.1 Applications of self-cleaning TiO_2. Reprinted from O. Carp *et al.*, Photoinduced reactivity of titanium dioxide, *Prog. Solid State Chem.*, **32**, 33–177, Copyright (2004) with permission from Elsevier.

Function	Material	Application
Cleaning easiness	Material for road	Tunnel lighting Tunnel wall
	Material for house	Kitchen parts, bathroom and interior furnishing
	Materials for electric and electronic devices	Computer display, electronic wires
	Daily necessities and consumer products	Tableware, kitchenware
Self-cleaning by rainfall	Materials for road	Traffic sign, lighting, soundproof wall, guardrail decorative laminate panel
	Materials for buildings	Exterior tiles, siding boards, curtain wall, painted steel plate, aluminium panel, building stone, crystallized glass, glass film, window, sash, screen door, gate door, sun parlous, veranda parts
	Materials for electric and electronic devices	Upper glass of a solar cell, insulator
	Materials for vehicles	Painting and coating of vehicles, outside of windows, headlight

(UV(A)), and dust is easily washed off, for example, by rainwater due to the spreading of the water on the superhydrophilic surface.[21] Table 6.1 summarizes the main self-cleaning applications of TiO_2.[22]

6.3.1 Introducing Porosity in TiO_2 Film

A mesoporous material, as characterized by IUPAC, is a porous material with a pore diameter of $2.0 \leq d \leq 50$ nm.[23] Usually, mesoporous metal oxides are prepared by template-based methods using soft templates (surfactants and block polymers),[24] and/or hard templates (porous silica, polystyrene spheres, porous carbon).[25] The importance of these types of materials stems from their high specific surface areas and pore volumes, as well as from the usually narrow pore size distributions which offer more active sites for catalytic reactions to take place.[26] Since the adsorption of the molecules (electron acceptor and donor) on the surface of the photocatalyst is essential in the photocatalytic systems, enhancing the surface area of the photocatalyst is expected to increase its photocatalytic activity. Moreover, mesoporous TiO_2 is an interesting material for photocatalytic applications not only because of its higher surface area but also due to its continuous particle framework, which may be beneficial compared to separate individual nanoparticles. The low number of studies that have been reported on ordered mesoporous TiO_2 as a photocatalyst is likely related to the difficulties in making it as an ordered material.[27] Moreover, the mechanical properties of coatings prepared using mesoporous metal oxides are relatively poor and need to be improved.

6.3.2 Modifying TiO_2 with Other Metal Oxides

Mixed oxide composite materials are sometimes more efficient photocatalysts than pure semiconductors.[28,29] This is due to the generation of new active sites as a result of interactions between the oxides. Furthermore, mixing of metal oxides may lead to an increased surface area, and to improved mechanical strength and thermal stability.[30]

SiO_2 exhibits high thermal stability, excellent mechanical strength and supports the creation of new catalytically active sites due to the interaction between TiO_2 and SiO_2.[31] A mixed metal oxide (TiO_2/SiO_2) can enhance the photocatalytic performance improving the ability of surface adsorption and increasing the amount of hydroxyl surface groups.[32] Simultaneously, SiO_2 acts as the carrier of TiO_2 leading to a large surface area as well as to a suitable porous structure.[33,34] On the other hand, it is well known that the coupling of two semiconductors,[30] in particular TiO_2 and ZnO,[35] is useful in achieving a more efficient separation of photo-generated electron–hole pairs. The electron transfers from the conduction band of ZnO to the conduction band of TiO_2 under illumination and, conversely, the hole transfers from the valence band of TiO_2 to that of ZnO, resulting in a decrease of the rate of electron–hole recombination. This, in turn, leads to an improvement in the photocatalytic activity.

6.4 Photoinduced Superhydrophilicity

The discovery of the phenomenon of photoinduced superhydrophilicity of TiO_2 thin films opens a lot of new aspects for industrial applications such as self-cleaning and anti-fogging. According to the photoinduced superhydrophilicity phenomenon, the TiO_2 surface becomes superhydrophilic with a water contact angle $<5^\circ$.[36] To this day, the reasons behind the photoinduced superhydrophilicity are still an open question. But the three most proposed ones are: (I) generation of surface vacancies, (II) photoinduced reconstruction of Ti–OH bonds, and (III) photocatalytic decomposition of organic adsorbents.

6.4.1 Generation of Surface Vacancies

When electron–hole pairs are generated upon irradiation with UV light, the holes diffuse to the surface and cause oxygen vacancies at the surface. This results in a reduction of Ti^{4+} to Ti^{3+}. These Ti^{3+} sites can cause water dissociation, resulting in adsorbed −OH species, which are hydrophilic.[20]

Hashimoto and co-authors further show that hydrophilicity closely depends on the crystal surface. The (110) surface shows better hydrophilic properties than the (001) surface since it has bridging oxygen atoms which are not found on the (001) surface.[37]

Although a lot of experimental results correspond with the proposed mechanism, there are also some results that disagree with this mechanism. White *et al.* studied the hydrophilic effect on the TiO_2 (110) surfaces prepared with or without oxygen vacancies. They pointed out that the hydrophilic property is not influenced whether or not oxygen vacancies (up to 14%) are present, which is obviously contradicting the current mechanism.[38]

6.4.2 Photo-induced Reconstruction of Ti–OH Bonds

The reconstruction of hydroxyl groups of the TiO_2 surface is attributable to photo generated holes. As is explained in eqn (6.4) and (6.5), the photo generated holes are trapped at surface lattice oxygen atoms, causing a dissociation of their bonds with titanium atoms, followed by the adsorption of water molecules, which leads to the appearance of hydroxyl groups on surface defects.[39]

$$\text{Ti–O–Ti} + h^+ + H_2O \rightarrow [\text{Ti–O}^\bullet \cdot \text{HO–Ti}] + H^+ \qquad (6.4)$$

$$[\text{Ti–O}^\bullet\ \text{HO–Ti}] + H_2O \rightarrow [\text{Ti–OH} \cdot \text{HO–Ti}] + {}^\bullet\text{OH} \qquad (6.5)$$

However, Anpo and co-workers showed that a vacuum can cause an immediate disappearance of the superhydrophilicity of TiO_2. The authors argued that these results cannot be explained by the newly formed OH groups during UV-illumination, because those cannot be removed by evacuation at room temperature.[40] Thus, the proposed reconstruction of the Ti–OH bond mechanism needs to be modified.

6.4.3 Photocatalytic Decomposition of Organic Adsorbents

The light-induced superhydrophilicity conversion was initially considered to be directly associated with photocatalytic activity. When a TiO_2 surface is contaminated, it turns hydrophobic (strictly, it has a water contact angle smaller than 90°) due to the adsorption of certain gas (organic) components contained in the air. After UV irradiation, the organic adsorbents are photocatalytically oxidized and removed from the TiO_2 surface.[41]

As a matter of fact, it cannot be considered that the removal and adsorption of organic contaminate is the only reason causing the wettability change of the TiO_2 surface. Additionally, a preferential adsorption of water molecules with dissociation on the photo-generated surface defective sites contributes to the formation of highly hydrophilic TiO_2 surfaces.[41]

Hennessy *et al.* found that the well-protected clean surface of rutile TiO_2 (110) still exhibits a large water contact angle,[42] consistent with previously reported measurements by the Hashimoto group.[37] A mechanochemical treatment such as an ultrasonic treatment or cleaning the surface with

NaOH should enhance hydrophilicity thus organic contaminants are decomposed in the same way as by photocatalysis. But the contact angle only decreases to values about 10–20° and no superhydrophilic state is reached.[41] This leads to the conclusion that photocatalytic decomposition of the adsorbed pollutants cannot be the only process to achieve superhydrophilicity.

6.5 Methods for Preparing Self-cleaning Coatings on Polymer Surfaces

Several routes to apply self-cleaning coatings onto surfaces are possible, such as physical or chemical vapour deposition. However, while these techniques exhibit many advantages, they are expensive and cumbersome.[43] In addition, the range and shape of materials that can be coated by these techniques is limited.[43] An alternative coating method is the sol–gel route allowing novel materials, such as organic–inorganic hybrids, to be deposited on various substrates from a solution at or near room temperature. Consequently, this technique is suitable for coating polymers, which usually have limited thermal stability.[43] It enables the coating of large surfaces and also the attainment of thin layers with good thermal and mechanical stabilities as well as high optical quality.[44] Furthermore, the sol–gel method is relatively simple and inexpensive.

To obtain a mechanically stable layer on a polymeric substrate *via* a sol–gel process, a pretreatment of the substrate is usually required. Impurities which adversely affect the wettability of the substrate by the coating solution must be removed. Frequently, an additional pretreatment is necessary to create reactive functional groups on the polymer surface to which the active inorganic material can be covalently bound.

However, the direct deposition of a photocatalytically active layer onto a polymer surface may induce photocatalytic degradation of the polymer resulting in an exfoliation of the photocatalytically active layer.[45] To avoid damage of the polymer surface, a photocatalytically inactive intermediate layer should be deposited to prevent direct contact between the polymer and the photocatalytically active layer. However, the role of the intermediate layer is not only to protect the polymer from the photocatalytically active layer but also to enhance the binding between the organic polymer and the inorganic photocatalytically active layer.[46] Figure 6.9 presents the deposition of different layers on the polymer surface.

6.6 The Sol–Gel Process

The sol–gel process can be defined as follows: the synthesis of an inorganic or organic network by a chemical reaction in solution at low temperature.[47] This solution, which is called a sol, is a colloidal suspension of solid particles in a liquid (a colloid is a suspension in which the dispersed phase is so

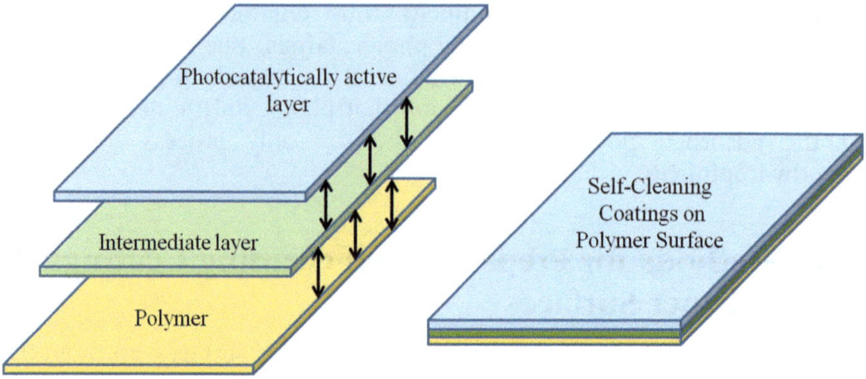

Figure 6.9 Schematic presentation of the deposition of different layers on the polymer surface.

Table 6.2 Applications of sol–gel methods.

Applications of sol–gel method	
Mechanical and chemical	Protection with hard coats, catalysts, anticorrosion films.
Optics and photonics	Solar cells, laser elements, waveguides, optical switching, antireflection coatings.
Electronics	Capacitors, non-volatile memories, transport electrodes.
Thermal	Refractory ceramics, fibres, low expansion ceramics.
Biomedical	Biocompatible films, entrapment of enzyme and cells, medical tests.

small (~1–1000 nm) that the gravitational force is negligible and interactions are dominated by short range forces, such as Van der Waals attraction and surface charges).[48] The continuous solid skeleton made of colloidal particles is called a gel.

Sol–gel synthesis is realized *via* various process paths. By employing this method, it is possible either to generate small (nano) particles to coat parts or surfaces or even to form three-dimensional structures.

The sol–gel process is used in many different applications. Table 6.2 summarizes the applications of the sol–gel method.[49]

In Figure 6.10, different nanostructures produced by sol–gel methods are shown. The starting point of the production route from the solution of precursors is always a sol, which evolves towards the formation of a gel containing both liquid and solid phases whose morphologies range from discrete particles to continuous polymer networks. Dried gels, either xerogels or aerogels (xerogels are dried by evaporation, whereas aerogels are dried by supercritical extraction of solvent), are distinguished by their enormous surface area and correspondingly small pore sizes.[49]

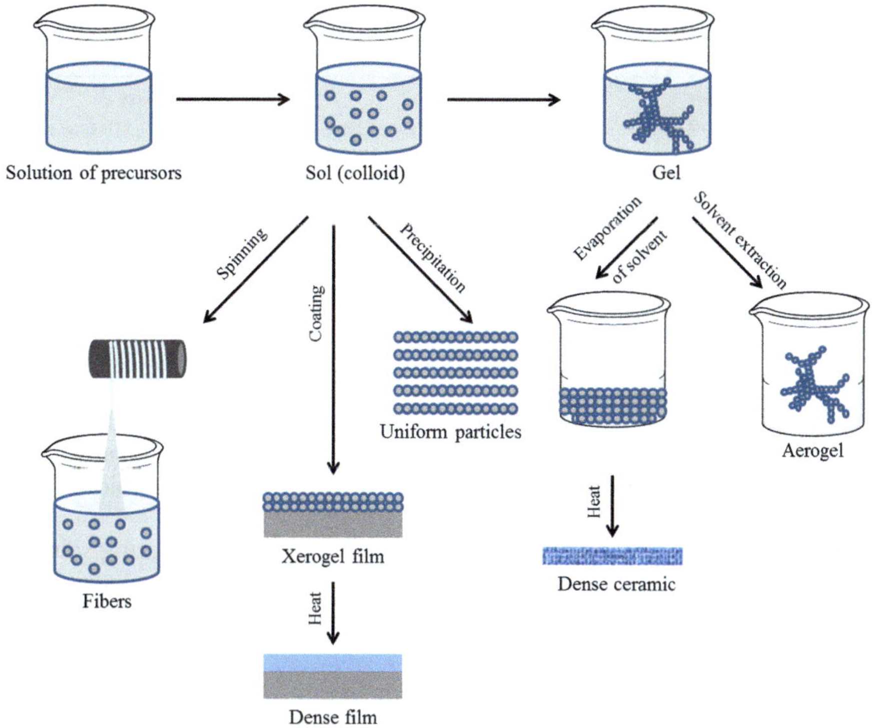

Figure 6.10 Sol–gel process for preparation of different materials and devices.

Upon removal of the solvent in the formation of a xerogel, the network is not completely linked and the resulting product has a porosity of about 50 percent. By a subsequent drying process at high temperatures, it is possible to obtain a solid ceramic with low porosity. Aerogels can also be produced directly from the gel stage by removing the solvent at high pressure and high temperature (supercritical extraction of solvent), thus the gel network formed retains its structure, and the resulting solid comprises up to 95 percent pores. It is also possible to produce fibers obtained from the sol, which is then sintered at high temperatures.[49]

In fact, one of the main sol–gel applications is the coating of surfaces. In this case, dip and spin coating procedures are used to deposit the sol onto the substrate to form the thin film. Actually, the following conditions have to be matched during the film formation:

- The solution must wet the substrate;
- The solution must remain stable during the aging process;
- The solution solidifies as a homogeneous transparent film;
- The solution should have some tendency towards crystallization into a stable high temperature phase;
- For multiple layers, the previous layer must be either insoluble or heat treated to make layers insoluble before subsequent depositions.

Table 6.3 Substances used in the preparation of sols and their roles.

	Substance	Role	Example
Basis	Alkoxide	Ti donor	Titanisopropylate
	Alcohol	Solvent	Propan-2-ol
	Water	Hydrolysis	
Additives	Chelator	Hydrolysis Controlling	Acetylacetone, acetic acid
	Acid, base	—	Acetic acid, HCl, HNO_3 NH_4OH
	Additional solvent	Drying control	Polyethyleneglycol (PEG), dimethylformamide (DMF)
	Organic binder	—	Hydroxypropylcellulose

Alkoxide solutions and aqueous solutions of different metal salts (chloride, nitrate, …) fulfil these conditions. Table 6.3 presents the substances used in the preparation of various sols and their respective roles in the coating process.[50]

In general, the precursors used for the preparation of the coating sols can be divided into two groups: metal–organic compounds, *e.g.*, metal alkoxides, and inorganic salts.[49]

With an alkoxide ($M(OR)_n$) as the precursor, the sol–gel reactions can be described as follows:

$$\text{Hydrolysis: -MOR} + H_2O \rightarrow \text{-MOH} + \text{ROH} \tag{6.6}$$

$$\text{Condensation: -MOH} + \text{ROM-} \rightarrow \text{-MOM-} + \text{ROH} \tag{6.7}$$

$$\text{-MOH} + \text{HOM-} \rightarrow \text{-MOM-} + H_2O \tag{6.8}$$

Eqn (6.9)–(6.15) describe the reaction mechanisms for acid (eqn (6.9)–(6.12)) and base (eqn (6.13)–(6.15)) catalyzed hydrolytic condensation of alkoxy silane. Titanium alkoxides hydrolyse and condensate following the same mechanisms.

Acid catalyzed

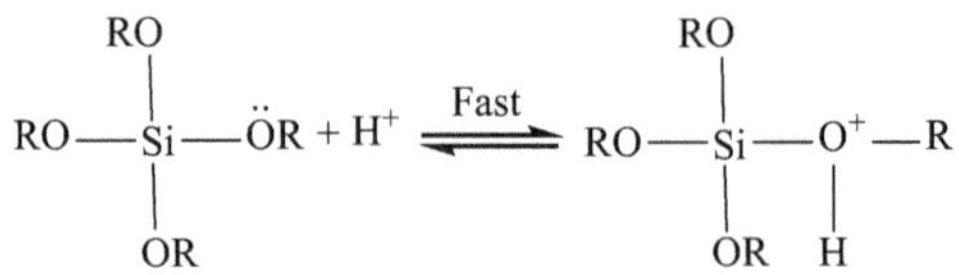

(6.9)

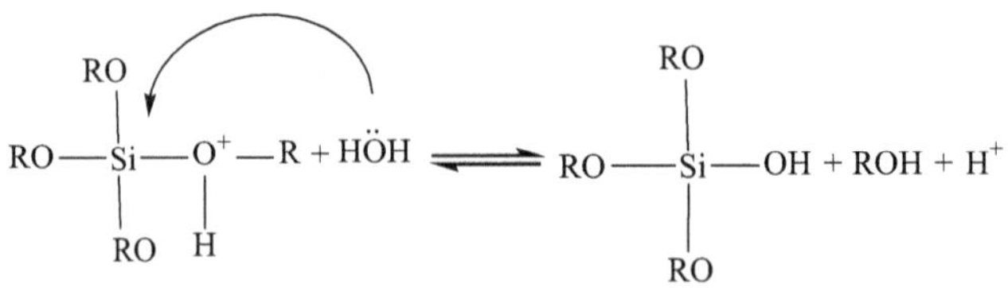

(6.10)

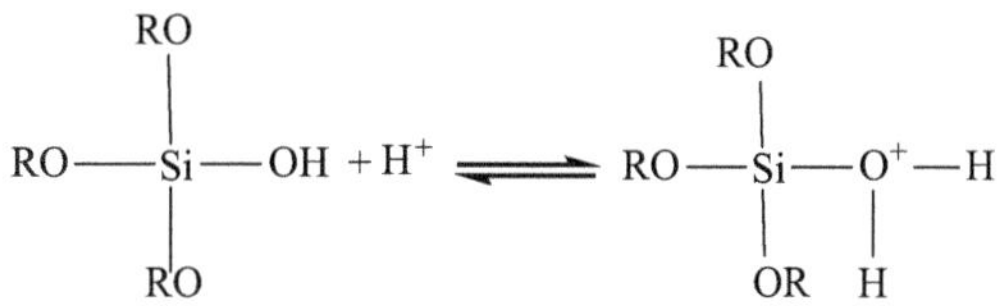

(6.11)

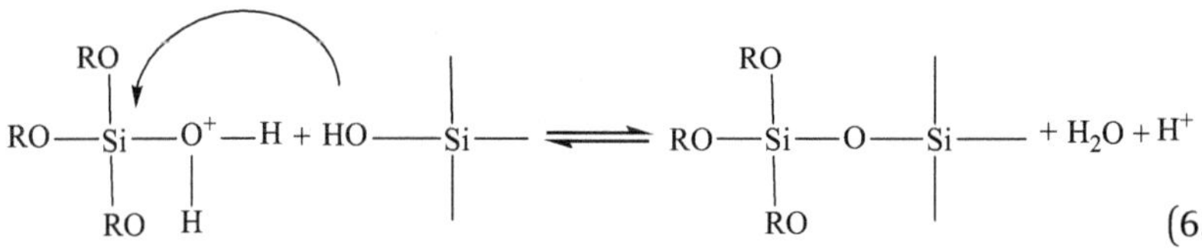

(6.12)

Base catalyzed

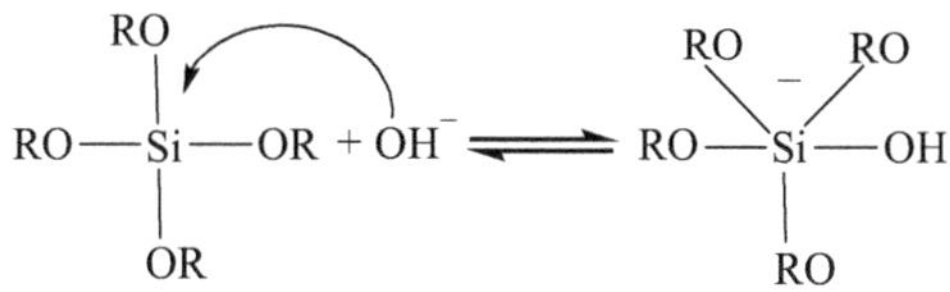

(6.13)

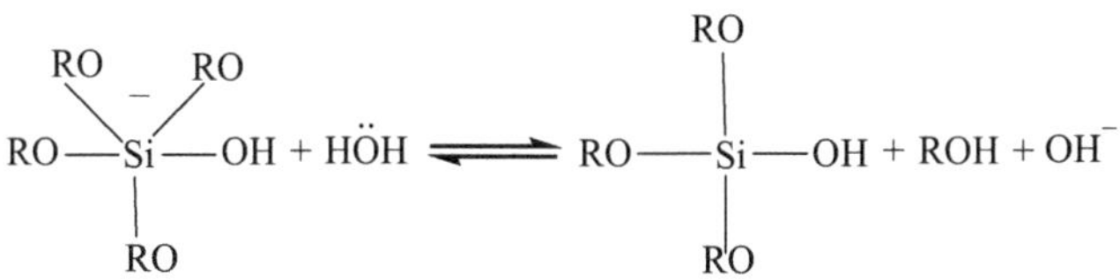

(6.14)

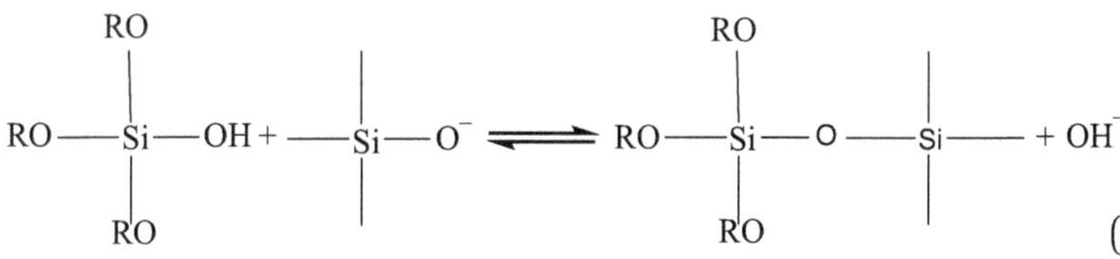

(6.15)

These reactions show that a gel is formed because of the condensation of partially hydrolyzed species into a three-dimensional polymeric network and any factors affecting these reactions are likely to impact on the properties of the resulting product. The pH of the sol plays a very important role on the further formed gel. Under acidic conditions, the hydrolysis reaction occurs at a faster rate than the condensation reaction and thus the resulting gel is weak due to the low linking between the long chains formed. However, by increasing the pH, the condensation reaction is accelerated relative to the hydrolysis reaction. Thus, the gel formed under basic conditions is highly branched (Figure 6.11).[29]

Figure 6.11 A cartoon showing linear weakly cross-linked and highly branched clusters, A: acid catalyzed B: base catalyzed hydrolysis.
Reproduced from ref. 60 with permission from the Royal Society of Chemistry.

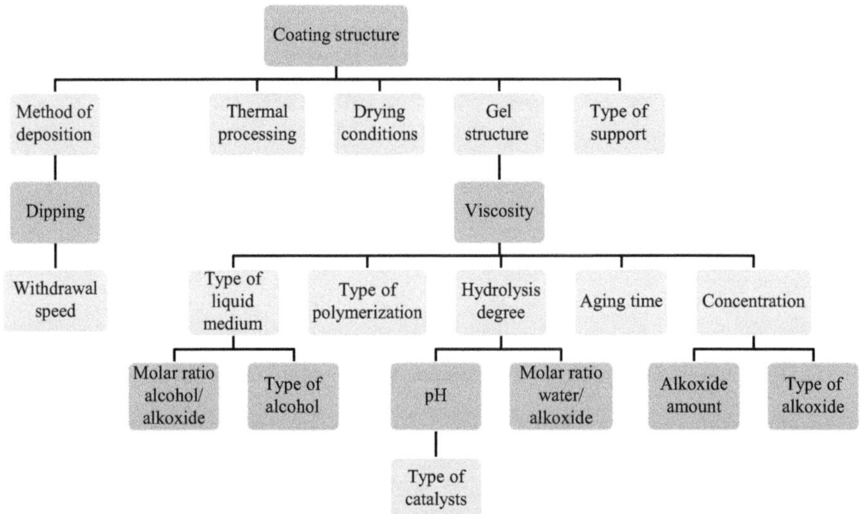

Figure 6.12 Factors that influence film structure.

Various other factors that can influence film structure are presented in Figure 6.12.[49]

6.7 The Coating Process

To deposit the materials prepared by the sol–gel technique on the required substrate (*e.g.*, on a glass or polymer surface), spin, spray, or dip coating processes are used.

6.7.1 Spin Coating

In general, spin coating is divided into four stages: deposition, spin up, spin off, and evaporation (Figure 6.13). The deposition, spin up, and spin off stages occur sequentially while the evaporation stage occurs throughout the process.

The turntable is accelerated to distribute the applied solution by centrifugal force. The remaining layer is thinned as a result of the high spinning speeds. This stage is followed by the drying of the applied layer. By rapid rotation, a uniform evaporation of the solvent used is made possible. By the evaporation stage or simply by "drying", the solvents are removed from the surface and the low volatile components of the solution remain on the surface of the substrate. The speed of rotation and the viscosity of the coating solution affect the thickness of the layer.[48]

The spin coating sol–gel synthesis is used to coat surfaces exhibiting a rotational symmetry (*e.g.*, lenses and eyeglass lenses). The disadvantages of spin coating are few, but they are becoming more important as the substrate size increases. Large substrates cannot be spun at a sufficiently high rate in order to allow the film to thin. The biggest disadvantage of spin coating is its lack of material efficiency. Typical spin coating processes utilize only 2–5% of the material dispensed onto the substrate, while the remaining 95–98% is flung off into the coating bowl and is disposed of.[51]

6.7.2 Spray Coating

Spray coating techniques are widely used in industry fields for coating complex shapes on various (curved and flat) polymeric substrates. In this process, very fine droplets are formed from the solution using atomizers or nebulizers. These fine droplets are then carried into the coating chamber with a carrier gas and deposited on the substrate by gravity or with an electrostatic field. The quality of the coating is determined by the size of the droplet, which can be reduced by decreasing the viscosity of the solution, increasing the atomizing pressure or using a Venturi nozzle (*cf.* Figure 6.14). Although the spray coating is a fast coating process that wastes less of the coating sols, the spray-coating method offers limited control of the uniformity of thickness.[52]

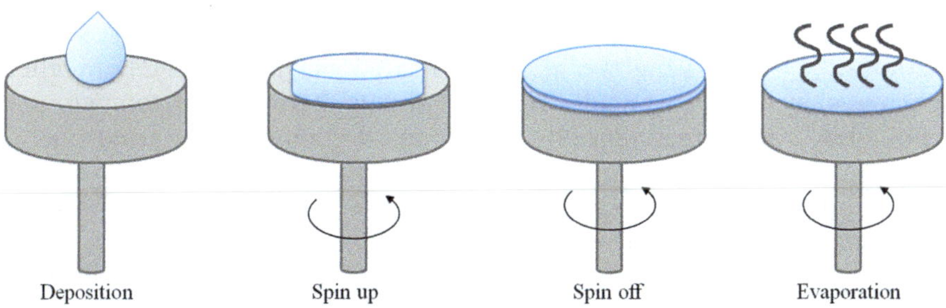

Figure 6.13 The steps of film deposition by a spin coating process.

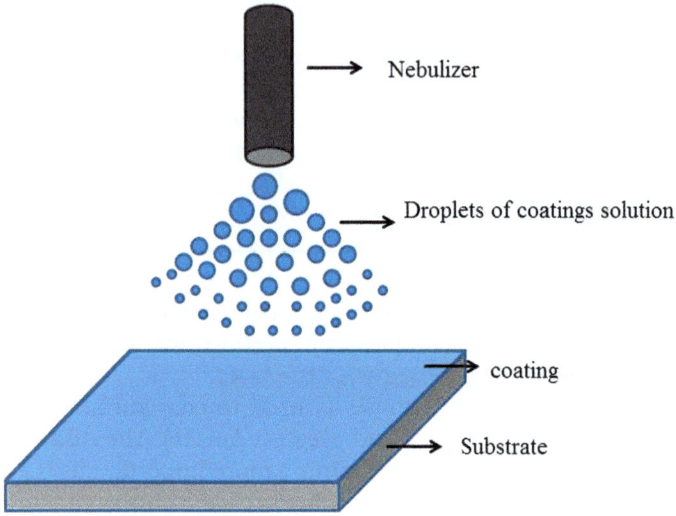

Figure 6.14 Film deposition by a spray coating process.

6.7.3 Dip Coating

In the dip coating process, the substrate is coated by immersing it in a liquid, *i.e.*, the sol, followed by its withdrawal at a well-defined withdrawal speed under controlled atmospheric conditions, such as temperature and humidity, due to their effect on the application or drying of the coating. During this process, the sol is converted to a film *via* aggregation, gelation, and drying (Figure 6.15). The thickness of the layer increases by increasing the pull rate and by the use of a sol with a higher viscosity or lower surface tension.[48] One of the advantages of this coating technique is its ability to be applied on flat panels, cylinders or complex geometry with quite a large surface.

6.8 Self-cleaning Coatings on Polycarbonate

TiO_2 and TiO_2–SiO_2 coatings on polycarbonate (PC) are well known but only a limited number of publications have focused on photocatalytically active self-cleaning TiO_2 and TiO_2–SiO_2 coatings on PC prepared by sol–gel methods.[16,44,45,53–59] Langlet *et al.* have prepared TiO_2 films on PC (and other substrates) by depositing acidic TiO_2 sols, drying them at room temperature followed by a heat treatment (90–140 °C) in an ethanol–water atmosphere. The optical transmission of the films in the visible range of the spectrum was found to be lower than that of bare polycarbonate. The films were photocatalytically active; the activity increased with increasing film thickness.[53,54] Langlet and co-workers also deposited a TiO_2 sol peptized under alkaline conditions.[44] The films prepared on PC were characterized directly

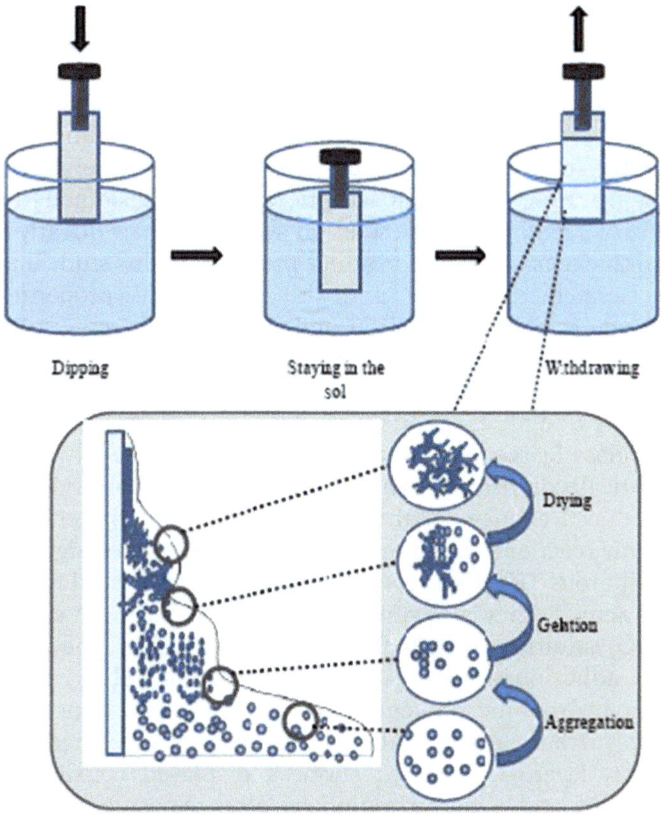

Figure 6.15 Steps of film deposition by the dip coating process.

after deposition and after a subsequent heat treatment in air at temperatures up to 140 °C. However, the transmission in the visible range of the spectrum was found to be lower than that of uncoated PC. The transmission was decreasing with increasing film thickness (about 70–370 nm) whereas the photocatalytic activity increased. The post-deposition heat treatment did not influence the photocatalytic activity for temperatures up to 110 °C. Kwon *et al.* have used different alkoxide precursors to prepare TiO_2 nano-crystalline thin films on polycarbonate, polymethyl methacrylate, glass, and aluminium *via* a sol–gel process. They observed that the films on PC exhibited poor adhesion and peeled off after a certain period of time. All films were found to photocatalytically decolorize an aqueous methylene blue solution under UV(A) irradiation.[56] Lam *et al.* studied the effect of pretreating the polymer sheets by NaOH etching and UV(C) irradiation on the mechanical stability of TiO_2 coatings on PC slides. TiO_2 nanoparticles were found to adhere more strongly on UV(C)-treated PC than on NaOH-etched PC whereas the TiO_2 films on NaOH-etched PC exhibited a higher photocatalytic activity than the films on UV(C)-treated PC in a methylene blue degradation test.[57] Yaghoubi *et al.* prepared TiO_2 films on a PC surface chemically pretreated to create hydrophilic groups on the surface. The films were prepared by dip-coating using an anatase sol with TiO_2 nanoparticles 30 nm in size. A pre-coat by peroxotitanium complexes was employed to improve adhesion and to inhibit substrate degradation. The TiO_2 coating reduced the transparency by 10–15%. The photocatalytic activity determined through decolorization of a methylene blue solution was found to be linearly dependent on the film thickness. The TiO_2 coatings prepared in this study improved the hardness, scratch resistance, and the mechanical properties as was demonstrated by nano-indentation and nano-scratch tests.[58] Fateh *et al.* have prepared transparent photocatalytically active mesoporous TiO_2 and Sachtleben Hombikat UV100 thin film coatings on PC precoated with SiO_2. The newly prepared photocatalyst films have been compared with Pilkington Activ sheet glass by the determination of their photonic efficiencies for the photocatalytic decolorization of a methylene blue solution and by measuring the light-induced changes of the surface wettability. The efficiency of the photocatalytic reactions was found to be about three times higher in the case of the mesoporous TiO_2 films compared with films prepared from UV-100 or Pilkington Activ.[16] They also published a systematic study concerning the effect of SiO_2 addition to TiO_2 films on the wettability, the photocatalytic activity, the adhesion strength, and the mechanical stability of the films.[45] Transparent hydrophilic photocatalytic TiO_2/SiO_2 thin films were used to coat the polycarbonate (PC) substrate which was precoated by an intermediate SiO_2 layer. The coated surfaces displayed considerable photocatalytic activity and superhydrophilicity after exposure to UV light. The addition of SiO_2 resulted in an improvement of the photocatalytic activity of the TiO_2 film reaching the highest value at a molar ratio of TiO_2–SiO_2 equal to 1 : 0.9. The prepared films exhibited a good stability against UV(A) irradiation. Moreover, transparent TiO_2/ZnO thin films on a SiO_2 interlayer were successfully deposited on the surface of polycarbonate to provide polymeric sheets with a self-cleaning, superhydrophilic, and photocatalytically active surface layer.[46] The experimental results showed that increasing the molar ratio of ZnO in TiO_2 coatings up to 5% yields maximum photonic efficiency values of 0.023%, as assessed by the photocatalytic degradation of methylene blue. Moreover, the superhydrophilic coating with a molar TiO_2–ZnO ratio of 1 : 0.05 exhibited the best self-cleaning properties combined with a good mechanical stability and a very good stability against UV irradiation. Matsuda *et al.* have prepared transparent TiO_2/SiO_2 films with a 1 : 5 molar ratio on various types of substrates, including PC, by dip-coating and post-treatment at temperatures <100 °C under ambient pressure. They observed that TiO_2 nanocrystals were precipitated mainly at the surface of the TiO_2–SiO_2 films during posttreatment in hot water, whereas the addition of poly(ethylene glycol) in the films led to the dispersion of TiO_2 nanocrystals in the whole of the films after the treatment. The films were found to be photocatalytically active in the methylene blue degradation test as well as in other test systems. The authors assumed that in the case of films where the TiO_2 are mainly precipitated at the surface, the residual SiO_2 under-layer is acting as a protective layer for an organic polymer substrate against photocatalytic degradation.[55] Horiuchi *et al.* have prepared Ti-containing mesoporous SiO_2 thin films on PC by spin-coating a sol containing metal ethoxides as the TiO_2 and SiO_2 source, and polyethylene stearyl ether as a structure-directing agent. After drying, the organic structure-directing agent was photocatalytically removed, resulting in superhydrophilic films after UV(A) irradiation for 4 to 5 days. The superhydrophilic state was maintained for several days under dark conditions.[59]

6.9 Concluding Remarks

As outlined above, techniques are available with which self-cleaning layers on polymeric substrates may be prepared. It has been shown that the products have the characteristics, *i.e.*, photocatalytic activity, long-lasting superhydrophilicity even in the dark, mechanical stability and sufficient optical properties, demanded by the users. But basic scientific work is still necessary to come to a deeper understanding of the mechanisms underlying the light-induced changes in wettability. Future progress in this area will result in new products with noticeably improved self-cleaning properties.

References

1. M. R. Hoffmann, S. T. Martin, W. Choi and D. W. Bahnemann, *Chem. Rev.*, 1995, **95**, 69.
2. I. P. Parkin and R. G. Palgrave, *J. Mater. Chem.*, 2005, **15**, 1689.
3. H. Irie and K. Hashimoto, *Handbook of Environmental Chemistry* Part M, 2005, vol. 2, p. 425.
4. A. Fujishima and X. T. Zhang, *C. R. Chim.*, 2006, **9**, 750.
5. A. Fujishima, X. Zhang and D. A. Tryk, *Surf. Sci. Rep.*, 2008, **63**, 515.
6. K. Nakata and A. Fujishima, *J. Photochem. Photobiol., C*, 2012, **13**, 169.
7. V. A. Ganesh, H. K. Raut, A. S. Nair and S. Ramakrishna, *J. Mater. Chem.*, 2011, **21**, 16304.
8. H. E. Çamurlu, Ö. Kesmez, E. Burunkaya, N. Kiraz, Z. Yeşil, M. Asiltürk and E. Arpaç, *Chem. Pap.*, 2012, **66**, 461.
9. L. Zhang, R. Dillert, D. Bahnemann and M. Vormoor, *Energy Environ. Sci.*, 2012, **5**, 7491.
10. S. Nishimoto and B. Bhushan, *RSC Adv.*, 2013, **3**, 671.
11. S. Banerjee, D. D. Dionysiou and S. C. Pillai, *Appl. Catal., B*, 2015, **176–177**, 396.
12. P. Gould, *Materialstoday*, 2003, 44.
13. A. Mills, A. Lepre, N. Elliott, S. Bhopal, I. P. Parkin and S. A. O'Neill, *J. Photochem. Photobiol., A*, 2003, **160**, 213.
14. M. Houmard, D. Riassetto, F. Roussel, A. Bourgeois, G. Berthomé, J. C. Joud and M. Langlet, *Appl. Surf. Sci.*, 2007, **254**, 1405.

15. J. Yu, X. Zhao, J. C. Yu, G. Zhong, J. Han and Q. Zhao, *J. Mater. Sci. Lett.*, 2001, **20**, 1745.
16. R. Fateh, A. A. Ismail, R. Dillert and D. W. Bahnemann, *J. Phys. Chem. C*, 2011, **115**, 10405.
17. J. M. Goddard and J. H. Hotchkiss, *Prog. Polym. Sci.*, 2007, **32**, 698.
18. B. J. Ryan and K. M. Poduska, *Am. J. Phys.*, 2008, **76**, 1074.
19. X. Feng and L. Jiang, *Adv. Mater.*, 2006, **18**, 3063.
20. R. Wang, N. Sakai, A. Fujishima, T. Watanabe and K. Hashimoto, *J. Phys. Chem. B*, 1999, **103**, 2188.
21. Z. Liu, X. Zhang, T. Murakami and A. Fujishima, *Sol. Energy Mater. Sol. Cells*, 2008, **92**, 1434.
22. O. Carp, C. L. Huisman and A. Reller, *Prog. Solid State Chem.*, 2004, **32**, 33.
23. A. Corma, *Chem. Rev.*, 1997, **97**, 2373.
24. W.-S. Chae, S.-W. Lee and Y.-R. Kim, *Chem. Mater.*, 2005, **17**, 3072.
25. K. Inumaru, M. Yasui, T. Kasahara, K. Yamaguchi, A. Yasuda and S. Yamanaka, *J. Mater. Chem.*, 2011, **21**, 12117.
26. C. Suwanchawalit and S. Wongnawa, *J. Nanopart. Res.*, 2010, **12**, 2895.
27. A. A. Ismail and D. W. Bahnemann, *J. Mater. Chem.*, 2011, **21**, 11686.
28. X. Fu, L. A. Clark, Q. Yang and M. A. Anderson, *Environ. Sci. Technol.*, 1996, **30**, 647.
29. T. Cetinkaya, L. Neuwirthová, K. M. Kutláková, V. Tomášek and H. Akbulut, *Appl. Surf. Sci.*, 2013, **279**, 384.
30. Y. Gnatyuk, N. Smirnova, A. Eremenko and V. Ilyin, *Adsorpt. Sci. Technol.*, 2005, **23**, 497.
31. A. Ennaoui, B. R. Sankapal, V. Skryshevsky and M. Lux-Steiner, *Sol. Energy Mater. Sol. Cells*, 2006, **90**, 1533.
32. L. Zhou, S. Yan, B. Tian, J. Zhang and M. Anpo, *Mater. Lett.*, 2006, **60**, 396.
33. P. Cheng, M. Zheng, Y. Jin, Q. Huang and M. Gu, *Mater. Lett.*, 2003, **57**, 2989.
34. R. Abd Aziz, N. Asyikin and I. Sopyan, *J. Inst. Eng. Malaysia*, 2009, **70**, 34.
35. S. Liao, H. Donggen, D. Yu, Y. Su and G. Yuan, *J. Photochem. Photobiol., A*, 2004, **168**, 7.
36. A. Mills and M. Crow, *Int. J. Photoenergy*, 2008, 470670, 6 pages.
37. A. Nakajima, S. Koizumi, T. Watanabe and K. Hashimoto, *J. Photochem. Photobiol., A*, 2001, **146**, 129.
38. J. M. White, J. Szanyi and M. A. Henderson, *J. Phys. Chem. B*, 2003, **107**, 9029.
39. N. Sakai, A. Fujishima, T. Watanabe and K. Hashimoto, *J. Phys. Chem. B*, 2003, **107**, 1028.
40. M. Takeuchi, K. Sakamoto, G. Martra, S. Coluccia and M. Anpo, *J. Phys. Chem. B*, 2005, **109**, 15422.
41. R.-D. Sun, A. Nakajima, A. Fujishima, T. Watanabe and K. Hashimoto, *J. Phys. Chem. B*, 2001, **105**, 1984.

42. D. C. Hennessy, M. Pierce, K.-C. Chang, S. Takakusagi, H. You and K. Uosaki, *Electrochim. Acta*, 2008, **53**, 6173.
43. J. D. Mackenzie and E. Bescher, *J. Sol–Gel Sci. Technol.*, 2003, **27**, 7.
44. M. Langlet, A. Kim, M. Audier, C. Guillard and J. Herrmann, *Thin Solid Films*, 2003, **429**, 13.
45. R. Fateh, R. Dillert and D. Bahnemann, *Langmuir*, 2013, **29**, 3730.
46. R. Fateh, R. Dillert and D. Bahnemann, *ACS Appl. Mater. Interfaces*, 2014, **6**, 2270.
47. H. Schmidt, *J. Non–Cryst. Solids*, 1988, **100**, 51.
48. C. J. Brinker and G. W. Scherer, *Sol-gel Science, The Physics and Chemistry of Sol-gel Processing*, Academic Press, *Boston*, 5th edn, 2002.
49. *The Sol-Gel Process, Uniformity, Polymers and Applications*, ed. R. E. Morris, Nova Science Publishers, New York, 2011.
50. S. Beling, *Nasschemische Abscheidung und Charakterisierung von TiO_2-Schichten für den Sensoreinsatz beim Gasnitrieren*, Shaker, Aachen, 2002.
51. N. Sahu, B. Parija and S. Panigrahi, *Indian J. Phys.*, 2009, **83**, 493.
52. B. Faure, G. Salazar-Alvarez, A. Ahniyaz, I. Villaluenga, G. Berriozabal, Y. R. De Miguel and L. Bergström, *Sci. Technol. Adv. Mater.*, 2013, **14**, 023001.
53. M. Langlet, A. Kim, M. Audier and J. M. Herrmann, *J. Sol–Gel Sci. Technol.*, 2002, **25**, 223.
54. M. Langlet, A. Kim, M. Audier, C. Guillard and J. M. Herrmann, *J. Mater. Sci.*, 2003, **38**, 3945.
55. A. Matsuda, T. Matoda, Y. Kotani, T. Kogure, M. Tatsumisago and T. Minami, *J. Sol–Gel Sci. Technol.*, 2003, **26**, 517.
56. C. H. Kwon, H. Shin, J. H. Kim, W. S. Choi and K. H. Yoon, *Mater. Chem. Phys.*, 2004, **86**, 78.
57. S. W. Lam, A. Soetanto and R. Amal, *J. Nanopart. Res.*, 2009, **11**, 1971.
58. H. Yaghoubi, N. Taghavinia and E. K. Alamdari, *Surf. Coat. Technol.*, 2010, **204**, 1562.
59. Y. Horiuchi, H. Ura, T. Kamegawa, K. Mori and H. Yamashita, *J. Mater. Chem.*, 2011, **21**, 236.
60. R. K. Nagarale, W. Shin and P. K. Singh, *Polym. Chem.*, 2009, **1**, 390.

CHAPTER 7

Nanostructured Self-cleaning Coating with Antireflection Properties

ZHAO ZHAO AND ZAICHENG SUN*

Beijing Key Laboratory for Green Catalysis and Separation, Department of Chemistry and Chemical Engineering, Beijing University of Technology, 100 Pingleyuan, Chaoyang, Beijing, P. R. China
*Email: sunzc@bjut.edu.cn

7.1 Introduction

Self-cleaning coatings have attracted extensive attention recently in terms of research and commercial applications, such as building glass, eyeglasses, and solar cell modules.[1–3] Generally, self-cleaning coatings can be classified into two categories: hydrophobic and hydrophilic. On a hydrophilic coating, water is made to spread out (water sheeting) over the surface, which carries away dirt and other impurities. Whereas using the hydrophobic technique, water droplets slide and roll over the surface, thereby cleaning them. In addition to the sheeting effect, metal oxides have an additional property of chemically breaking down organic dirt deposits by a sunlight-assisted cleaning mechanism, *i.e.* photocatalytic effect. For the fabrication of self-cleaning surfaces by coating methods, therefore, reflection issues will be encountered unavoidably. In practical applications, reducing the reflection and improving the transmission or absorption of light from wide angles of incidence in a broad wavelength range are crucial for enhancing the performance of optical, optoelectronic, and electro-optical devices. Inspired

RSC Smart Materials No. 21
Self-cleaning Coatings: Structure, Fabrication and Application
Edited by Junhui He

Published by the Royal Society of Chemistry, www.rsc.org

by the unique multi-scale structures of an insect's compound eyes, large quantities of nano-arrays with a sub-wavelength structure have been introduced into surfaces, which exhibit promising broadband antireflective and self-cleaning properties simultaneously. Among all the kinds of nanostructure self-cleaning coatings, semiconductor materials with promising band gap structures have been widely investigated during recent decades, especially in the arena of energy and environmental applications, due to their advantage of good photocatalytic activity, easy processing and low cost. This chapter mainly deals with the basics of antireflection (AR), followed by superhydrophobic/superhydrophobic, photocatalysis-based self-cleaning materials. Furthermore, recent advancements in nano-structured self-cleaning coatings with antireflection properties, including inorganic semiconductors and polymer surfaces, and multifunctional coating materials, have also been analyzed and discussed.

7.2 Principle of Nano-structure Antireflection

7.2.1 Basic Concept of Antireflection

Incident light is irradiated on the boundary between two materials with different refractive indices. According to the Fresnel equation,[4] a fundamental phenomenon occurring is optical reflection;[5] a certain fraction of the incident light is reflected, while the remainder is absorbed or transmitted.

$$R=\frac{n_\mathrm{f}^2(n_\mathrm{s}-n_0)^2\cos^2\frac{1}{2}x+(n_\mathrm{f}^2-n_0n_\mathrm{s})^2\sin^2\frac{1}{2}x}{n_\mathrm{f}^2(n_\mathrm{s}+n_0)^2\cos^2\frac{1}{2}x+(n_\mathrm{f}^2+n_0n_\mathrm{s})^2\sin^2\frac{1}{2}x} \tag{7.1}$$

where, n_f is the refractive index of the thin film, n_s is the refractive index of the substrate, n_0 is the refractive index of the incident medium, x is the optical path, $x=2\pi nd/\lambda$, and d is the physical thickness of the thin film.

If the optical path $x=(2m-1)\,\lambda/2$, $m=1, 2, 3\ldots$ odd multiple of half-wave, then $\cos x=-1$, $\sin^2(x/2)=1$, $\cos^2(x/2)=0$, the equation can be presented in the following simplified form:

$$R=\left(\frac{n_\mathrm{f}^2-n_0n_\mathrm{s}}{n_\mathrm{f}^2+n_0n_\mathrm{s}}\right)^2 \tag{7.2}$$

Zero reflectance can be obtained when $n_\mathrm{f}=\sqrt{n_\mathrm{s}n_0}$ and $x=(2m-1)\lambda/2$, $m=1$, 2, 3…

To achieve the maximal light transmission, it has to fulfill the following two assumptions:

(a) The reflected waves have the same intensity and one wave is reflected per interface. Other optical interactions such as scattering, absorption *etc.* are negligible.

(b) The optical path length must be chosen for the reflected wave to interfere destructively. The thickness of the film d is an odd multiple of $\lambda/4$.

By assuming one of the two materials is air ($n_0 = 1$), R can be presented as a simplified form of the Fresnel equation:

$$R = \left(\frac{n_f^2 - n_s}{n_f^2 + n_s}\right)^2 \tag{7.3}$$

Zero reflectance can be obtained when $n_s = \sqrt{n_f}$ according to eqn (7.3). For a typical transparent dielectric material, n_s of the glass is about 1.5. Therefore, the desired refractive index of the coating film, n_f can be calculated as 1.23. Nevertheless, it is still impractical to achieve exactly zero reflectance from a single-layer AR coating. As the thickness of the film d needs to meet the odd multiple of $\lambda/4$, the downside of this method is that:

(a) The wavelength and incident angle of incident light are strictly limited.
(b) It is difficult to achieve broadband antireflective.

7.2.2 The Basis of Multiple Layers on the Substrate

The above analysis was done on the basis of a single layer on the substrate. As-obtained antireflection properties are also based on a single wavelength light. For multi-layered AR coating, the basic principle remains the same; the only difference lies in the mathematical model, which relies on vector analysis of the individual reflected rays. As shown in Figure 7.1(b), the reflected light from the interface ij between the adjacent layers i and j meets the equation:

$$R_{ij} = |R_{ij}|\exp[-2(\delta_i + \delta_j)]$$

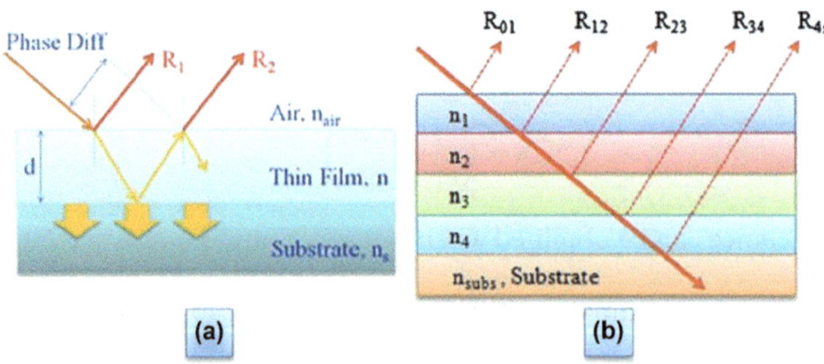

Figure 7.1 Propagation of light rays through (a) a single layer film on a substrate ($n_s > n$) and (b) a multilayer film on a substrate.
Reproduced from ref. 5 with permission from the Royal Society of Chemistry.

where, $|R_{ij}| = |(n_i - n_j)/(n_i + n_j)|$, $\delta_i = 2\pi n_i d_i \cos\theta_i/\lambda$ is the phase thickness of each layer (δ_i is the angle of refraction, d_i is the physical thickness of the layer).[6]

Hence, the resultant reflection vector of the entire coating is given by:

$$R_{\text{sum}} = R_{01} + R_{12} + R_{23} + \ldots + R_{\text{ns}}$$

By choosing a suitable reflective index and thickness of each layer, a minimum R_{sum} can be achieved, and a wide spectral region AR coating can be obtained. However, due to strictly selected expensive materials, and the complexity of technology preparation, multilayer AR coating is still expensive for some applications, and intrinsically, this technique is unsuitable for large area coating.

7.2.3 Gradient Refractive Index Coating

Since the application of layered AR coatings is limited, gradient refractive index AR coatings have been proposed as a desirable choice.[6–11] As shown in Figure 7.2, normal homogeneous layers have a step-changed refractive index varying from the incident medium to the substrate. Alternatively, inhomogeneous layers with a refractive index that varies monotonically throughout the thickness, can also achieve high transmittance over a broadband range.[12] An additional advantage of gradient refractive index antireflection films is that their optical properties are quite insensitive to film thickness.

Typical expressions for the continuous gradient RI with linear, cubic and quintic profiles are presented below:[13]

$$\text{Linear index profile: } n = n_0 + (n_s - n_0)t, \quad 0 \le t \le 1$$

$$\text{Cubic index profile: } n = (n_s - n_0)(3t^2 - 2t^3)$$

$$\text{Quintic index profile: } n = n_0 + (n_s - n_0)t(10t^3 - 15t^4 + 6t^5)$$

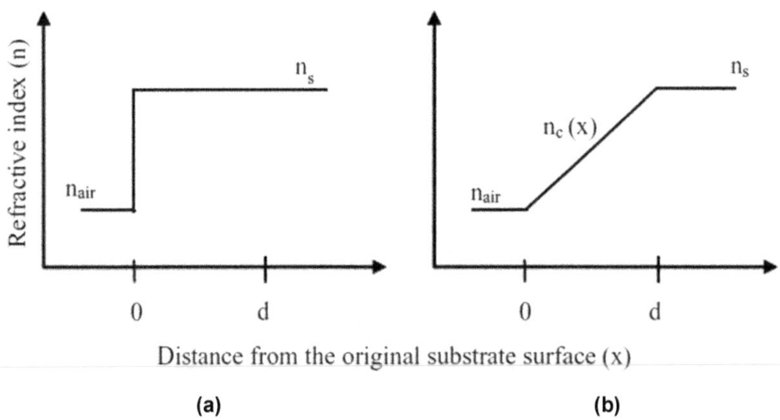

Figure 7.2 Representation of a coating exhibiting linear inhomogeneity (b) as compared to that exhibiting a step change (a) in refractive index; from ref 15. Copy with permission from John Wiley & Sons Inc.

where n_0 and n_s are the RIs of the incident and substrate media, and t is the thickness of the graded interface region. Gradient refractive index antireflection films are usually produced by incorporating porosity, which is performed by using chemical etch/leach processes.[7,14]

7.2.4 The Basis of Antireflection Based on Nanostructure

AR coatings based on micro- and nanostructure arrays inspired by the evolution of "Biomimetics", by a "moth's eye", or bionic structures have led to an alternative way of reducing reflectance.[16] Highly magnified SEM images reveal the well-organized nanostructure patterns as shown in Figure 7.3(b).

Incident light on objects mainly undergoes three processes: reflection, scattering and partial absorption. However, when light is incident on micro- and nanostructure arrays, there are two different interactions depending on the characteristic dimension of the nanostructure patterns and the wavelength of the incident light:[5]

(a) When the AR units have dimensions less than the light wavelength, light interacts with the AR surface wholly. If the AR surface has a gradient refractive index, light rays tend to bend progressively (Figure 7.4(a)).
(b) When the wavelength of light is much smaller than the dimensions of the nanostructure pattern, if the spacing and depth between the nanostructures in the textured surface are in the same dimensional constraints of light wavelength, light rays would get trapped in the crevices resulting in multiple internal reflections (Figure 7.4(b)).

Inspired by nature, people begun to directly prepare antireflective bionic structures on the materials' surfaces.[18–22] AR coatings based on micro- and

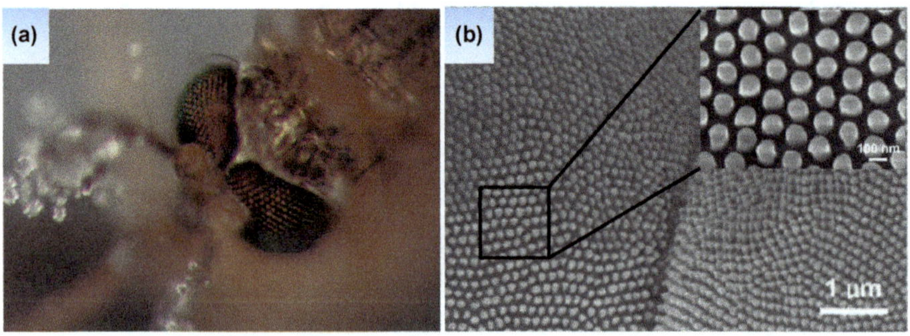

Figure 7.3 (a) A photograph of an insect's compound eyes.[17] (b) SEM images of a single mosquito eye.
Reproduced from ref. 17 with permission from John Wiley and Sons. Copyright © 2007 WILEY-VCH Verlag GmbH & Co. KGaA, Weinheim.

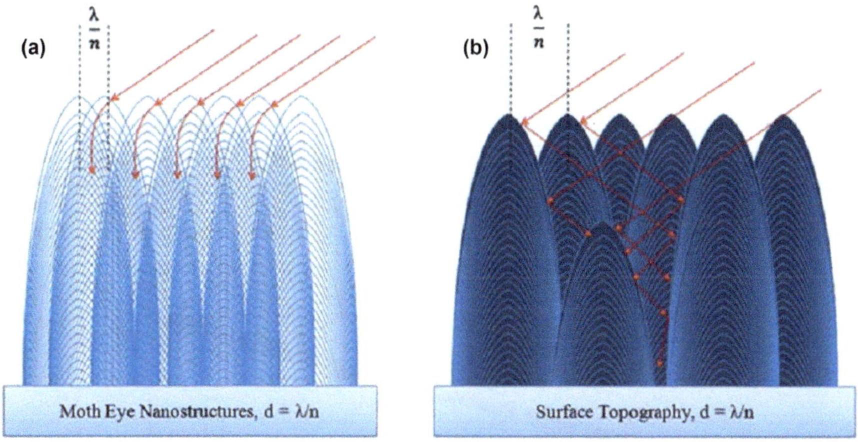

Figure 7.4 (a) Light interacting with the whole rough surface due to comparable dimensions. (b) Light undergoing multiple internal reflections through a nanostructure pattern.
Reproduced from ref. 5 with permission from the Royal Society of Chemistry.

nanostructure arrays overcome several shortcomings of the multiple layers method:

(1) Lowered mismatches of thermodynamic and adhesion coefficient between different materials, enhanced stability at the same time.
(2) Reduced loss of reflected light within the wide scope of the spectrum.
(3) Low cost and easy process control for applications, making it suitable for large area coating.

7.3 Principle of Nano-structure Self-cleaning

7.3.1 Superhydrophobicity

For special wettability,[23] a "superhydrophobic" surface should exhibit a water advancing contact angle larger than 150° and low contact angle hysteresis.[24,25] Such hydrophobic materials can be extensively viewed in nature.[26] Plant leaf surfaces, such as lotus, are representative models for the unusual wetting characteristics of superhydrophobicity and self-cleaning, which have been investigated and documented.[27,28] Due to the papillose epidermal cells being covered with epicuticular wax crystalloids,[29] particulate contamination would be completely removed in the case of rain. When a water droplet rolls off the rough surface, dirt particles with a small contact area with the rough surface lead to minimized adhesion, so they are easily adsorbed to the water droplet, and removed from the surface (Figure 7.5). This biological cleaning mechanism was described as the "lotus effect" by Neinhuis *et al.* in 1997.[30] It is well-known that the wettability of a solid

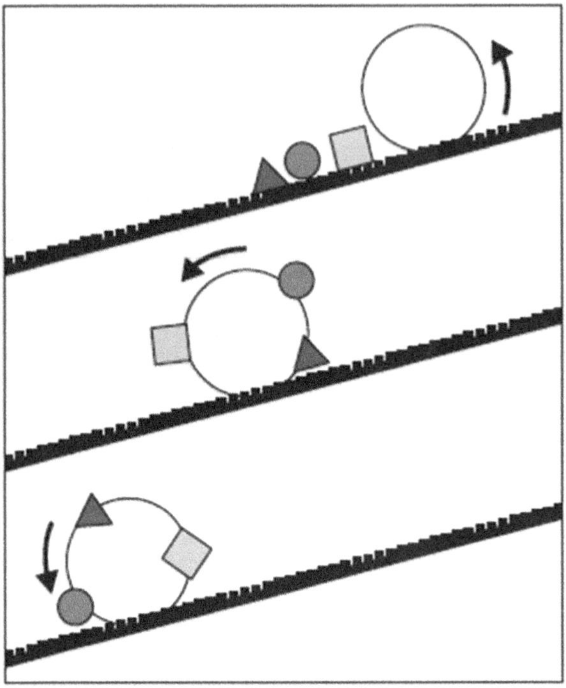

Figure 7.5 Schematic diagram showing the self-cleaning process on a rough surface.[31] Reproduced with permission from Springer.

Figure 7.6 The contact angle (θ) of a hydrophilic and superhydrophilic surface. Adapted from P. Gould, Smart, clean surfaces, *Mater. Today*, **6**, 44–48, Copyright 2003, with permission from Elsevier.[37]

surface is generally governed by both the chemical properties and the surface microstructures. In addition, hydrophobicity is usually determined by the roughness factor, which is frequently controlled by the use of surface nanostructure. Hence, superhydrophobic surfaces tend to attract considerable attention from researchers.

7.3.2 Superhydrophilicity

Relative to the potential applications of superhydrophobic surfaces, researchers also focus on superhydrophilic surfaces, of which the wettability surface is normally high, and the water contact angle tends to be less than 5° (as shown in Figure 7.6).[32–36] When a water drop falls on the

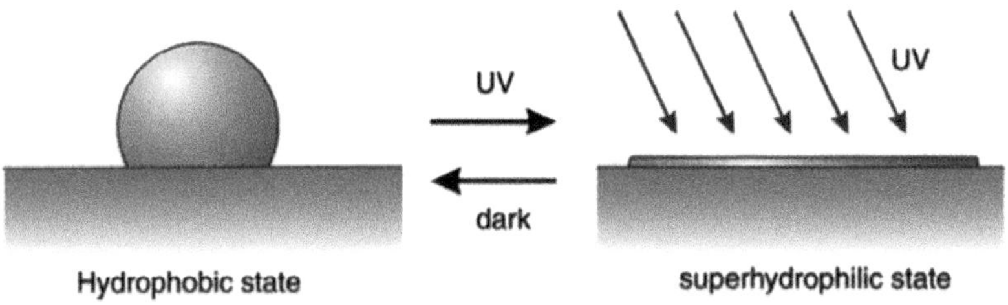

Figure 7.7 A schematic diagram of the effect of UV irradiation on the contact angle of water on photocatalytic surfaces.
Reproduced from Y. Takata *et al.*, Selective oxidation of CO in hydrogen-rich stream over Cu–Ce catalyst promoted with transition metals, *Int. J. Hydrogen Energy*, **30**, 209–220, Copyright 2005, with permission from Elsevier.[42]

superhydrophilic surface, it could form a continuous or near continuous water film, which carries away dirt and other impurities. And scattering is suppressed and optical transmission is enhanced.[37]

Although superhydrophilic and superhydrophobic are opposites, reversible switching between superhydrophilicity and superhydrophobicity can sometimes occur in response to external stimuli (*e.g.* temperature,[38,39] light,[39] and pH[40]) as shown in Figure 7.7. These wettability-changed surfaces that respond to stimuli instantaneously have received much attention due to their significant potential foreground in fundamental research and industry. However, mechanical stability is a crucial problem for the practical applications of responsive surfaces with a superhydrophobic to superhydrophilic wetting transition.[41]

7.3.3 Photocatalysis-induced Self-cleaning

Materials used in hydrophilic surfaces are typically photocatalytic semiconductor materials (*e.g.* TiO_2, ZnO, V_2O_5 and so on).[43–45] These surfaces utilized such materials as self-cleaning coatings, by which the mechanism happens in two different ways: photocatalysis, a process in which the organic dirt molecules get decomposed using chemical energy generated by sunlight, and the sheeting of water rather than droplets, which makes the surface superhydrophilic (contact angle less than 5° or the time for complete wetting by small droplets of water has been observed to be less than 0.5 s), thereby chemically carrying away the organic dirt adsorbed onto the surfaces. Among all of photocatalytic semiconductor materials, TiO_2 has become the best choice of potential candidates for self-cleaning coatings, and hydrophilic self-cleaning surfaces in general, because of its favorable physical and chemical properties, low cost, stability, non-toxicity (as a bulk material, the toxicity of nano TiO_2 is contended[46]) and high abundance.[47]

The TiO_2 surface becomes superhydrophilic *via* UV light irradiation, because photo-generated h^+ oxidizes the lattice oxygen at the surface of the

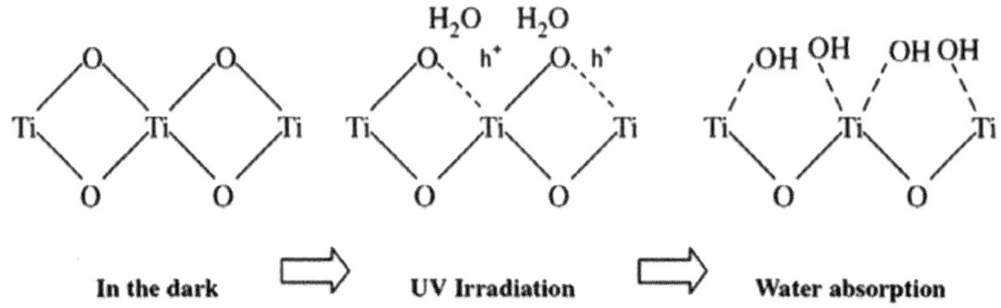

Figure 7.8 The mechanism for photoinduced hydrophilicity of TiO_2 by the formation of surface hydroxide groups.
Reproduced from J. Chen and C.-s. Poon, Photocatalytic construction and building material: From fundamentals to applications, *Build. Environ.*, 44, 1899–1906, Copyright 2009, with permission from Elsevier.[49]

material resulting in oxygen vacancies so that the bond can be interrupted by the adsorbed water molecule.[48] This causes the formation of surface hydroxide groups, which dramatically decrease the static contact angle of a TiO_2 surface as shown in Figure 7.8. In the dark, the hydroxyl groups gradually desorb from the surface in the form of H_2O_2 or $H_2O + O_2$ and the contact angle gradually reverts to its initial value.

The mechanisms of the photocatalysic self-cleaning processes that occur on titania surfaces are illustrated in Figure 7.9. Photocatalysis can be generally categorized into three processes:

(1) A pair of charged carriers (an electron, e^-, and a hole, h^+) excited by absorbing of light with energy equal to or greater than its band gap energy;
(2) Recombination of these charge carriers in the bulk or on the surface;
(3) Migration from bulk to the surface.

Therefore, holes produce hydroxyl radicals causing the oxidization of adsorbed organic molecules, while electrons eventually reduce atmospheric dioxygen to give the superoxide radicals or hydrogen peroxide, which quickly attack nearby organic molecules. The process in which the adsorbate molecule gets photoexcited and interacts with the ground state catalyst substrate is known as a catalyzed photoreaction. If the initial photoexcitation takes place in the catalyst substrate and transfers an electron or energy into a ground state molecule, the process is referred to as a sensitized photoreaction. Moreover, if the initial photoexcitation takes place in the catalyst substrate and transfers an electron or energy into a ground state molecule, the process is referred to as a sensitized photoreaction.[50]

Hydroxyl groups that form on the surfaces not only contribute to superhydrophilicity, but also to the formation of hydroxyl radicals that play an important role in the decomposition of organic compounds. The self-cleaning performance of photocatalytic surfaces is, therefore, two fold (as shown in Figure 7.10): many organic molecules are decomposed to CO_2 and

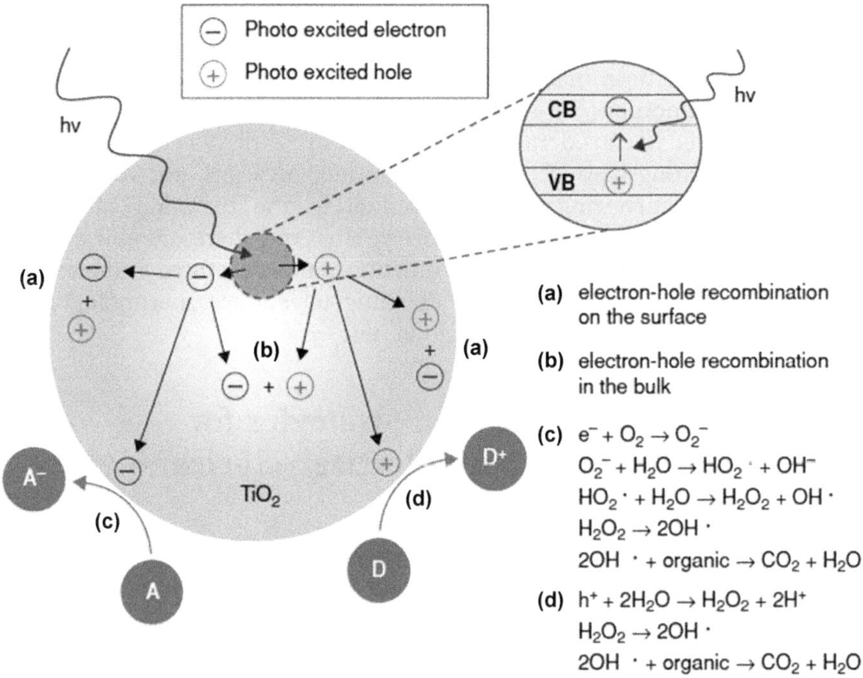

Figure 7.9 The semiconductor (TiO_2) undergoes photoexcitation upon illumination with UV light. The electron and the hole that result can follow one of several pathways.
Reproduced from ref. 51 with permission from the Royal Society of Chemistry.

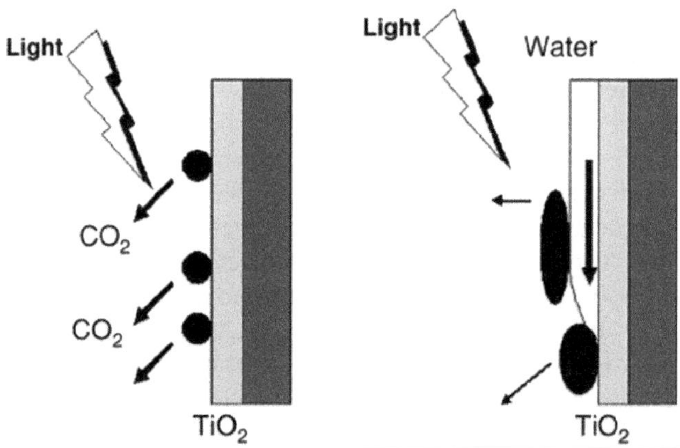

Figure 7.10 Schematic diagram of the decontamination process occurring on the superhydrophilic self-cleaning surface.
Reproduced from A. Fujishima *et al.*, TiO_2 photocatalysis and related surface phenomena, *Surf. Sci. Rep.*, **63**, 515–582, Copyright 2008 with permission from Elsevier.[53]

H_2O when irradiated with UV light, and, when water contacts the surface, it forms "sheets" and carries away dirt and other contaminants.[52]

The advancement in nanotechnology with the use of nanoparticles,[54] nanowires,[55] nanotubes,[56] and nanobelts,[57] typically in the range of 1–100 nm, has led to considerable development in the photocatalysis-induced self-cleaning coating because of its large surface area per unit volume and enhanced light harvesting, thereby facilitating the diffusion of photo-generated holes and electrons, respectively. However, there are still a lot of challenges that require resolving by utilizing nano-structured self-cleaning coatings, especially robustness that depends on surface morphology and optical transparency for the glazing industries.

7.4 Progress in Fabrication Strategies for Nanostructured Antireflective Self-cleaning Coatings

People have proposed many methods to achieve nano-structured self-cleaning coatings with antireflection properties.[58–61] The different methods of fabrication strategies can be divided into two distinct categories: bottom-up and top-down design methods.

7.4.1 Bottom-up Design Method

The chemical synthesis method is a simple and reliable thin film deposition technique, which can achieve an antireflective nanostructured surface without utilizing expensive processing equipment or a lithography mask and can be applied to a large area and scaled up for mass production. Recently, the availability of chemical nanofabrication technology has enabled the engineering of materials with desired AR characteristics.[58,59,62] ZnO nanostructures have brought exciting possibilities for next-generation AR coatings (ARCs) to suppress the Fresnel reflection effectively, as shown in Figure 7.11. Critical features related to the ARC, such as broadband, omnidirectional, and polarization-independent characteristics, have been examined to demonstrate ZnO nanorods array (NRA) layers as an excellent ARC for unpolarized and polarized light. The specular and diffuse reflection observed in NRA layers suggests the importance of the effect of the nanostructured surface. Through measuring the angular distribution of light scattered at nanostructured surfaces, it was found that the reduction of total reflectance mainly results from the significant suppression of specular reflection.

Although ZnO NRA layers with two-dimensional structures meet the requirements for a very low average refractive index, the chemical stability of ZnO has impeded its further application as an AR coating that can be dissolved in both acidic and basic solutions.[64] To improve the stability of the ZnO nanorod arrays, a few groups have reported the preparation of $ZnO@TiO_2$ core/shell nanowires *via* an extra sol–gel route or hydrothermal

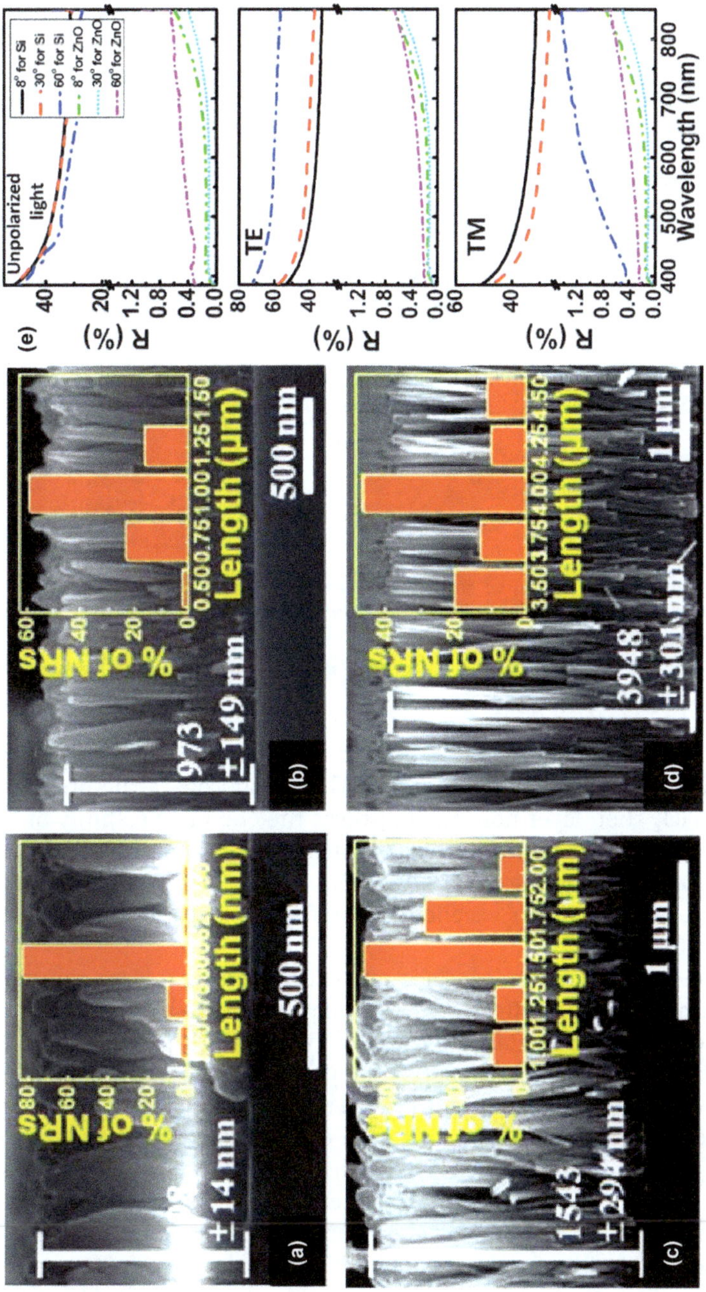

Figure 7.11 Cross-sectional SEM images of ZnO nanorod arrays with different average standard deviation of length distribution: (a) 498 ± 14 nm, (b) 973 ± 149 nm, (c) 1543 ± 294 nm, and (d) 3948 ± 301 nm. (e) The insets indicate the length distribution of the NRAs. The spectral reflectance as a function of AOI for polished Si and 3948 nm long ZnO NRAs/Si over a range from 385 nm to 850 nm.

Reprinted from ref. 63 with permission from the Royal Society of Chemistry.

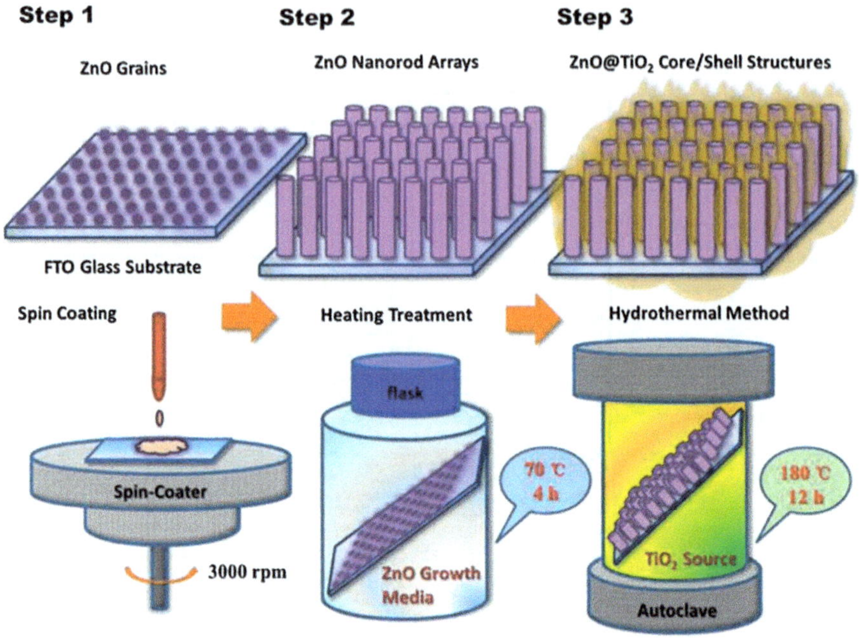

Figure 7.12 Schematic diagram of the preparation procedure for ZnO@TiO_2 core/shell nanorod array films based on an FTO glass substrate using a hydrothermal method.
Reprinted from ref. 65 with permission from the Royal Society of Chemistry.

methods, as shown in Figure 7.12. In addition, ZnO nanorod arrays provide relatively lower surface areas compared with mesoporous structures. The mesoporous TiO_2 shell layer not only improves the chemical stability of the coating, but also imparts improved hydrophilicity and photocatalytic activity upon the ZnO@TiO_2 coating layer. In addition, the ZnO@TiO_2 nanorod array structure results in an excellent antireflective effect, which affords better transmittance.

The surface free energy and surface geometrical structure are the most important parameters that affect the wettability of the surface.[66] Since the successful preparation of anatase TiO_2 crystals with a high percentage of exposed {001} facets by Lu *et al.* in 2008,[67] the photocatalytic performance has undergone a significant improvement when TiO_2 nanostructures possess this well-defined crystal facet,[68–70] the fabrication of such as nanostructured self-cleaning TiO_2 thin films with orientated high reactive facets is preferentially desirable. Recently, TiO_2 films with 100% exposed {001} facets were synthesized *via* a hydrothermal route as shown in Figure 7.13.

Besides the high surface free energy of high index facets, different surfaces of the TiO_2 also show discriminant photoinduced hydrophilicity.[72] Figure 7.14 shows the atom arrangement in the (110), (100), and (001) surfaces of TiO_2 (rutile). Bridging oxygens, which are higher in position and

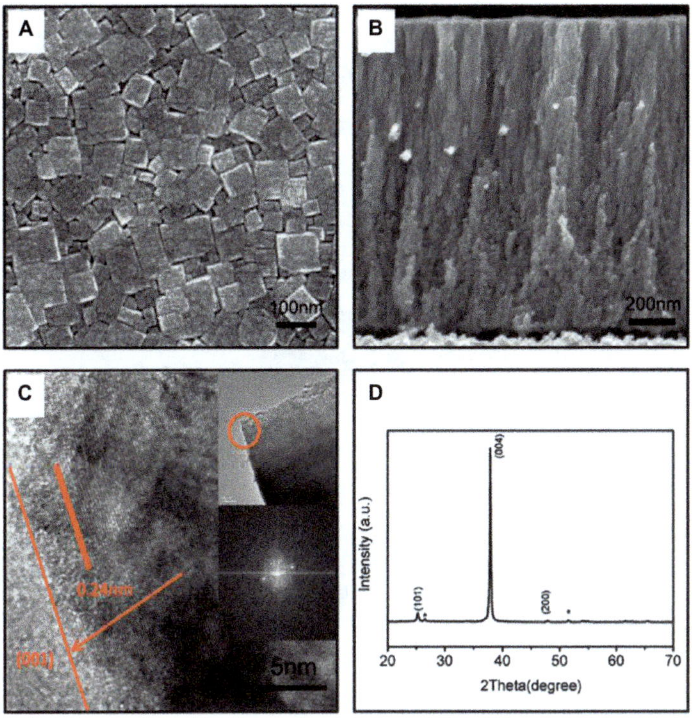

Figure 7.13 Electron microscopy and XRD characterization of TiO_2 nanowire arrays with nearly 100% exposed {001} facets.
Reprinted from ref. 71 with permission from the Royal Society of Chemistry.

energetically more reactive than their surrounding atoms, exist on both the (110) and (100) planes. However, the (001) surface differs from the (110) and (100) surfaces in that all the surface Ti cations are four-fold coordinated with two oxygen atoms within the surface plane and two in the plane below. Hence the (001) surface lacks bridged oxygen atoms and, therefore, it is inert under UV illumination. XPS measurements showed that there are a few Ti^{3+} signals in the Ti_{2p} spectrum on surfaces (110) and (100) due to the removal of bridged oxygen. However, none of the Ti^{3+} signals was observed from surface (001) as no bridged oxygen was attached onto the (001) surface. Therefore, UV illumination could induce some oxygen vacancies on the TiO_2 surface, which may be one of the reasons for the photoinduced hydrophilicity.[73]

Because TiO_2 has a large refractive index ($n \sim 2.5$–2.7),[74] 20% reflectivity at the air–TiO_2 interface would make glass lose too much transmittance. For windows and solar cells applications, high transparency is highly desired. To obtain low surface reflection, one general method is to lower the refractive index of the film through a composite of low refractive index material (such as SiO_2, Figure 7.15).

Nanoparticle size and film thickness are of considerable importance not only for transparency but also for photocatalytic activity. An optimum value

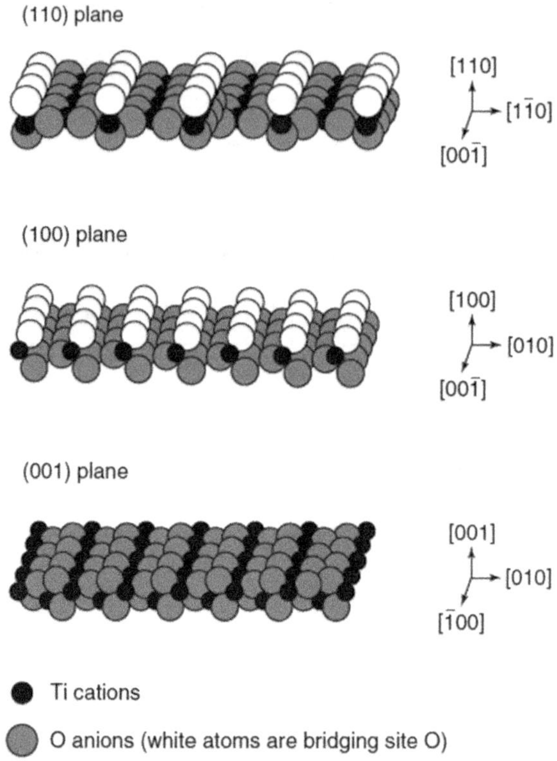

Figure 7.14 Schematic illustration of the atomic alignments on ideal TiO_2 (110), (100), and (001) single-crystal faces.
Reproduced with permission from R. Wang, N. Sakai, A. Fujishima, T. Watanabe and K. Hashimoto, *J. Phys. Chem. B*, 1999, **103**, 2188–2219.[73]

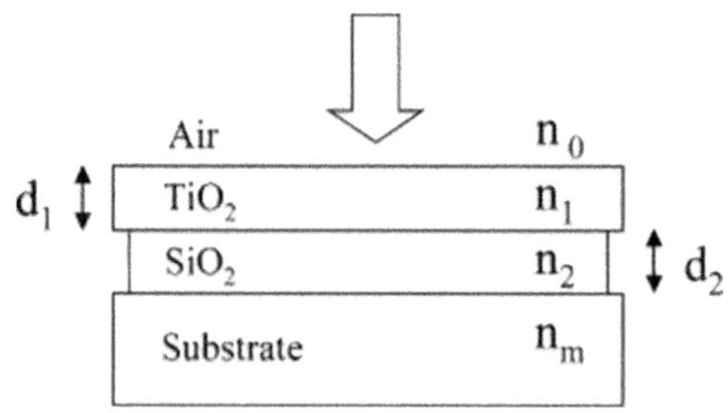

Figure 7.15 Scheme of a double-layered antireflective coating showing the various parameters.
Reprinted with permission from X. Zhang, A. Fujishima, M. Jin, A. V. Emeline and T. Murakami, *J. Phys. Chem. B*, 2006, **110**, 25142–25148.[75]

balance exists between maximum superhydrophilicity and photocatalytic activity, as shown in Figure 7.16. The smaller component units have larger surface areas and optical transparency, and will also have higher rates of electron–hole recombination. Thicker films (up to a few microns in thickness) have shown improved photocatalytic activity, but optical clarity and durability are poor.[51,76]

Although two-dimensional nanostructured coatings can achieve excellent antireflective performance, it is still difficult to realize broad spectrum antireflectance, which requires preparation technology and equipment. Three-dimensional nanostructured coatings can effectively make up for the inadequacy of two-dimensional antireflective structures. The lotus leaf's surface is a good example of how combining a micro-/nano-hierarchical structure with a low surface energy wax layer exhibits a high water contact angle (CA) and low CA hysteresis.[78,79] Xiong *et al.* have fabricated ZnO-based hierarchical structures with nanorods on microcraters by a two-step method with a combination of top-down and bottom-up approaches. A layer of a *c*-axis aligned, transparent, conductive ZnO thin film was deposited by pulsed DC sputtering on a Corning glass substrate. The microcraters were created with anisotropic etching on the as-deposited ZnO thin film. ZnO nanorods were then synthesized onto the etched film by means of metal organic chemical vapor deposition. The resulting hierarchical film exhibits a high water contact angle ($>160°$) with a low contact angle hysteresis ($2°$) and low reflection over a wide spectral range, as shown in Figure 7.17.

Sun's group has also fabricated the hierarchical nanostructure of a coaxial ZnO@TiO_2 nanorod array decorated with a new polymetallic sulfide solid-solution. This novel nanostructure combines the merits of all components and meets the requirements of photocatalysis-induced self-cleaning applications (shown in Figure 7.18).

7.4.2 Top-down Design Method

Etching techniques can prepare structures of several nanometers with high accuracy and resolution, which are described as a selective dissolution or ablation of a surface or substrate. The versatility of etching processing is demonstrated for the preparation of nanotextured surfaces with excellent antireflective performance.[82]

The rapid and large-scale preparation of antireflective nanostructures is highly desired in practical applications and production. A templating technique is widely used for fabricating large-scale nanostructured ARCs. Jiang's group reported bioinspired monolayer silica colloidal crystals as a template for fabricating broadband moth-eye ARCs on single crystalline silicon substrates. The resulting subwavelength-structured ARCs exhibit superior broadband antireflective performance to commercial SiN_x coatings. A schematic illustration of the templating procedures for making wafer-scale silicon nipple arrays is shown in Figure 7.19.

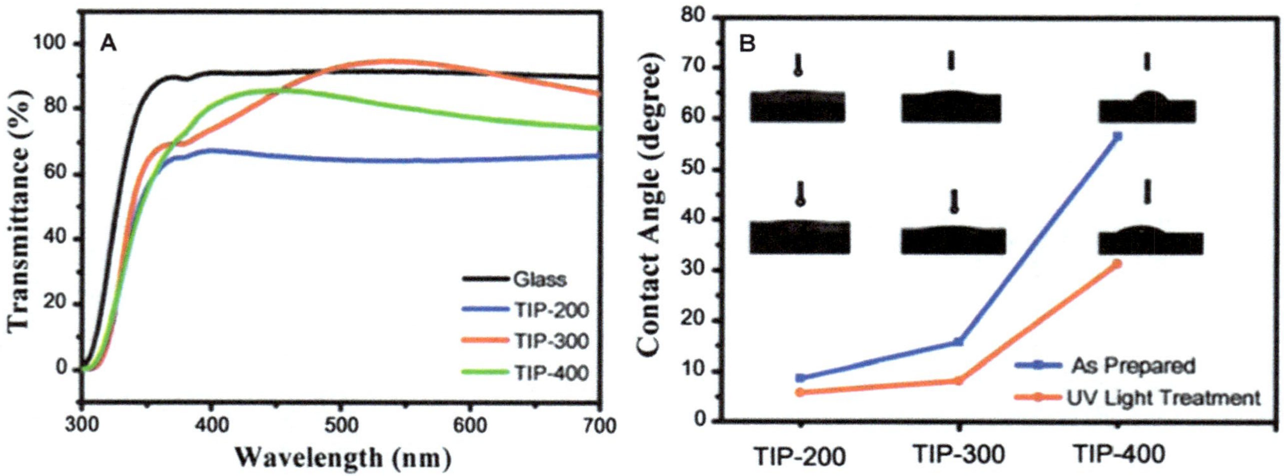

Figure 7.16 (A) The transparency of the glass substrate and thin films of TiO_2 seeds with different thicknesses of TiO_2 nanosheets. (B) The contact angle of the TiO_2 nanosheet thin films coated with different thicknesses. The insets show the contact angle optical images.
Reprinted from ref. 77 with permission from the Royal Society of Chemistry.

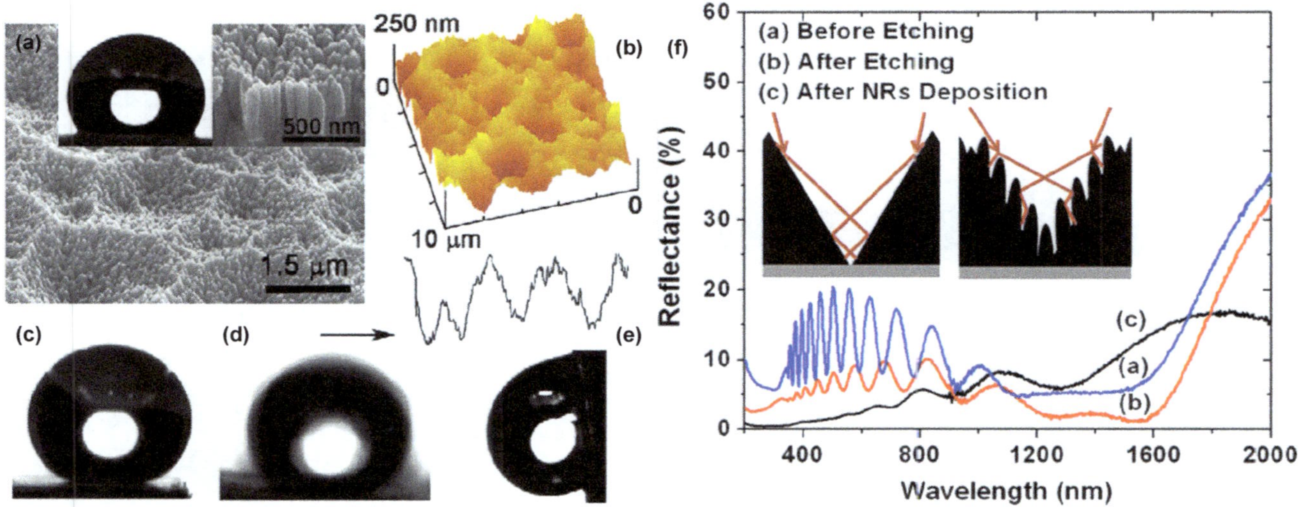

Figure 7.17 (a) FESEM image (45° view), inset shows a droplet of water (left) and the cross sectional FESEM image (right). (b) AFM image with cross sectional profile. (c) Static CA of water after stearic acid modification. (d) Water droplet on surface-modified etched AZO film with NRs rolls off the surface when tilted by 5°. (e) Water droplet on surface-modified etched AZO film without NRs does not move even when tilted by 90°. (f) Reflection spectra of the various AZO films.
Reprinted from J. Xiong *et al.*, Biomimetic hierarchical ZnO structure with superhydrophobic and antireflective properties, *J. Colloid Interface Sci.*, **350**, 355–347, Copyright 2010 with permission from Elsevier.[80]

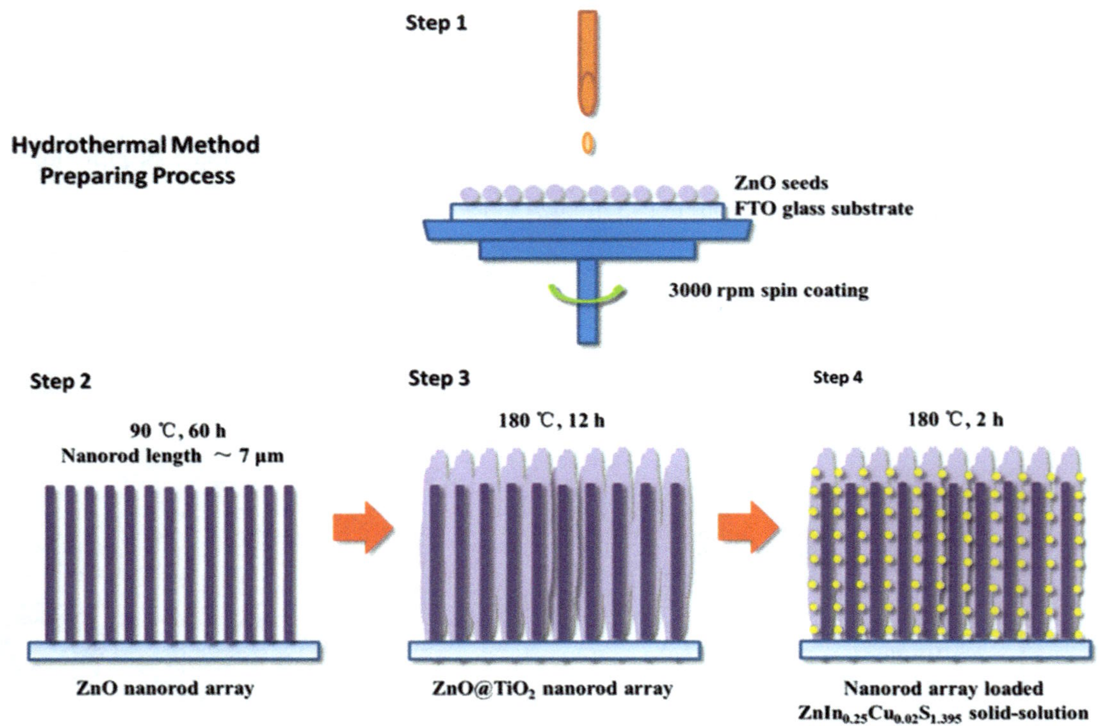

Figure 7.18 Hydrothermal fabrication process of the ZnO@TiO_2 nanorod array film loaded with a polymetallic sulfide solid-solution. Reprinted from ref. 81 with permission from the Royal Society of Chemistry.

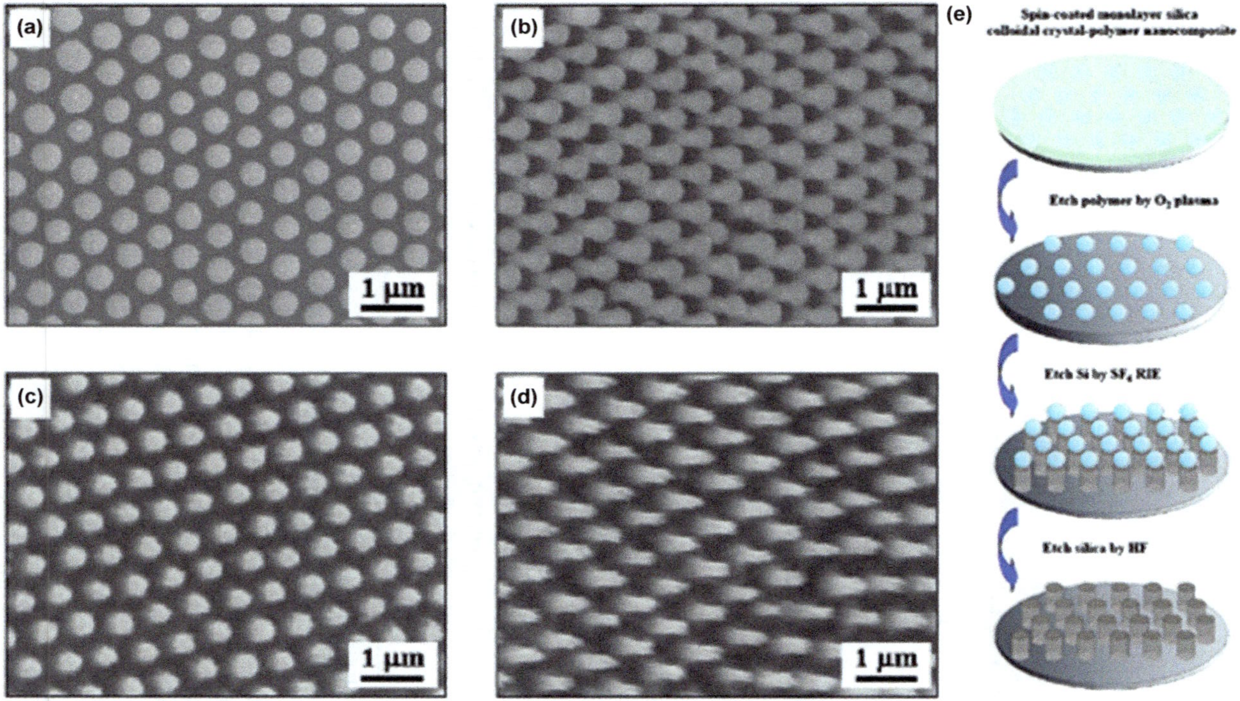

Figure 7.19 (a) SEM image of a spin-coated monolayer non-close-packed colloidal crystal consisting of 360 nm silica spheres. (b) Tilted 20° SEM image showing templating silica array and underneath silicon nipples etched by SF6 RIE for 9 min. (c) Templated silicon nipple array after 6 min SF6 etch. (d) Silicon nipples after 9 min SF6 etch. (e) Schematic illustration of the templating procedures for fabricating moth-eye ARCs on silicon substrates.
Reprinted from C.-H. Sun, P. Jiang and B. Jiang, *Appl. Phys. Lett.*, 2008, **92**, 061112 with the permission of AIP Publishing.[83]

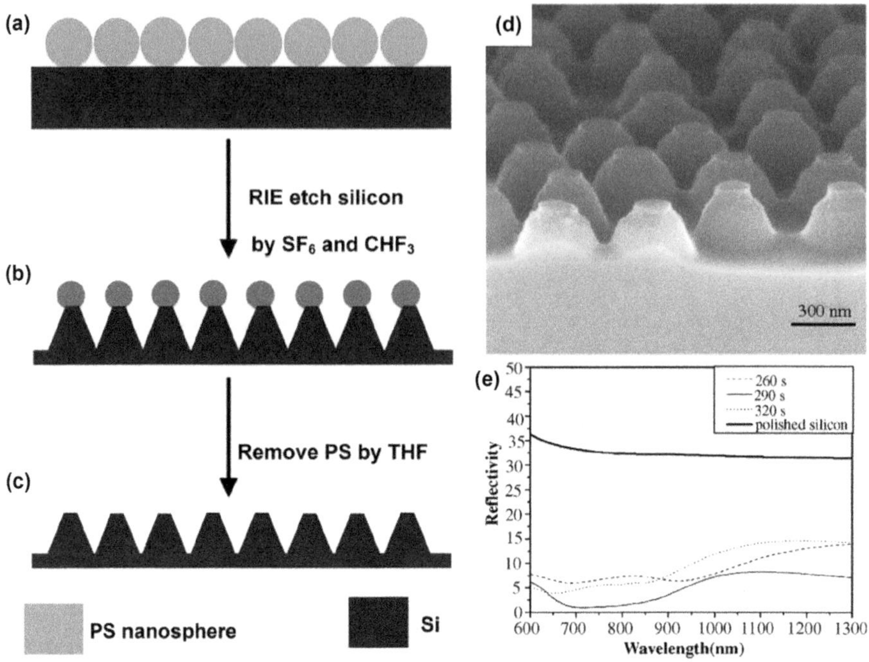

Figure 7.20 Schematic illustration of the procedure for creating antireflective subwavelength arrays on Si (a)–(c). Cross-sectional SEM images of Si nanopillar arrays (d). Specular optical reflection at normal incidence for a flat polished Si wafer (bold solid line) and Si nanopillar arrays etched for 260 s (dashed line), 290 s (solid line), and 320 s (dotted line), respectively (e).
Reprinted from ref. 88 with permission from John Wiley and Sons. Copyright © 2008 Wiley-VCH Verlag GmbH & Co. KGaA, Weinheim.

An etching technique combined with a self-assembly process is more desirable than direct-print lithographic approaches for practical applications due to its simplicity and low cost. Several techniques based on self-assembly masks, such as anodic porous aluminium,[84,85] spin-coated spheres,[86] and evaporated Ag islands,[87] have been demonstrated. Chi's group demonstrated how Si nanopillar arrays with antireflective behavior are created based on the combination of self-assembly of polymer spheres into two-dimensional arrays and reactive-ion etching (RIE), illustrated in Figure 7.20.

The practical applications of etching methods are highly restricted by the high cost of the masks or complications with the procedures. A structure with low reflectivity is the focus of research, otherwise the order of the structure on the surface has less of an impact on antireflective performance. Hence, utilizing a chaotic mask can effectively simplify and reduce the cost of preparation, which provides a possibility for its application. Wang's group used metal island films as masks to generate large-area antireflective subwavelength structures SWSs. These island films can be easily formed by

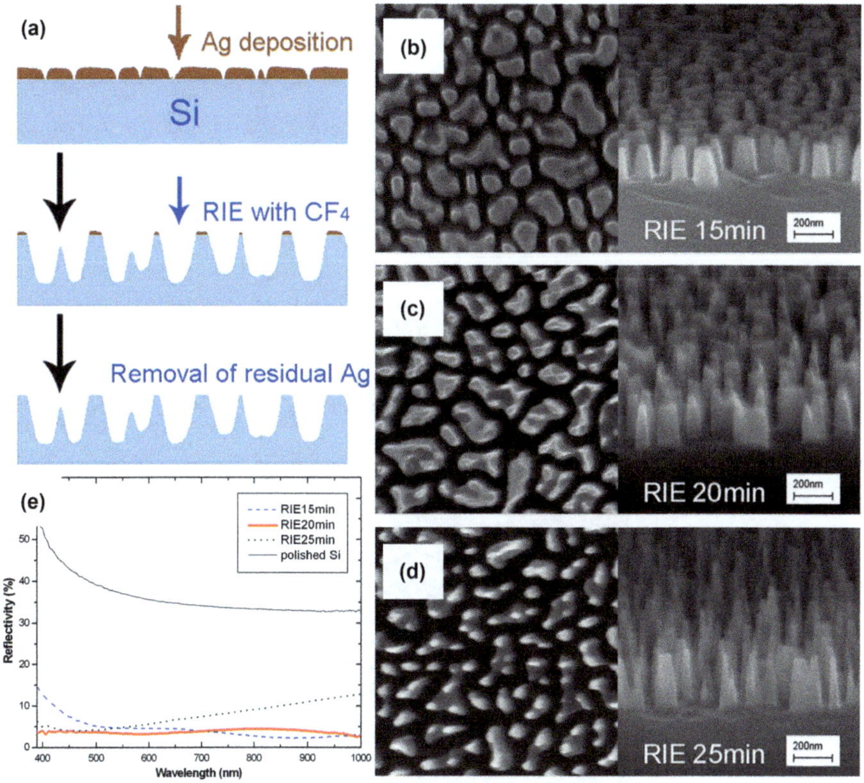

Figure 7.21 Schematic diagram for the fabrication procedure (a). Top view (left column) and side view (right column) SEM images for the Si SWSs with RIE times (b) 15 min, (c) 20 min, and (d) 25 min. Hemispherical reflectivity for the Si SWS surfaces in the wave band of 390–1000 nm. A measured reflectivity for polished Si is also shown for comparison (e). Reprinted from S. Wang, X. Z. Yu and H. T. Fan, *Appl. Phys. Lett.*, 2007, **91**, 06110587 with the permission of AIP Publishing.

thermal dewetting of continuous films over a large area. As shown in Figure 7.21, Ag islands with dimensions ranging from 50 to 400 nm were fabricated by a single step of film deposition under 300 °C. In the process, nucleation, growth, and combination of small Ag clusters are attributed to the formation of larger islands. The growth mode is different from that in thermal dewetting of continuous films. Si SWSs with a low reflectivity of 5% were achieved by following a RIE process.

Currently, a non-mask technique has been one of the most extensively used top-down techniques in the fabrication of antireflective and self-cleaning coatings due to its several advantages, such as uncomplicated technology (the mask does not need to be removed during the procedure), less material is required and the processing cost is decreased. Recently, "black silicon" with a very low reflectance has been produced by plasma immersion ion implantation (PIII), which is usually used to dope semiconductors (shown in Figure 7.22).

Figure 7.22 SEM images of unique microstructures of black silicon. (a) Top view, (b) side view (viewed at 30° to the normal), (c) the formation mechanism of the needle-like structure by the PIII process.
Reprinted from Y. Xia *et al.*, A novel method to produce black silicon for solar cells, *Sol. Energy*, **85**, 1574–1578, Copyright 2011 with permission from Elsevier.[89]

The black silicon exhibits a unique microstructure with a needle-like structure. The average reflectance of the black silicon can be reduced to 1.79% over the wavelength ranging from 300 nm to 1100 nm.

7.5 Conclusion and Outlook

In this chapter, we have presented the principle and fabrication methods of photocatalytic self-cleaning antireflective nano-coatings and their developments are also reviewed. These promising self-cleaning nanostructured coatings can be exploited in a range of applications and advantage can be taken of the beneficial side effects for mankind and the environment. However, the stability, operational lifetime and safety of living organisms have to be highly required and warranted to offer commercial products to attract the confidence of the marketplace, which could help preserve natural resources consumed during the washing and cleaning processes eventually.

References

1. R. Blossey, *Nat. Mater.*, 2003, **2**, 301–306.
2. C.-A. Lin, M.-L. Tsai, W.-R. Wei, K.-Y. Lai and J.-H. He, *ACS Nano*, 2015, DOI: 10.1021/acsnano.5b05564.
3. Y. Yuan, Y. Chen, W. Chen and R. Hong, *Sol. Energy*, 2015, **118**, 222–231.
4. A. Vašíček, *Optics of Thin Films*, North-Holland Publisher Co., 1960.
5. H. K. Raut, V. A. Ganesh, A. S. Nair and S. Ramakrishna, *Energy Environ. Sci.*, 2011, **4**, 3779–3804.
6. H. A. Macleod, *Thin-film Optical Filters*, CRC Press, 2010.
7. M. J. Minot, *J. Opt. Soc. Am.*, 1976, **66**, 515–519.
8. S. Chattopadhyay, Y. Huang, Y.-J. Jen, A. Ganguly, K. Chen and L. Chen, *Mater. Sci. Eng., R*, 2010, **69**, 1–35.
9. D. Chen, *Sol. Energy Mater. Solar Cells*, 2001, **68**, 313–336.
10. D. Poitras and J. Dobrowolski, *Appl. Opt.*, 2004, **43**, 1286–1295.
11. R. Jacobsson, *Prog. Opt.*, 1966, **5**, 247–286.
12. H. Fabricius, *Appl. Opt.*, 1992, **31**, 5191–5196.
13. W. H. Southwell, *Opt. Lett.*, 1983, **8**, 584–586.
14. C. Striemer and P. Fauchet, *Appl. Phys. Lett.*, 2002, **81**, 2980–2982.
15. W. A. Daoud, *Self-cleaning Materials and Surfaces: A Nanotechnology Approach*, John Wiley & Sons, 2013.
16. W. Miller, G. Bernard and J. Allen, *Science*, 1968, **162**, 760–767.
17. X. Gao, X. Yan, X. Yao, L. Xu, K. Zhang, J. Zhang, B. Yang and L. Jiang, *Adv. Mater.*, 2007, **19**, 2213.
18. R. Brunner, A. Deparnay, M. Helgert, M. Burkhardt, T. Lohmüller and J. P. Spatz, Optical Engineering + Applications, 2008.
19. Y. Li, X. J. Huang, S. H. Heo, C. C. Li, Y. K. Choi, W. P. Cai and S. O. Cho, *Langmuir*, 2007, **23**, 2169–2174.
20. Q. Xie, J. Xu, L. Feng, L. Jiang, W. Tang, X. Luo and C. C. Han, *Adv. Mater.*, 2004, **16**, 302–305.

21. S. Niu, B. Li, Z. Mu, M. Yang, J. Zhang, Z. Han and L. Ren, *J. Bionic Eng.*, 2015, **12**, 170–189.
22. H.-Y. Guo, Q. Li, H.-P. Zhao, K. Zhou and X.-Q. Feng, *RSC Adv.*, 2015, **5**, 66901–66926.
23. T. Sun, L. Feng, X. Gao and L. Jiang, *Acc. Chem. Res.*, 2005, **38**, 644–652.
24. A. Tuteja, W. Choi, M. Ma, J. M. Mabry, S. A. Mazzella, G. C. Rutledge, G. H. McKinley and R. E. Cohen, *Science*, 2007, **318**, 1618–1622.
25. M. Miwa, A. Nakajima, A. Fujishima, K. Hashimoto and T. Watanabe, *Langmuir*, 2000, **16**, 5754–5760.
26. B. Bhushan and Y. C. Jung, *Prog. Mater. Sci.*, 2011, **56**, 1–108.
27. W. E. Ward, *J. Aesth. Art Crit.*, 1952, **11**, 135–146.
28. L. Feng, S. Li, Y. Li, H. Li, L. Zhang, J. Zhai, Y. Song, B. Liu, L. Jiang and D. Zhu, *Adv. Mater.*, 2002, **14**, 1857–1860.
29. M. Sun, C. Luo, L. Xu, H. Ji, Q. Ouyang, D. Yu and Y. Chen, *Langmuir*, 2005, **21**, 8978–8981.
30. C. Neinhuis and W. Barthlott, *Ann. Bot.*, 1997, **79**, 667–677.
31. W. Barthlott and C. Neinhuis, *Planta*, 1997, **202**, 1–8.
32. M. Ma and R. M. Hill, *Curr. Opin. Colloid Interface Sci.*, 2006, **11**, 193–202.
33. J. Drelich, E. Chibowski, D. D. Meng and K. Terpilowski, *Soft Matter*, 2011, **7**, 9804–9828.
34. C. Euvananont, C. Junin, K. Inpor, P. Limthongkul and C. Thanachayanont, *Ceram. Int.*, 2008, **34**, 1067–1071.
35. P. Patel, C. K. Choi and D. D. Meng, *JALA(1998–2010)*, 2010, **15**, 114–119.
36. K. Mittal, *Advances in Contact Angle, Wettability and Adhesion, Volume Two*, John Wiley & Sons, 2015.
37. P. Gould, *Materials Today*, 2003, **6**, 44–48.
38. X. Zhang, Y. Guo, P. Zhang, Z. Wu and Z. Zhang, *ACS Appl. Mater. Interfaces*, 2012, **4**, 1742–1746.
39. T. Sun, G. Wang, L. Feng, B. Liu, Y. Ma, L. Jiang and D. Zhu, *Angew. Chem., Int. Ed.*, 2004, **43**, 357–360.
40. F. Xia, H. Ge, Y. Hou, T. Sun, L. Chen, G. Zhang and L. Jiang, *Adv. Mater.*, 2007, **19**, 2520–2524.
41. F. Xia and L. Jiang, *Adv. Mater*, 2008, **20**, 2842–2858.
42. Y. Takata, S. Hidaka, J. Cao, T. Nakamura, H. Yamamoto, M. Masuda and T. Ito, *Energy*, 2005, **30**, 209–220.
43. A. Fujishima, T. N. Rao and D. A. Tryk, *J. Photochem. Photobiol., C*, 2000, **1**, 1–21.
44. X. Feng, L. Feng, M. Jin, J. Zhai, L. Jiang and D. Zhu, *J. Am. Chem. Soc.*, 2004, **126**, 62–63.
45. M. Miyauchi, A. Nakajima, T. Watanabe and K. Hashimoto, *Chem. Mater.*, 2002, **14**, 2812–2816.
46. A. Nel, T. Xia, L. Mädler and N. Li, *Science*, 2006, **311**, 622–627.
47. K. Hashimoto, H. Irie and A. Fujishima, *Jpn. J. Appl. Phys.*, 2005, **44**, 8269.
48. N. Sakai, A. Fujishima, T. Watanabe and K. Hashimoto, *J. Phys. Chem. B*, 2003, **107**, 1028–1035.

49. J. Chen and C.-s. Poon, *Build. Environ.*, 2009, **44**, 1899–1906.
50. A. L. Linsebigler, G. Lu and J. T. Yates Jr., *Chem. Rev.*, 1995, **95**, 735–758.
51. I. P. Parkin and R. G. Palgrave, *J. Mater. Chem.*, 2005, **15**, 1689–1695.
52. M. J. Hanus and A. T. Harris, *Prog. Mater. Sci.*, 2013, **58**, 1056–1102.
53. A. Fujishima, X. Zhang and D. A. Tryk, *Surf. Sci. Rep.*, 2008, **63**, 515–582.
54. X.-T. Zhang, O. Sato, M. Taguchi, Y. Einaga, T. Murakami and A. Fujishima, *Chem. Mater.*, 2005, **17**, 696–700.
55. Y.-A. Dai, H.-C. Chang, K.-Y. Lai, C.-A. Lin, R.-J. Chung, G.-R. Lin and J.-H. He, *J. Mater. Chem.*, 2010, **20**, 10924–10930.
56. G. K. Mor, M. A. Carvalho, O. K. Varghese, M. V. Pishko and C. A. Grimes, *J. Mater. Res.*, 2004, **19**, 628–634.
57. W. Zhou, H. Liu, J. Wang, D. Liu, G. Du and J. Cui, *ACS Appl. Mater. Interfaces*, 2010, **2**, 2385–2392.
58. R. Prado, G. Beobide, A. Marcaide, J. Goikoetxea and A. Aranzabe, *Sol. Energy Mater. Sol. Cells*, 2010, **94**, 1081–1088.
59. K. Askar, B. M. Phillips, Y. Fang, B. Choi, N. Gozubenli, P. Jiang and B. Jiang, *Colloids Surf. A*, 2013, **439**, 84–100.
60. J. C. Thies, E. Currie and G. J. Meijers and K. Gan, *U.S. Patent*, 7 892 606[P], 2011.
61. L. Yao and J. He, *Prog. Mater. Sci.*, 2014, **61**, 94–143.
62. S. Nishimoto and B. Bhushan, *Rsc Adv.*, 2013, **3**, 671–690.
63. Y.-C. Chao, C.-Y. Chen, C.-A. Lin, Y.-A. Dai and J.-H. He, *J. Mater. Chem.*, 2010, **20**, 8134–8138.
64. R. Comparelli, E. Fanizza, M. Curri, P. Cozzoli, G. Mascolo and A. Agostiano, *Appl. Catal., B*, 2005, **60**, 1–11.
65. R. Wang, H. Tan, Z. Zhao, G. Zhang, L. Song, W. Dong and Z. Sun, *J. Mater. Chem. A*, 2014, **2**, 7313–7318.
66. Z. Yoshimitsu, A. Nakajima, T. Watanabe and K. Hashimoto, *Langmuir*, 2002, **18**, 5818–5822.
67. H. G. Yang, C. H. Sun, S. Z. Qiao, J. Zou, G. Liu, S. C. Smith, H. M. Cheng and G. Q. Lu, *Nature*, 2008, **453**, 638–641.
68. J. Pan, G. Liu, G. Q. M. Lu and H. M. Cheng, *Angew. Chem., Int. Ed.*, 2011, **50**, 2133–2137.
69. X. Zhao, W. Jin, J. Cai, J. Ye, Z. Li, Y. Ma, J. Xie and L. Qi, *Adv. Funct. Mater.*, 2011, **21**, 3554–3563.
70. X. Han, Q. Kuang, M. Jin, Z. Xie and L. Zheng, *J. Am. Chem. Soc.*, 2009, **131**, 3152–3153.
71. Z. Zhao, H. Tan, H. Zhao, D. Li, M. Zheng, P. Du, G. Zhang, D. Qu, Z. Sun and H. Fan, *Chem. Commun.*, 2013, **49**, 8958–8960.
72. T. Watanabe, A. Nakajima, R. Wang, M. Minabe, S. Koizumi, A. Fujishima and K. Hashimoto, *Thin Solid Films*, 1999, **351**, 260–263.
73. R. Wang, N. Sakai, A. Fujishima, T. Watanabe and K. Hashimoto, *J. Phys. Chem. B*, 1999, **103**, 2188–2194.
74. H. K. Pulker, G. Paesold and E. Ritter, *Appl. Opt.*, 1976, **15**, 2986–2991.
75. X. Zhang, A. Fujishima, M. Jin, A. V. Emeline and T. Murakami, *J. Phys. Chem. B*, 2006, **110**, 25142–25148.

76. V. A. Ganesh, H. K. Raut, A. S. Nair and S. Ramakrishna, *J. Mater. Chem.*, 2011, **21**, 16304–16322.
77. F. Wang, G. Zhang, Z. Zhao, H. Tan, W. Yu, X. Zhang and Z. Sun, , *RSC Adv.*, 2015, **5**, 9861–9864.
78. X. Tang, J. Dong and X. Li, *J. Colloid Interface Sci.*, 2008, **325**, 223–227.
79. W. Hou and Q. Wang, *J. Colloid Interface Science*, 2009, **333**, 400–403.
80. J. Xiong, S. N. Das, B. Shin, J. P. Kar, J. H. Choi and J.-M. Myoung, *J. Colloid Interface Sci.*, 2010, **350**, 344–347.
81. R. Wang, X. Xu, Y. Zhang, Z. Chang, Z. Sun and W. Dong, *Nanoscale*, 2015, **7**, 11082–11092.
82. J. W. Arnold, T. L. Brewer and S. Punyakumleard, *U.S. Patent*, 4 910 122[P], 1990.
83. C.-H. Sun, P. Jiang and B. Jiang, *Appl. Phys. Lett.*, 2008, **92**, 061112.
84. Y. Kanamori, K. Hane, H. Sai and H. Yugami, *Appl. Phys. Lett.*, 2001, **78**, 142–143.
85. H. Sai, H. Fujii, K. Arafune, Y. Ohshita, M. Yamaguchi, Y. Kanamori and H. Yugami, *Appl. Phys. Lett.*, 2006, **88**, 201116.
86. C.-H. Sun, W.-L. Min, N. C. Linn, P. Jiang and B. Jiang, *Appl. Phys. Lett.*, 2007, **91**, 231105.
87. S. Wang, X. Z. Yu and H. T. Fan, *Appl. Phys. Lett.*, 2007, **91**, 061105.
88. H. Xu, N. Lu, D. Qi, J. Hao, L. Gao, B. Zhang and L. Chi, *Small*, 2008, **4**, 1972–1975.
89. Y. Xia, B. Liu, J. Liu, Z. Shen and C. Li, *Sol. Energy*, 2011, **85**, 1574–1578.

CHAPTER 8

Antireflection and Self-cleaning Coatings: Principle, Fabrication and Application

LIN YAO AND JUNHUI HE*

Functional Nanomaterials Laboratory, Center for Micro/Nanomaterials and Technology and Key Laboratory of Photochemical Conversion and Optoelectronic Materials, Technical Institute of Physics and Chemistry, Chinese Academy of Sciences, Zhongguancundonglu 29, Haidianqu, Beijing 100190, China
*Email: jhhe@mail.ipc.ac.cn

8.1 Introduction

Transparent surface coatings with suitable optical path differences can suppress the reflection of substrates. Such coatings are usually called "antireflective coatings" (ARCs). There exist many antireflective surfaces in nature. Taking some diurnal butterflies as an example, they use arrays of non-close-packed nipples (about 300 nm) as an antireflective structure (ARS) to reduce reflection from their compound eyes.[1,2] The ARS on the cornea can increase light transmission and suppress reflection losses at the interfaces. Besides these excellent optical properties, the ARS arrays also exhibit self-cleaning capability because of the high fraction of air trapped between arrays.[3,4]

The antireflection phenomenon provides enormous inspiration for scientists to mimic it for many important applications. Antireflection technology has been widely used in some high-precision optical components, solar cells, flat panel displays and light emitting diode lighting to increase

RSC Smart Materials No. 21
Self-cleaning Coatings: Structure, Fabrication and Application
Edited by Junhui He

Published by the Royal Society of Chemistry, www.rsc.org

the transmittance of incident light.[5–8] For solar cells, due to reflection at the air/glass interface of the packaging glass and scattering by accumulated dust on outdoor panels, part of the incident energy on solar modules is lost. On one hand, ARC on the glass can help alleviate reflection in systems; on the other hand, the self-cleaning property can, to some extent, solve the dust accumulation problem.[5] Tseng *et al.* systematically studied the antireflection and light trapping effects. Their results showed a 76.9% enhancement of short-circuit current density compared with that of bare silicon due to suppression of surface reflection.[9]

It is common knowledge that smooth surfaces shine more than rough ones. The very basic idea of antireflection is that roughness is necessary to reduce reflection of surfaces. The reflection or optical disturbance will be zero if the medium for light propagation does not change or two media have the same refractive index (RI).[10] Therefore, many materials with a micro/nano structure have the potential to fabricate ARC, including silicon, silica, titania, zirconia, zinc oxide, cobalt oxide, tin oxide, carbon, poly(ethylene terephthalate) (PET), polystyrene (PS) and gallium nitride.[11–17] To date, two kinds of approaches are available for fabricating ARCs. One is coating porous or multilayered films on the surface of devices, and the other is fabricating sub-wavelength ARS directly on the substrate.[18–21] The corresponding fabrication routes can be classified into bottom-up and top-down modes. The bottom-up technique usually uses nanoparticles as building block to form ARCs. The top-down approach relies on etching or lithography *etc.* techniques performed with or without masks. Recently, the two-step method with a combination of top-down and bottom-up approaches has also attracted much attention.[11]

When contamination or fogging occurs on ARCs, however, their optical properties deteriorate dramatically. Contaminants accumulate and water molecules condense on the surface, leading to scattering and reflection of light. This problem may be solved by creation of a surface that has special wettability (superhydrophilicity or superhydrophobicity) and a photocatalytic property. ARCs with a self-cleaning property have been developed rapidly in recent years from window glass to various devices.[22]

In this chapter, we aim to provide recent developments in antireflective and self-cleaning surfaces, with particular emphasis on silicon and fused silica materials, since they are both commonly used in many functional devices. In the first section, the basic concept and principle of antireflection and self-cleaning are briefly described. In the next section, the fabrication pathways and their mechanisms are discussed. Then, we introduce the latest typical progress in self-cleaning ARCs over the last five years. In the last section, some applications of the self-cleaning ARCs surfaces are introduced. A large number of research articles have reported the fabrication of ARCs or superhydrophobic/superhydrophilic surfaces within the past few years. This chapter intends to give readers both an integrated tour of antireflective and self-cleaning coatings fabrication technologies and a basic realization of their great application prospects.

8.2 Theoretical Aspects of Antireflection and Self-cleaning

8.2.1 Principle of Antireflection

ARCs can effectively enhance the transmission of light. The Fresnel equation offers the basic mathematical model of the reflection of coatings.[23] We learn that RI, which quantifies the speed of light in the current medium with respect to that in a vacuum, is the major parameter for characterizing the transmission of light in the medium.

For single-layer coatings, the Fresnel equation, which offers the basic preliminary mathematical model of reflection and refraction, has the following two assumptions:

(1) The reflected waves have the same intensity and one wave is reflected per interface;
(2) Other optical interactions such as scattering, absorption, *etc.* are negligible.

As illustrated in Figure 8.1(a), there would be no reflection if there was a destructive interference between light reflected from the coating–substrate and the air–coating interfaces. Therefore, n_c for an ideal homogeneous ARC has to meet the following two conditions:

(1) $n_c = (n_a n_s)^{1/2}$, where n_a and n_s are the refractive indices of the air and the substrate, respectively.
(2) $d = \lambda/4n_c$, where d is the thickness of the coating and λ is the wavelength of the incident light.

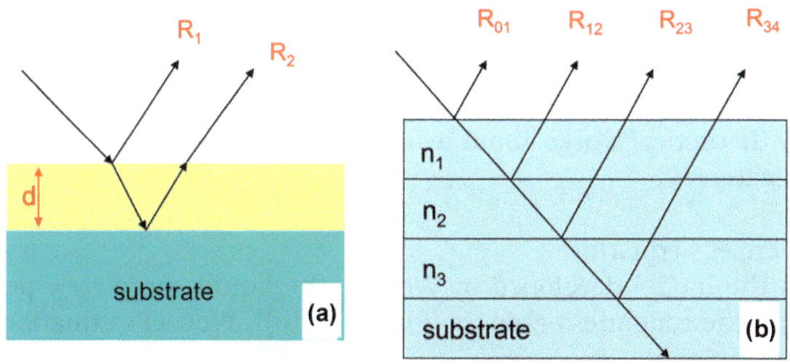

Figure 8.1 Propagation of light rays through (a) a single-layer coating and (b) multilayer coating on a substrate ($n_s > n_n$, n_s and n_n are the refractive indices of the substrate and coatings, respectively).[228]
Reprinted from L. Yao and J. He, Recent progress in antireflection and self-cleaning technology – From surface engineering to fiunctional surfaces, *Prog. Mater. Sci.*, **61**, 94–143. Copyright 2014 with permission from Elsevier.

On the basis of Fresnel's theory, the antireflection property is also dependent on the s and p-polarization of the light.[23] The s-polarization and p-polarization have the electric field perpendicular and parallel to the incidence plane, respectively. Sunlight shows an appreciable degree of polarization when light reflects at shallow angles due to the fact that light reflecting at shallow angles has the p-polarized light reflecting to the maximum.

For multi-layer coatings, the mathematical model has some differences. As shown in Figure 8.1(b), the reflected light from the interface ij between the adjacent layers i and j is:[24]

$$R_{ij} = |R_{\mathrm{mn}}| \exp[-2(\delta_i + \delta_j)]$$

in which, $|R_{\mathrm{mn}}| = [(n_i - n_j)/(n_i + n_j)]$, $\delta_i = 2\pi n_i d_i \cos\theta_i/\lambda$ (θ_i is the angle of refraction, d_i is the physical thickness of the layer).

For the entire coating:

$$R_{\mathrm{sum}} = R_{01} + R_{12} + R_{23} + \ldots + R_{\mathrm{ns}}$$

By adjusting the reflective index and the thickness of each layer, a minimum R_{sum} can be obtained, and an ARC can be achieved.

8.2.2 Principle of Self-cleaning

8.2.2.1 Special Wettability-induced Self-cleaning

Wettability is an important property of a solid surface, and contact angle (θ) has been commonly used to characterize the surface wettability.

For a perfectly smooth and chemically homogeneous solid surface, the contact angle of liquid is given by Young's equation[25]:

$$\cos\theta = (\gamma_{\mathrm{SV}} - \gamma_{\mathrm{SL}})/\gamma_{\mathrm{LV}}$$

where, γ_{SV}, γ_{SL}, and γ_{LV} are the interfacial tensions of the solid–vapor, solid–liquid and liquid–vapor interfaces, respectively (Figure 8.2(a)).

For a realistic solid surface, which is rough and chemically heterogeneous, the early theoretical works about its contact angle were done by Wenzel and Cassie–Baxter.

I. Wenzel's equation[26]

   In the model developed by Wenzel, the liquid completely penetrates into the roughness grooves (Figure 8.2(b)). Wenzel's equation is:

   $$\cos\theta_{\mathrm{w}} = r\cos\theta$$

   where θ_{w} is the contact angle on a rough surface, θ is the Young's contact angle on a similar smooth surface, and r is the surface roughness factor, defined as the ratio of the actual surface area to the projected surface area (for a perfectly smooth surface: $r = 1$, and for a rough surface: $r > 1$).

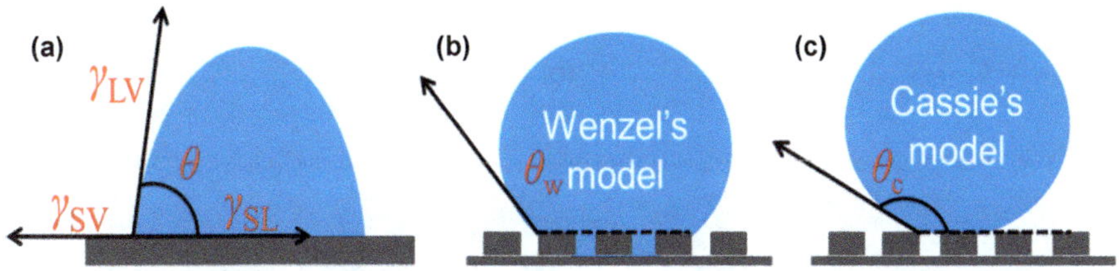

Figure 8.2 (a) A liquid droplet on a smooth surface. (b) A liquid droplet in the Wenzel state. (c) A liquid droplet in the Cassie state.[228]
Reprinted from L. Yao and J. He, Recent progress in antireflection and self-cleaning technology – From surface engineering to functional surfaces, *Prog. Mater. Sci.*, **61**, 94–143. Copyright 2014 with permission from Elsevier.

The above equation points out that when θ is $<90°$, the wetting could be enhanced by r; when θ is $>90°$, the wetting is lessened by r.

II. Cassie–Baxter equation[27]
The Cassie–Baxter state considers that under some roughness conditions, air bubbles may be trapped when $\theta > 90°$ (Figure 8.2(c)). In this case, the liquid–surface interface actually consists of two phases, *i.e.*, the liquid–solid interface and the liquid–vapor interface, respectively. Therefore, the apparent contact angle is the sum of all the contributions of the different phases:

$$\cos\theta_c = f_1 \cos\theta_1 + f_2 \cos\theta_2$$

where θ_c is the apparent contact angle, f_1 and f_2 are the surface fractions of phase 1 and phase 2, respectively; and θ_1 and θ_2 are the contact angles on phase 1 and phase 2, respectively.

If one of the surfaces is the air–liquid interface and f is defined as the fraction of the solid surface that is wetted by the liquid, the air fraction can be expressed as $(1-f)$. With $\theta = 180°$ for air, the resulting apparent contact angle can be calculated by the following equation:

$$\cos\theta_c = f\cos\theta + (1-f)\cos 180° = f\cos\theta + f - 1$$

In the above equations, the contact angles are all static. However, owing to chemical heterogeneity, surface roughness and surface reorganization, different contact angles can coexist along the contact line upon contact with a probe liquid.[28] As shown in Figure 8.3(a), the contact angle of the droplet at the front is greater than that at the back, and they correspond to the advancing and receding contact angle, respectively. The difference between the advancing and receding contact angles is defined as the contact angle hysteresis, which is a measurement of how well a drop of liquid sticks to the solid surface (Figure 8.3(b)).

There are two kinds of surface with special wettability: a superhydrophobic surface (water contact angle $>150°$ and low contact angle

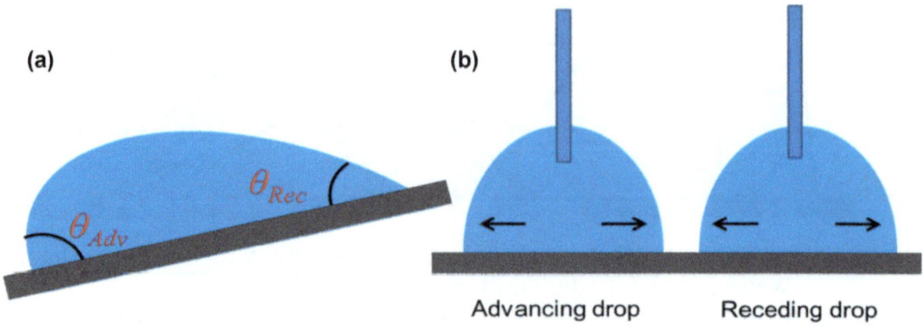

Figure 8.3 (a) Schematic illustration of contact angle hysteresis on a tilted surface, (b) advancing and receding contact angles on a flat surface *via* increasing and decreasing the volume of droplet, respectively.[228]
Reprinted from L. Yao and J. He, Recent progress in antireflection and self-cleaning technology – From surface engineering to fiunctional surfaces, *Prog. Mater. Sci.*, **61**, 94–143. Copyright 2014 with permission from Elsevier.

hysteresis) and superhydrophilic surface (water contact angle $<5^\circ$ within 0.5 s or less). A self-cleaning superhydrophobic surface should always have a low contact angle hysteresis. The leaf of the lotus plant is a representative model for a superhydrophobic and self-cleaning surface. Due to the papillose epidermal cells being covered with wax crystals (Figure 8.4(a)), the leaf surface is very rough and water-repelling. Therefore, water droplets on the surface are in the Cassie–Baxter state, making them highly spherical and able to roll off easily. Usually, contaminants are larger than the cellular structure of the leaves, leaving the particles resting on the tips of the surface structure. When a water droplet rolls over the contaminants, dirt particles are adsorbed to the water droplet, and moved away from the surface (Figure 8.4(b)). On such intrinsic hierarchically structured surfaces, both the contact area and adhesion to the surface are dramatically reduced.

Besides, special wettability also endows the surfaces with an antifogging function. Fogging occurs when water vapor condenses on the surface owing to temporary changes in humidity, temperature, or convection.[30] If the surface has a rather low contact angle, it could form a continuous or near-continuous film, scattering is suppressed and optical transmission is enhanced. On the other hand, if the surface has quite a high contact angle and low contact angle hysteresis, the droplets may slide from the surface rapidly. Thus it may be difficult for fog to cover the surface. Superhydrophobic or superhydrophilic coatings are in line with the above requirements and have shown good prospects for practical applications.[22]

8.2.2.2 Photocatalysis-induced Self-cleaning

Since Frank and Bard first discovered that TiO_2 can decompose cyanide in water in 1977,[31] TiO_2 has been an ideal photocatalyst in decomposing organic contamination for environmental applications. TiO_2 has a strong

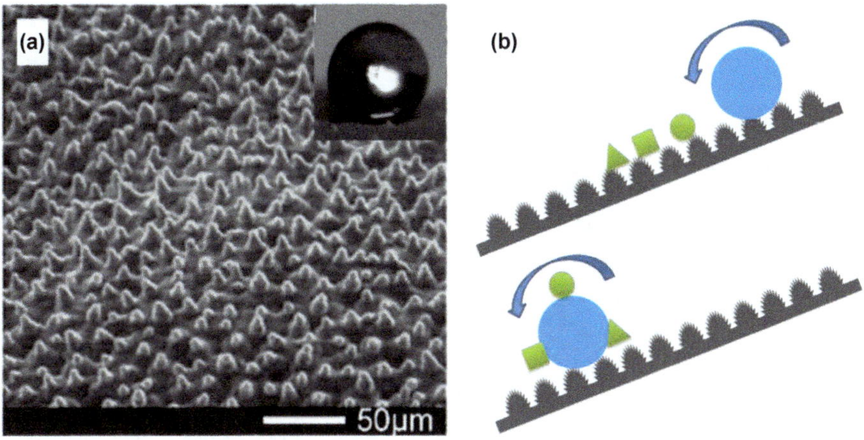

Figure 8.4 (a) SEM image of a lotus leaf showing epidermal cells, and (b) diagram showing the self-cleaning process on a rough surface.
Reprinted with permission from M. H. Sun, C. X. Luo, L. P. Xu, H. Ji, O. Y. Qi, D. P. Yu *et al.*, Artificial lotus leaf by nanocasting, *Langmuir*, 2005, **21**, 8978–8981.[29] Copyright (2005) American Chemical Society.

oxidation power under UV illumination, and various organics accumulating on TiO_2 thin films could be oxidized into CO_2, H_2O, NO_3^- or other simple basic products.[32–34] Moreover, under UV illumination, the surface of TiO_2 becomes superhydrophilic with a water contact angle close to 0°, making the surface easily cleaned by water. Due to these two characteristics, TiO_2-coated substrates can achieve an outstanding self-cleaning property when they are exposed outdoors, as shown in Figure 8.5.

I. Mechanism of photocatalysis
The redox potential for TiO_2 photo-generated holes is +2.53 V, which can produce hydroxyl radicals ($^{\bullet}OH$). The redox potential for conduction band electrons is −0.52 V, which is enough to reduce dioxygen to superoxide, or to hydrogen peroxide.[35] The basic processes for an electron–hole pair reducing or oxidizing an adsorbate and forming a singly oxidized electron donor and singly reduced electron acceptor can be summarized as Figure 8.6.[36]

Previous results have shown that many liquid and solid organic compounds can be photocatalytically decomposed into nontoxic gas-phase products. The decomposition of organic compounds (alkanes, alcohols, ketones, carboxylic acids, *etc.*) can be summarized as Scheme 8.1 with alcohols as an example.[37,38] As reported by Schwitzgebel *et al.*,[39] long-chain organic compounds (*e.g.*, C_{16} and C_{17} acids) and even polymers can also be decomposed by chains breaking.

II. Mechanism of photo-induced superhydrophilicity
A TiO_2 surface becomes superhydrophilic *via* UV light irradiation; however, the reduced water contact angle (WCA) gradually reverts to

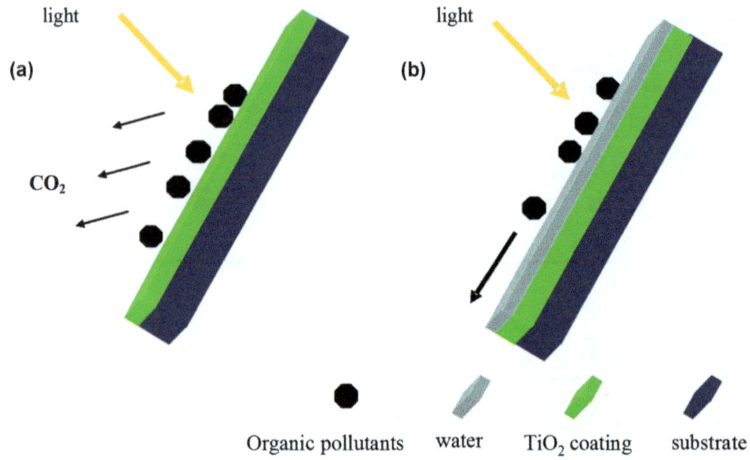

Figure 8.5 Schematic diagram of the decontamination process occurring on TiO_2-coated substrates. (a) Photocatalyic degradation of organics, (b) removal of contaminants by water sheet due to the superhydrophilicity of the TiO_2 surface.[228]
Reprinted from L. Yao and J. He, Recent progress in antireflection and self-cleaning technology – From surface engineering to fiunctional surfaces, *Prog. Mater. Sci.*, **61**, 94–143. Copyright 2014 with permission from Elsevier.

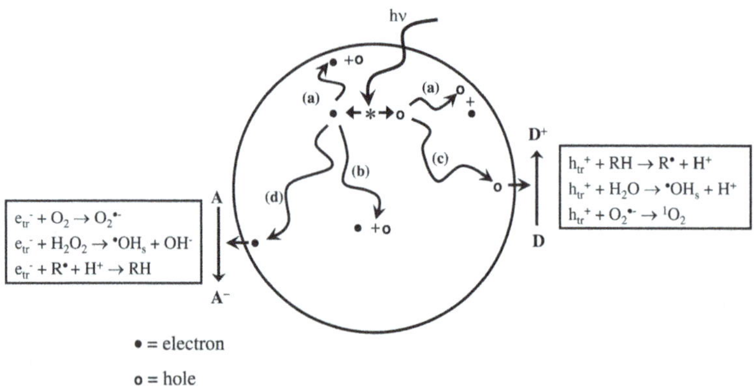

Figure 8.6 The processes occurring on a bare TiO_2 particle after UV excitation. Reprinted from Fujishima *et al.*, A., TiO_2 photocatalysis and related surface phenomena, *Surf. Sci. Rep.*, **63**, 515–582. Copyright 2008 with permission from Elsevier.[36]

$$RCH_2COOH + h^+ (\cdot OH) \rightarrow RCH_2\cdot + CO_2 + H^+ (H_2O)$$

$$RCH_2\cdot + O_2 \rightarrow RCH_2OO\cdot$$

$$RCH_2OO\cdot + \cdot OOH \rightarrow RCH_2OOOOH$$

$$RCH_2OOOOH \rightarrow RCHO + O_2 + H_2O$$

Scheme 8.1 Processes proposed for decomposition of organic compounds, *e.g.*, alcohols, on the surface of TiO_2.[37]

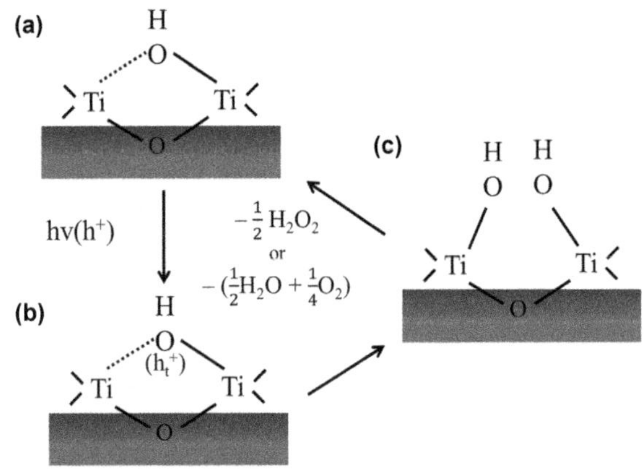

Figure 8.7 Surface structural models for the TiO_2 surface. (a) (Before UV irradiation) the OH group is bound to oxygen vacancy, (b) (at the transition state) the photogenerated hole is trapped at the lattice oxygen, and (c) (after UV irradiation) new OH groups are formed.
Reprinted from N. Sakai, A. Fujishima, T. Watanabe and K. Hashimoto, Quantitative evaluation of the photoinduced hydrophilic conversion properties of TiO_2 thin film surfaces by the reciprocal of contact angle, *J. Phys. Chem. B.*, 2003, **107**, 1028–1035. Copyright (2003) American Chemical Society. 40.

the original WCA in the dark. Fujishima's group[35,36,40,41] proposed a mechanism for a photoinduced wettability conversion of the TiO_2 surface, which has been widely accepted by scientists. The mechanism was proposed on the basis of the reconstruction of surface hydroxyl groups under UV irradiation in air. The density of surface hydroxyl groups is strongly correlated with the reciprocal of WCA. As illustrated in Figures 8.6 and 8.7, molecular oxygen captures the photoexcited electrons, while the holes diffuse to the TiO_2 surface, being trapped at lattice oxygen atoms. Afterwards, the hole weakens the binding energy between the Ti atom and the lattice oxygen, and the bond is interrupted by one adsorbed water molecule, forming new hydroxyl groups. In the dark, the hydroxyl groups gradually desorb from the surface in the form of H_2O_2 or $H_2O + O_2$. Compared to the OH groups bound to oxygen vacancies, the OH groups produced by UV irradiation are thermodynamically less stable because of the different desorption temperature.[40] As a result, the reconstruction between the stable OH groups and the thermodynamically metastable OH groups changes the surface energy of the TiO_2 and switches the surface between superhydrophilicity and hydrophobicity.

8.3 Fabrication Strategies and Methods

For nearly two centuries, researchers have been seeking novel methods to increase light transmission in optical systems as well as to eliminate unwanted reflection. Meanwhile, self-cleaning surfaces in nature have increasingly attracted the attention of scientists due to their unique properties. The technology of fabricating antireflective surfaces with highly efficient and durable self-cleaning properties has an extensive range of applications from glasses, displays, and solar panels to window optical lenses. In the following section are discussed different chemical and physical methods for fabricating ARCs and how to form rough surfaces with special surface energy.

8.3.1 Fabrication of Antireflective Surfaces

ARCs are a basic way to increase light transmission. Varied methods have been used to generate ARCs, which can be classified into both bottom-up and top-down approaches. Accurate control over the morphology of ARCs is crucial in order to achieve ideal antireflective surfaces. Techniques combining both bottom-up and top-down approaches may pave an effective way toward making the dream of application in industry come true.

8.3.1.1 Bottom-up Approach

(1) Layer-by-layer assembly

In 1966, Iler[42] demonstrated a novel method for creating multilayers of inorganic colloidal particles, which is now known as layer-by-layer (LbL) assembly. It was one of the first studies to disclose the successful assembly of nanoparticle thin films. Unfortunately, Iler's work did not attract much attention at the time. After twenty five years, Decher and co-workers,[43] using LbL processing, fabricated multilayer thin films from oppositely charged polyelectrolytes. Their work has led to a great many studies about films for various potential applications based on LbL methods. Now LbL has been proven a simple and effective way to fabricate various coatings with tailored chemical compositions and structures. Since then, much more attention has been paid to the LbL assembly technique, whose building blocks could be linear polyelectrolytes,[44] nanoparticles[45] or nanowires with low RI,[46] co-polymer micelles,[47] proteins,[48] and so forth. LbL assembly can be driven by multiple weak interactions, including electrostatic interactions, hydrogen bonds, halogen bonds, coordination bonds, charge–transfer interactions, biospecific interactions, guest–host interactions, cation–dipole interactions, and the combined interaction of the above forces, *etc.*[49] An outline of the LbL assembly technique conducted *via* electrostatic interaction is illustrated in Figure 8.8.[50]

The LbL assembly technique has several advantages in the preparation of ARCs. First, as mentioned above, varied materials (building blocks) can be used to fabricate ARCs and the LbL processing is substrate-independent. Second, it is possible to deposit ARCs of large areas on non-flat surfaces by the LbL assembly technique. Third, it is easy for the LbL assembly technique to control the film thickness, and

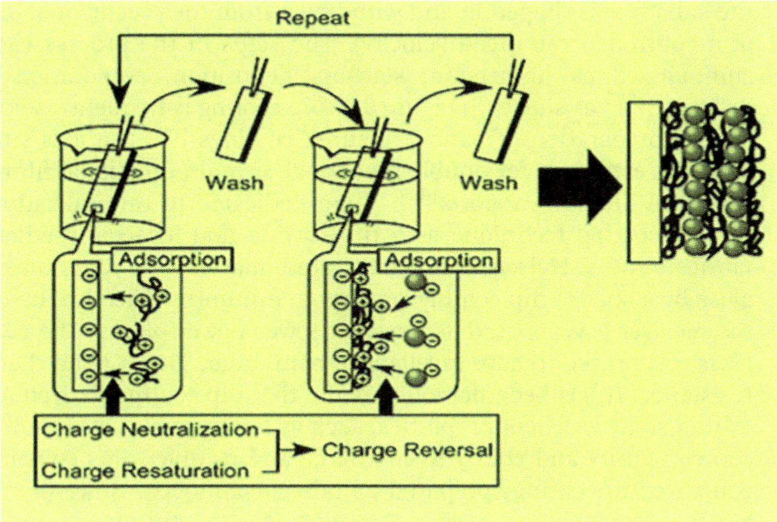

Figure 8.8 Outline of LbL assembly through electrostatic interaction. Reproduced from ref. 50 with permission from the PCCP Owner Societies.

therefore, to tailor the wavelength of maximum transmittance simply by changing the number of film deposition cycles.[51,52,53]

The groundbreaking work of applying the LbL assembly technique to the fabrication of ARCs was carried out by Rubner and co-workers.[44] They described an aqueous-based LbL process controlled at the molecular level that may coat surfaces of almost any material. High-efficiency antireflection coatings were created from phase-separated polyelectrolyte multilayer films that underwent a reversible pH-induced swelling transition. After that, many LbL-assembled high-performance ARCs have been fabricated. He's group has done a large amount of work to explore simple, economical but efficient ways for the preparation of ARCs by the LbL technique. They designed and synthesized silica nanoparticles of varied structures, such as solid silica nanoparticles,[54] monodisperse raspberry-like PS@SiO_2 composite nanoparticles,[55] nonspherical mesoporous silica nanoparticles,[56] hollow mesoporous silica nanospheres,[45] hierarchically mesoporous silica nanoparticles[57] and so forth, and used these multifarious silica nanoparticles as building blocks for LbL assembly. Coatings with ideal antireflectivity and special wettability were achieved by tailoring their porosity, thickness and surface roughness based on the Fresnel equation, the quarter-wavelength design and the requirements for special wettability.

(2) Dip-coating

Dip-coating has advantages such as simplicity, controllability, reliability and reproducibility. The process is generally described as that the substrate is dipped in and withdrawn from the precursor solution at a controlled rate (feed velocity). The steps in the process can be summarized as immersion, start-up, deposition, evaporation and drainage,[58,59] as shown in Figure 8.9. Dip-coating is frequently selected for the preparation of nano-structured coatings because it is simple and fast, especially for double-functional or multi-functional films.

Several literature reports[14,60–63] are dedicated to the utilization of the dip-coating technique as a time-saving and low-cost method to fabricate ARCs. Helsch *et al.*[63] fabricated quarter-wave ARCs on silica glass by a sol–gel dip coating process to minimize reflective losses of the receiver cover materials for solar tower power plants. The coated glass was proved to have excellent transmittance (98.6%) and thermal resistance. It has been demonstrated[62] that dip-coating-derived ARCs exhibit some excellent properties, such as homogeneity, stoichiometry control, purity and ability to coat large and complex area substrates, compared to coatings prepared by other techniques. However, there are also some inadequacies. For instance, the thickness and uniformity of fabricated coatings are sensitive to substrates and flow conditions such as a liquid bath and gas over-head.[59] Meanwhile, it is still a challenge for scientists to construct roughness-controlled films by dip-coating due to the flowing of sol–gel liquid during the preparation process.

(3) Spin-coating

Spin-coating is widely used in the microelectronics industry, where it is used to deposit uniform coatings of organic materials or to uniformly distribute particulate matter on a flat surface.[64] In short, in the procedure, an excess amount of precursor solution is placed on the substrate, which is then rotated at high speed in order to spread

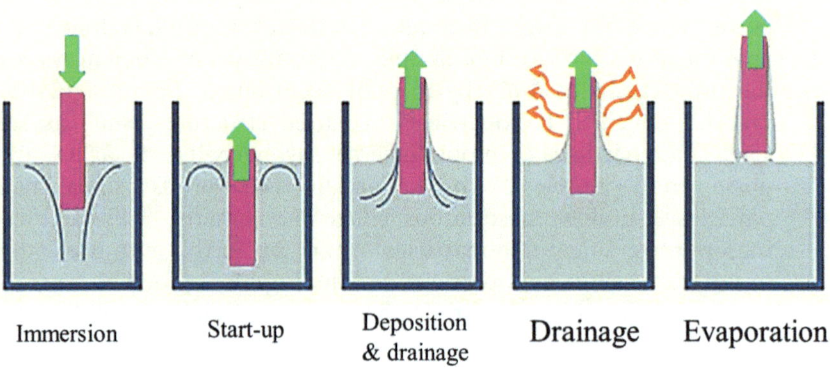

Figure 8.9 Steps involved in dip-coating.[228]
Reprinted from L. Yao and J. He, Recent progress in antireflection and self-cleaning technology – From surface engineering to fiunctional surfaces, *Prog. Mater. Sci.*, **61**, 94–143. Copyright 2014 with permission from Elsevier.

the fluid by centrifugal force. The film continues to thin slowly until disjoining pressure effects cause the film to reach an equilibrium thickness or until it turns solid-like due to a dramatic rise in viscosity resulting from solvent evaporation. The thin film finally forms due to solvent evaporation.[66] The key process and typical operation are illustrated in Figure 8.10.[65]

(4) Spray-coating

Spray-coating is quite a facile method, which is not specific to a particular substrate and can be easily applied to a large surface area.[67] As spray-coating does not typically require other complicated and costly processes, it is commercially feasible for a wide range of applications. The general process of spray-coating is shown in Figure 8.11(a). This method is simple, fast, material efficient, easy to use, carried out in a non-vacuum system and permits films with properties required for large area applications to be obtained.[68] Among the diversified spraying techniques, thermal spraying has become an excellent solution to the processing limitations of polymer/ceramic nanocomposites. Coating materials available for thermal spraying include metals, ceramics, alloys, plastics and composites. Usually, they are fed in powder or wire form, heated to a molten or semimolten state and accelerated towards substrates in the form of micrometer/nanometer scale particles. Thermal spray generally is a process in which a material is heated, accelerated, and propelled by a high-temperature jet through a confining nozzle toward a surface.[69,70] Coating quality is determined by its porosity, oxide content, hardness, bond strength and surface roughness. He's group[71] used a spray-coating method to construct coatings on slide glasses with synthesized silica nanoparticles 20 nm in size. When the number of spray-coating cycles was three, the transmittance of coating was greatly improved in the visible range. Recently, Zhang's group[72] developed a simple technique to fabricate superhydrophobic coatings by spraying microparticle/polymer suspensions to produce a robust polymer composite coating. Both the surface microstructures and corresponding superhydrophobicity depend on calcium stearate formed on the coating surfaces. The method for preparing a regenerative superhydrophobic coating has the advantage of simplicity in fabrication, easy availability of the materials, and applicability to a large area surface. Besides, multilayer coatings with multiple components can be fabricated by a spray-coating method as shown in Figure 8.11(b). By continuous spraying of polyanion and polycation solutions onto a charged surface, a uniform film that grows consecutively with time can be obtained.[73]

(5) Chemical vapor deposition (CVD)

CVD is a widely-used technology for the preparation of inorganic thin films. It can produce single layer, multilayer, composite, nanostructured, and functionally graded coating materials with well controlled dimension and unique structure at relatively low processing

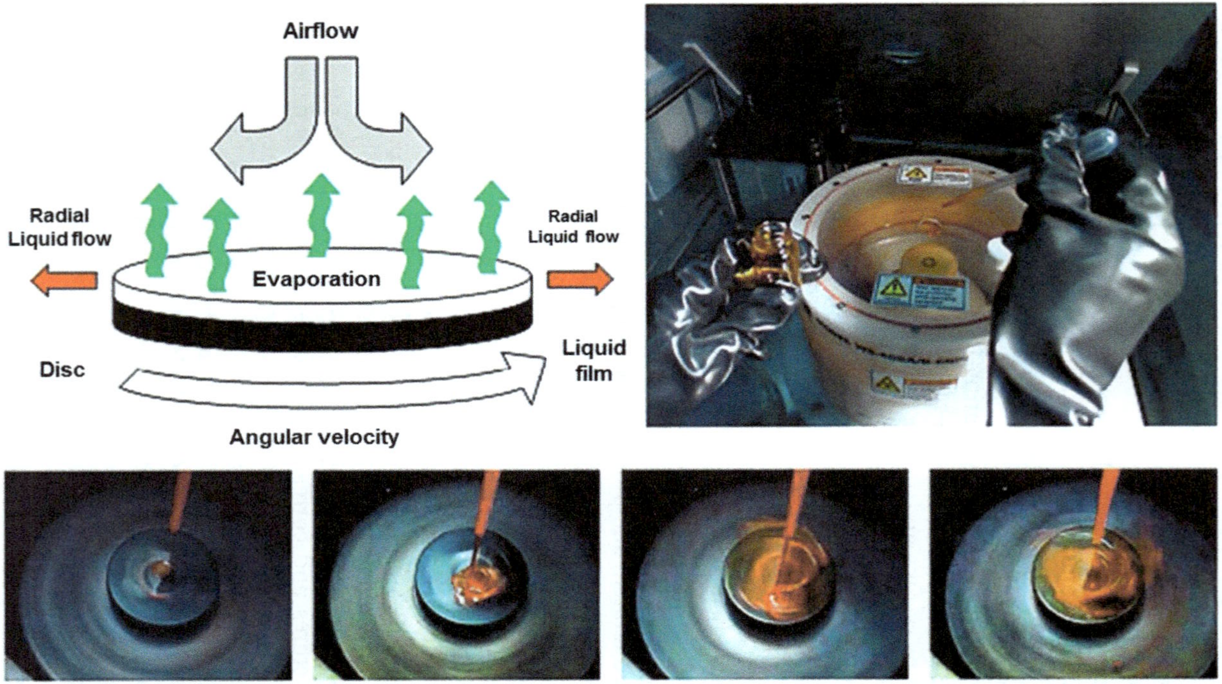

Figure 8.10 Schematic illustration of spin-coating (top left) along with a photograph of a typical spin-coating operation in a glovebox environment (top right) and high-speed images showing application of solution to a rotating substrate and film formation. The images were recorded at 300 images s^{-1} (below). The timing of the images (from left to right) after impact of the first drop is: $t = 17$, 100, 137 and 180 ms.
Reprinted from F. C. Krebs, Fabrication and processing of polymer solar cells: A review of printing and coating techniques, *Sol. Energy Mater. Sol. Cells*, **93**, 394–412. Copyright 2009 with permission from Elsevier.[65]

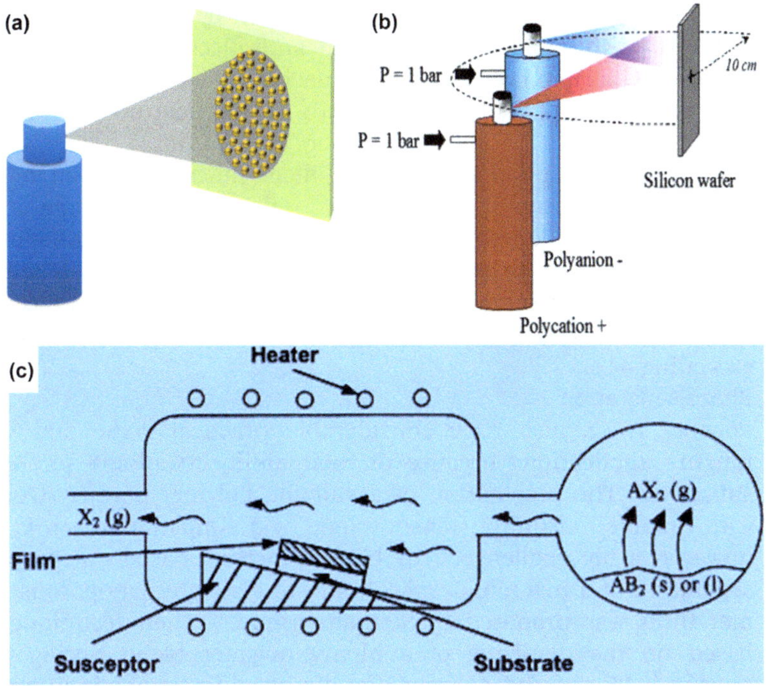

Figure 8.11 (a) Schematic diagram of the spray process, (b) continuous simultaneous film formation of a multilayer coating by spraying. Reprinted with permission from C. H. Porcel, A. Izquierdo, V. Ball, G. Decher, J. C. Voegel and P. Schaaf, Ultrathin coatings and (poly(glutamic acid)/polyallylamine) films deposited by continuous and simultaneous spraying. *Langmuir*, 2005, **21**, 800–802.[73] Copyright (2005) American Chemical Society. (c) Schematic diagram of chemical vapor deposition. Reprinted from K. L. Choy, Chemical Vapour deposition of coatings, *Prog. Mater. Sci.*, **48**, 57–170. Copyright 2003, with permission from Elsevier.[74]

temperatures.[74] CVD essentially involves the process of dissociation of molecules of gaseous reactants, which subsequently react chemically to form various structures when activated by heat, light, or plasma discharge. Mostly solid-state, stable products are formed as a result of the chemical reactions.[75] Basically, the process of CVD at a surface to generate a thin and solid film can be described as Figure 8.11(c).[74] In addition to the conventional CVD method, *i.e.*, so-called thermal activated CVD (TACVD),[76–78] many variants of the CVD method are used to fabricate ARCs, such as plasma enhanced CVD (PECVD),[79] photo-assisted CVD (PACVD),[80] hot-wire CVD (HWCVD)[81,82] and so on.

PECVD has been one of the most industrially preferred technologies for ARCs. It is a glow discharge CVD, which uses electron energy (plasma) as the activation method to enable deposition to occur at a low temperature and at a reasonable rate. By this technique, the precursor molecules can be decomposed by pulsed microwave-induced plasma and finally form an antireflective layer (TiO_2/SiO_2) and/or anti-scratch layer (SiO_2).[83] Remache and co-workers[84] designed porous silicon/PECVD SiO_x ARCs with nice optical properties. They first prepared mesoporous silicon layers on the front surface of the n^+ emitter of a $n^+ - p$ mono-crystalline silicon junction. The single layer ARCs achieved around 9% of effective reflectivity on junction $n^+ - p$ solar cells at wavelengths of 400 nm to 1000 nm. Then they deposited silicon oxide layers by PECVD on mesoporous silicon layers, leading to an effective reflectivity reduction to 3.8%. The multilayer coating was predicted to give an enhancement of the photogenerated current exceeding 44%.

(6) Phase-separation
Polymer ARCs also have considerable practical value and wide-ranging applications because of their ability to adhere to flexible substrates. The preparation of a suitable polymer porous structure with tunable maximum transmittance by a simple approach is still, however, a big challenge.[85] In 1999, a method based on the phase separation of a macromolecular liquid to generate nanoporous polymer films was proposed by Steiner's group.[86] Their technique was based on the demixing of a binary polymer blend during spin-coating.[87] PS and PMMA were first dissolved in tetrahydrofuran and spin-cast onto both sides of a microscope slide. During spin-coating, phase separation set in, and after evaporation of the solvent, lateral phase morphology was obtained. Then the glass slide was exposed to cyclohexane, which selectively dissolves PS. Finally, porous films with high optical transmission were obtained. A major advantage of the approach is its versatility. By using appropriate combinations of polymers and solvents, various monolayer and multilayers could be created, such as a copolymer PS-*b*-poly (dimethylsiloxane) (PS-*b*-PDMS) surface,[88] porous PS/poly (vinylpyrrolidone) (PS/PVP) films and so on.[15]

8.3.1.2 Top-down Fabrication

(1) Etching
The etching technique can be described as a selective dissolution or ablation of a surface or substrate. Basically, etching can be divided into two categories: wet etching and dry etching.

Wet etching can be described as a process using a chemical solution for dissolution with advantages such as precise selective attack and fast etching rate. Usually, the solution is water-based because water is corrosive to glass to some extent. Though the dissolution mechanism remains unclear, there are some reactions that are widely accepted to occur during dissolution: (1) hydration and hydrolysis; (2) ion-exchange; (3) network reconstruction. In most glass structures, the voids between oxygen and hydration are not

$$\text{Si–O–Si} + H_2O \leftrightarrow \text{Si–OH} + \text{OH–Si}$$

$$\text{Si–O–Si(OH)}_3 + OH^- \rightarrow \text{Si–O–Si(OH)}_4^- \rightarrow \text{Si–O}^- + \text{Si(OH)}_4$$

$$\text{Si–OR} + H_3O^+ \rightarrow \text{Si–OH} + R^+ + H_2O$$

$$\text{Si–OR} + H_2O \rightarrow \text{Si–OH} + R^+ + OH^-$$

Scheme 8.2 Chemical reactions in the etching mechanism.[90]

large enough to allow water molecules to penetrate into the surface, so the hydrolysis reaction may be accompanied by a network dissolution by releasing water soluble species such as $Si(OH)_4$ into solution and leaving large voids for further reaction, as shown in Scheme 8.2. By dehydration, the silanol groups (Si–OH) can condense into a Si–O–Si network, as shown in the reverse reaction of Scheme 8.2, which leads to a reconstruction of the glass surface, and form a porous material.[89,90]

He's group[91] reported a method that uses water to create a hierarchical superhydrophobic surface. However, the process temperature is relatively high (>160 °C). Recently, Qi *et al.*[92] demonstrated a simple method for creating hierarchical pyramidal structures on a whole silicon wafer using alkali etching at a low temperature (65 °C). The fabrication procedure for creating hierarchical structures is shown in Figure 8.12(a). There were three main steps: (a) fabrication of pyramidal structures on a silicon surface with KOH etching; (b) electroless deposition of a thin discontinuous layer of Ag nanoparticles on the pyramidal structures in HF/$AgNO_3$ solution; (c) generation of hierarchical structures by Ag-assisted etching and removal of Ag nanoparticles in nitric acid. Their research provided a cost-effective and facile approach for producing superhydrophobic and antireflective silicon surfaces with a hierarchical structure (Figure 8.12(b) and (c)).

Dry etching is currently used in the semiconductor industry due to its unique ability over wet etching to carry out anisotropic and directional etching or to create high aspect ratio structures (*e.g.* deep *via* holes or capacitor trenches). Typically, bombardment of ions (plasma of reactive gas such as fluorocarbons, oxygen, chlorine, boron trichloride) is used to dislodge portions of the material from the exposed surface with masked patterns. Other benefits of dry etching are reduced chemical hazard treatment and contamination, and ease of process automation and tool clustering.[94]

The versatility of plasma processing is demonstrated for the preparation of nanotextured surfaces with superior wettability properties. Leem[95] reported the fabrication of broadband and wide-angle antireflective subwavelength structures on a silicon substrate by inductively coupled plasma etching using Au nanopatterns as etching masks. Palumbo *et al.*[93] reported an effective plasma processing for preparing both superhydrophobic and superhydrophilic surfaces from

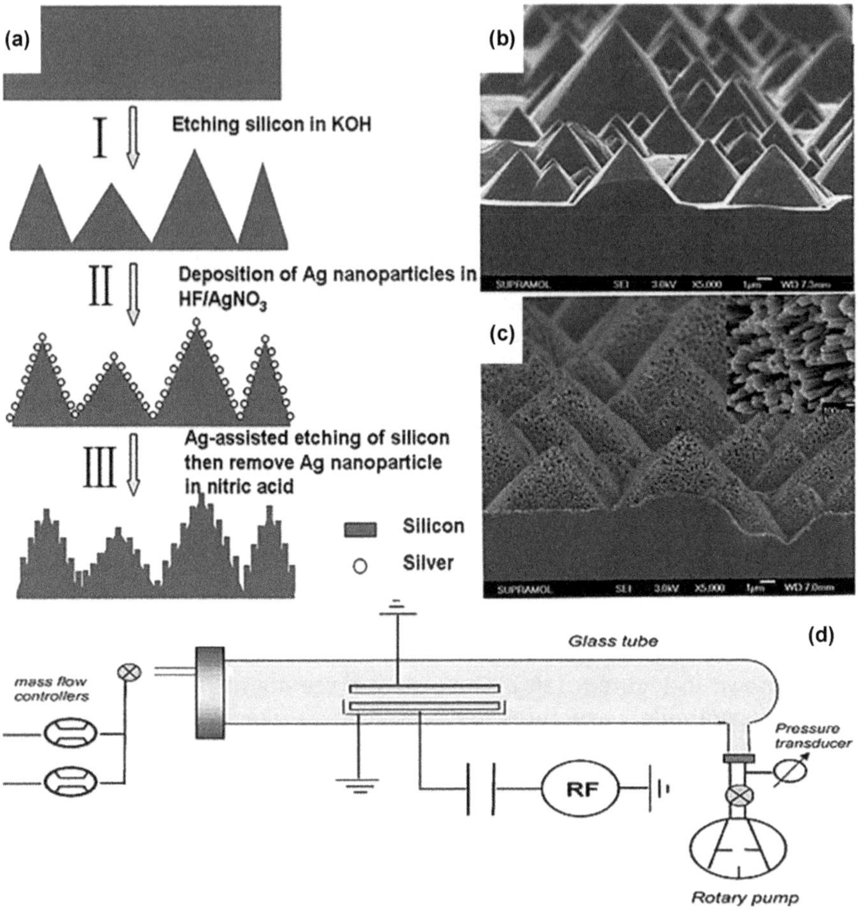

Figure 8.12 (a) Fabrication procedure for creating hierarchical structures on a silicon surface. (b) Cross-sectional SEM photograph of silicon pyramids created with KOH etching. (c) Cross-sectional SEM photograph of hierarchical structures generated with Ag-assisted etching. Inset: magnified SEM image. Reprinted with permission from D. Qi, N. Lu, H. Xu, B. Yang, C. Huang, M. Xu, *et al.*, Simple approach to wafer-scale self-cleaning antireflective silicon surfaces, *Langmuir*, 2009, **25**, 7769–7772.[92] Copyright (2009) American Chemical Society. (d) Scheme of the reactor for plasma texturing. Reprinted from ref. 93 with permission from John Wiley and Sons. Copyright © 2011 WILEY-VCH Verlag GmbH & Co. KGaA, Weinheim.

polycarbonate (PC). The polymer nanostructure size can be controlled by tuning the experimental plasma parameters during oxygen etching. The scheme of the reactor for plasma texturing is shown in Figure 8.13(d).

(2) Lithography

The semiconductor industry has been pushing high-precision nanoscale lithography to a rapid development. The previous fabrication

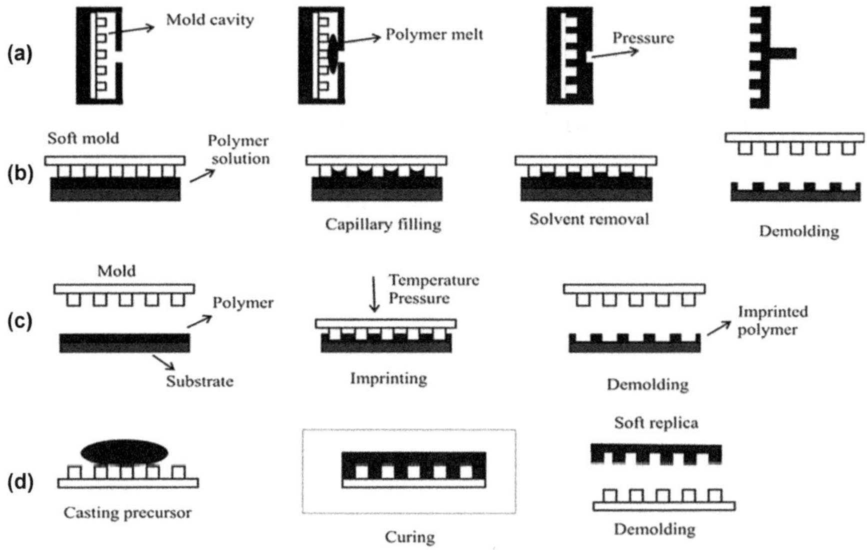

Figure 8.13 Different alternative lithographic processes. (a) Injection molding, (b) solvent-assisted molding, (c) hot embossing, and (d) soft lithography.[102,103]
Reproduced with permission from A. del Campo and E. Arzt, Fabrication approaches for generating complex micro-and nanopatterns on polymeric surfaces, *Chem. Rev.*, 2008, **108**, 911–945. Copyright 2008 American Chemical Society.

strategy relied on costly e-beam lithography[96] and photolithography.[97,98] Recently, unconventional nanofabrication techniques such as colloidal lithography,[99,100] nanoimprint lithography,[101] soft lithography[7] based on molding, embossing and printing were explored. The fundamentals of some lithography processes are shown in Figure 8.13.[102,103] These techniques have emerged due to high throughput and cost-effective process capabilities, and have been applied to the fabrication of ARCs, microchips and micro-electromechanical devices. These techniques always resort to expensive tools such as those used in deep-UV projection lithography and electron-beam lithography.

Currently, photolithography has been one of the most widely used top-dpwn techniques in the fabrication of antireflective and self-cleaning coatings[100,104,105] due to its several advantages, such as full wafer processing and short exposure time. However, drawbacks still exist such as the limited resolution and the limited possibilities of working with pre-existing topography or curved substrates, which still need further study.

8.3.2 Fabrication of Self-cleaning Surfaces

Techniques to produce self-cleaning surfaces can be generally classified into two categories: (a) making a rough surface; (b) modifying a rough surface with a material of low surface energy (for superhydrophobic surfaces) or high surface energy (for superhydrophilic surfaces).[106]

8.3.2.1 Fabrication of Superhydrophilic Surfaces

Superhydrophilic surfaces can induce "complete spreading" of water, which would have a broad range of applications, including self-cleaning, antifogging, water harvesting and in the biomedical field.[107,108] On one hand, the roughness of a surface plays a key role in controlling its wettability. On the other hand, superhydrophilicity can be achieved through photo-induced surface reactions by some semiconductors such as TiO_2, ZnO, WO_3 and V_2O_5.[109–112]

Nano-scale porous structures and some microstructures are known to exhibit superhydrophilicity. However, film porosity also plays an important role in obtaining superhydrophilicity. Wetting can be controlled by changing porosity and/or roughness. There are three types of rough surface: regular (designed) surface, irregular (random) surface, and hierarchically rough surface.[113] Films of nanoparticles are often deposited on substrates by LbL,[54,114] spin-coating,[115] sputtering,[116] hydrothermal synthesis[117] and so on. Microstructures can grow from polymerization,[118] lithographic[105] and electrochemical techniques.[119] Photoinduced surface reactions are important for achieving super-wetting properties, and a large number of methods were used to prepare superhydrophilic coatings with light-sensitive materials, such as drop casting,[111,120] peroxotitanate-complex deposition and liquid phase deposition,[121,122] physical vapor deposition,[123] atomic layer deposition[124–126] and so on. To avoid duplication, we just describe a few novel preparation methods in this section.

(1) Liquid phase deposition

Liquid phase deposition (LPD) is a process based on controlled hydrolysis and has been developed as a low temperature route to obtain several oxide films, including TiO_2.[127] LPD processes are environmentally friendly because they use low-cost and low energy-consuming equipment and water as a solvent.[128] Deposition rate and crystal orientation can be controlled by an appropriate choice of experimental parameters such as pH, temperature, surface characteristics and so on. Organic-polymer-core/TiO_2-shell composite particles and TiO_2 hollow spheres could be fabricated at a low deposition temperature.[129,130] Since the TiO_2 is successfully patterned at the micro and nano scale, it is very attractive in superhydrophilic applications.

In recent years, some research groups have successfully prepared porous TiO_2 films with a rough surface structure by combining the LPD method with a LbL self-assembly process,[114,131] and also extended it to films of other metal oxides such as FeOOH.[132] The combination of different technologies enhances its appeal for industrial applications due to its scalability and effectiveness in coating surfaces with complex morphologies.

(2) Physical vapor deposition (PVD)
Deposition of coatings by PVD techniques, in which a material is vaporized and then condensed onto a substrate to form a thin coating, has been widely used in many industrial sectors such as integrated circuits. Common PVD methods include thermal deposition, ion plating, ion implantation, sputtering, laser vaporization, and laser surface alloying.[133] TiO_2 thin films with a thickness ranging from 300 to 500 nm were prepared by PVD in an electron evaporation system.[123] To control the porosity and the surface roughness, evaporation was carried out at different glancing angles with respect to the evaporation source. The top views of TiO_2 thin films prepared by PVD and the effect of UV irradiation on the wetting angle on the films are shown in Figure 8.14. The control of deposition conditions by PVD has given thin films where not only the crystallographic structure but also the microstructure and porosity are controlled. Illumination, surface roughness, surface contamination, type of crystal planes at the surface, *etc.* play varied roles in the UV-induced wetting behavior of the films. Superhydrophilic TiO_2 thin films can also be fabricated by frequency magnetron sputtering deposition.[134,135] The temporal evolution of the water contact angle was found to depend on the total gas pressure, which influences the crystalline structure of the films.

(3) Hydrothermal treatment
Hosono *et al.* reported a fabrication method of TiO_2 nanosheet films by heat treatment.[136] The resulting anatase TiO_2 nanosheet films showed superhydrophilic properties without the need for any UV irradiation treatment. An aqueous solution of urea and the titanium sheet were placed in a Teflon-lined autoclave, which was heated at 100 °C for 4 days. After the hydrothermal treatment, the sheet was then transformed to TiO_2 by heating at 400 or 600 °C for 30 min in air. Figure 8.15(a) and (b) shows the presence of perpendicular TiO_2 nanosheets on the titanium sheet, and the perpendicular nanosheet thickness is around 2.5 nm. The high concentration of defects or dangling bonds is a result of the specific morphology and is responsible for the superhydrophilicity without prior UV irradiation (Figure 8.15(c)). Hydrothermal treatment also applies to silica glasses. He's group[91] fabricated a superhydrophilic surface by simple hydrothermal treatment of slide glass at 180 °C for 24 h. After reaction, a predominant three-dimensionally interconnected porous structure (the upper left corner in Figure 8.15(d)) was observed together with minor network structures (the lower right corner in Figure 8.15(d)), and the coating thickness was about 4 μm (Figure 8.15(e)). The rough surface exhibited a WCA of *ca.* 0° in less than 40 ms.

Figure 8.14 (a)–(e) Top views of TiO_2 thin films prepared by PVD. (f) Dependence of wetting angle on time. In (f), the filled symbols refer to the decay in wetting angle during illumination and the empty symbols refer to the recovery of wetting angle when the illuminated thin films were kept in the dark. The inset in (f) shows an enlarged timescale evolution of wetting angle for the PVD thin films under illumination.[123]
Reprinted from V. Rico *et al.*, Wetting angles and photocatalytic activities of illuminated TiO_2 thin films, *Catal. Today*, **143**, 347–354. Copyright 2009, with permission from Elsevier.

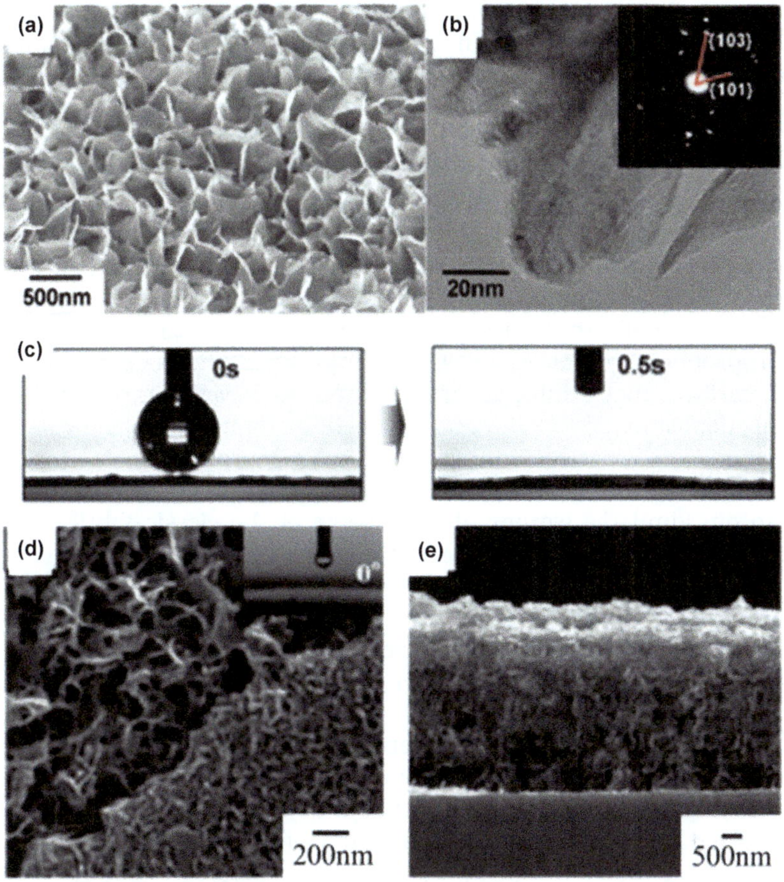

Figure 8.15 SEM (a) and TEM (b) images of a perpendicular TiO_2 nanosheet film fabricated *via* hydrothermal synthesis and heating in air at 400 °C for 30 min. (c) Measurement of the contact angle of water on the perpendicular TiO_2 nanosheet film. Reproduced with permission from E. Hosono, H. Matsuda, I. Honma, M. Ichihara and H. Zhou, Synthesis of a perpendicular TiO_2 nanosheet film with the superhydrophilic property without UV irradiation, *Langmuir*, 2007, **23**, 7447–7450.[136] Copyright (2007) American Chemical Society. (d) SEM image of coating prepared at 180 °C for 24 h. (e) Cross-sectional image of the coating in (d). Reproduced with permission from X. Liu and J. He. One-step hydrothermal creation of hierarchical microstructures toward superhydrophilic and superhydrophobic surfaces, *Langmuir*, 2009, **25**, 11822–11826.[91] Copyright (2007) American Chemical Society.

8.3.2.2 Fabrication of Superhydrophobic Surfaces

Superhydrophobic surfaces have a water contact angle greater than 150° and low contact angle hysteresis (lower than 5°). Because of their unique water-repellent property, these surfaces have attracted significant attention over the last two decades, which can essentially influence a lot of cutting-edge topics, including surface structures and fluid motivation, in engineering and biotech researches.[137–139] Learning from nature, we know that the low surface energy of –CF_3 groups, silyl compounds, or fluorocarbons is not the only condition to obtain superhydrophobicity.[140,141] Regardless of the material used and structure (particles, rod arrays, or pores) formed on the surface, suitable roughness in combination with low surface energy is required. Until now, procedures of roughening the surface followed by hydrophobization or transforming low-surface-energy materials into rough surfaces have been commonly used to produce superhydrophobic surfaces.[142] Since Tsujii *et al.*[143] first both theoretically and experimentally studied biomimetic superhydrophobic surfaces in 1996, various smart methods for attaining rough surfaces that exhibit superhydrophobicity have been reported, such as LbL deposition,[144,145] dip-coating,[146] lithography,[7,147] chemical vapor deposition,[148,149] electrospinning,[150–152] sol–gel processing,[153,154] plasma treatment,[155,156] one-pot reaction,[157,158] electrochemical deposition[159,160] and so on. Similarly, we only discuss some of the classic fabrications of superhydrophobic films in this section.

(1) Electrospinning

Electrospinning is a simple and versatile technique for continuous production of micro/nanofibers from polymers.[161,162] The electrospinning process is carried out by a high electric field between a syringe containing viscous polymer solution and a conducting substrate. To form uniform fibers, the molecular weight of the polymer and the concentration of the solution should be carefully considered and adjusted. By electrospinning a mixture of polymer and hydrophobic material or introducing secondary structures, uniform superhydrophobic fibers can be obtained.[161,163,164] Figure 8.16 shows a coaxial electrospinning operation, in which a coaxial nozzle consists of a central tube surrounded by a concentric annular tube.[165] Steckl's group produced core–sheath-structured nano/microfibers that combine different properties from individual core and sheath materials by coaxial electrospinning. The first Teflon electrospun fibers were reported using coaxial electrospinning with Teflon AF sheath and poly (ε-caprolactone) core materials. Superhydrophobic and oleophobic membranes were successfully produced by these core/sheath fibers. In addition, the combination of nanoparticles and electrospinning process is effective to increase surface area and porosity for superhydrophobic surfaces.

In 2009, the He group[151] reported the preparation of large area superhydrophobic titania membranes by electrospinning, followed by calcination and surface modification with 1*H*,1*H*,2*H*,2*H*-perfluorooctyltriethoxysilane (POTS). Varied surface structures were fabricated by combination of electrospinning and hydrothermal treatment. It is important to control precursor aging time (t_{aging}) and

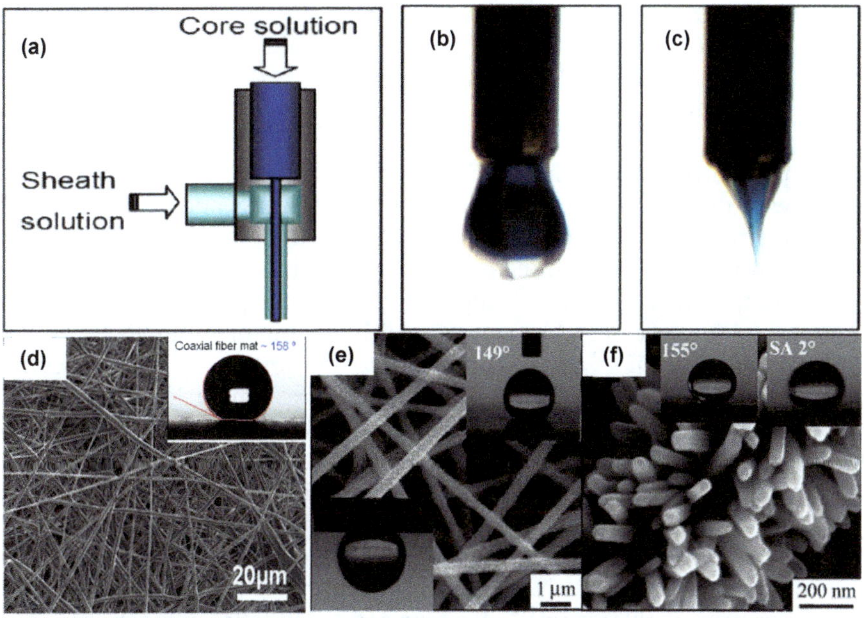

Figure 8.16 Coaxial electrospinning operation: (a) diagram of the coaxial nozzle; (b) core–sheath droplet without bias; and (c) Taylor cone and coaxial jet formation at 12.5 kV; (d) a coaxially electrospun PCL/Teflon AF membrane and contact angle measurement (2 μL water droplet). Reproduced with permission from D. Han and A. J. Steckl, Superhydrophobic and oleophobic fibers by coaxial electrospinning, *Langmuir*, 2009, **25**, 9454–9462.[165] Copyright (2009) American Chemical Society. (e) SEM image of fibers prepared using a precursor solution aged for a t_{aging} of 24 h and a $t_{electrospinning}$ of 90 min, insets are contact angles on the membrane; (f) SEM image of titania nanomembrane after hydrothermal treatment for 120 min. Insets are corresponding contact angle and sliding angle (t_{aging} = 24 h and membrane pore size = 0.786 μm). Reproduced with permission from H. Tang, H. Wang and J. He, Superhydrophobic titania membranes of different adhesive forces fabricated by electrospinning, *J. Phys. Chem. C*, 2009, **113**, 14220–14224.[151] Copyright (2009) American Chemical Society.

electrospinning time ($t_{electrospinning}$) well. Very interesting is that when the titania membrane was obtained by using a $t_{electrospinning}$ of 90 min followed by treatment with POTS, water droplets on the surface appeared spherical and adsorbed firmly onto the substrate even when the substrate was turned upside down (Figure 8.16(e)). However, if the titania membrane was subjected to hydrothermal treatment in a $Ti(OBu)_4$ acidic solution for 120 min, the water contact angle became 155° and the sliding angle remarkably changed to 2°. It truly meets the conditions of superhydrophobic self-cleaning. As shown in Figure 8.16(f), nanoparticles grew along the perpendicular direction of fibers and finally became rod-like crystals after 120 min of hydrothermal treatment. The formation of the hierarchical micronanostructure resulted in a thick air layer and kept the water droplet from contact with the fibers.

(2) Templating

The template-based method is an imprint-related way to prepare superhydrophobic surfaces.[166] Templating involves the use of a master with desired features, replication of the features and subsequent lifting off the replica or dissolution of the templates.[139] Many materials can be used as a template such as natural lotus leaves,[167,168] candle soot,[169] insect wings,[170] reptile skins,[170] master prepared by lithographic processes[171] or commercial inorganic membranes.[172]

Sheng *et al.*[173] have demonstrated fabrication of nanostructured superhydrophobic polymeric surfaces by a simple (one-step) reproductive method of anodic aluminium oxide (AAO) template extrusion. By tuning the diameter of the AAO template and the pressure to extrude, high-density polyethylene nanofiber surfaces with different nanometer roughness were prepared. The water contact angle test showed that the surface was superhydrophobic and sliding angles were very important for the dynamic wettability of the superhydrophobic surfaces. Park's group[174] also reported a synthetic route for ordering a set of Au nanoparticles on the vertically aligned conducting polymer for superhydrophobic surfaces with low water flow friction. By mimicking the lotus leaves, they combined two scale roughness composed of nanorods and nanoparticles, which is very favorable for the preparation of the superhydrophobic structure.

(3) Hydrophobization

After the surface coating has been roughened, most of them still do not exhibit superhydrophobicity. It is necessary to lower the surface energy of the roughened surfaces. The commonly used reactive molecules for low-surface-energy modification can be classified mainly into four categories: fluorinated molecules, alkyl molecules, nonfluorinated polymer and silicon/silane compounds.[142]

Due to the extremely low surface free energy and simple reaction of the silane group with the hydroxyl group, fluoroalkylsilanes are the most commonly used chemicals.[175,176] However, most fluoroalkylsilanes are expensive and potentially harmful to human health and the environment. Therefore, Wang's group used stearic acid to modify TiO_2 or SiO_2 particles on roughened cotton textiles to make them superhydrophobic.[177–179] Polymers with low surface energy can also be used for hydrophobization. Ramaratnam[180] developed a permanent nanocoating to create ultrahydrophobic fibers/textiles with excellent water repellency and self-cleaning ability, which contained 29 wt% styrene and 1.4 wt% reactive maleic anhydride groups on a PET fabric covered with epoxidized silica nanoparticles. The coating exhibited excellent water repellency and self-cleaning ability.

8.3.3 Progress in Antireflective Self-cleaning Coatings

In recent decades, surfaces with either an antireflective or self-cleaning property have been studied intensively. Scientists have made significant progress and a few excellent reviews have been published to give an integrated introduction.[4,23,181–185] Coatings with practical significance that combine both antireflective and self-cleaning properties still need, however, vigorous investigation and a consolidated summary. In this section, we introduce the latest progress in antireflective self-cleaning coatings.

Silicon and glass are the most important materials for solar cells and other optical devices. Unfortunately, due to their high refractive indices, more than 30% of incident light is reflected back from the surface of crystalline silicon and nearly 10% of light is lost from the air/glass interface.[186] In addition to effective improvement of the transmittance and wettability of substrates, low cost and simplification of the fabrication method are also very important for industrial applications. He's group did lots of work on antireflective self-cleaning coatings.[19,45,56,187–192] The building blocks used for the LbL assembly are from solid SiO_2 nanoparticles of different sizes (Figure 8.17(a)),[19] nonspherical mesoporous silica nanoparticles (Figure 8.17(b)),[56] hierarchically mesoporous silica nanoparticles (Figure 8.17(c))[187] to hollow mesoporous silica nanospheres (Figure 8.17(d)).[45]

Recently, Jiang's group developed several simple colloidal templating approaches for fabricating ARCs on both glass and silicon substrates.[193–196] All these approaches are based on a simple and scalable spin-coating technique to produce colloidal crystals with nonclose-packed structures. In 2008, they developed a simple and scalable templating technique for making ARCs on silicon and glass substrates, and investigated the antireflective and nonwetting properties of templated pillar arrays with high aspect ratio (up to *ca.* 10).[197] The outline of the templating procedure for patterning subwavelength pillar arrays on a silicon substrate is shown in Figure 8.18(a).

They first used a spin-coating technique to generate nonclose-packed colloidal layers on silicon wafers. After coating, the wafer and glass exhibited much lower reflection, which were below 2.5% and 0.5%, respectively. Figure 8.18(b) and (c) show water drop profiles on fluorosilane-modified silicon and glass pillar arrays, respectively. Both coatings were superhydrophobic and the WCA was *ca.* 172° for the former and *ca.* 160° for the latter. They believe that this technique is promising for developing self-cleaning ARCs for a large variety of technological applications and for industrial manufacture.

TiO_2-based coatings can be easily applied to transparent substrates such as plastics to provide a self-cleaning function. However, the coatings always increase the surface reflection of the transparent substrate because of the large RI of TiO_2, and the reflection of TiO_2 coated substrates can be as high as 20%.[199] Fujishima *et al.*[21] fabricated PET films with a moth-eye-like surface, which were coated with TiO_2 particles. The films exhibited a high

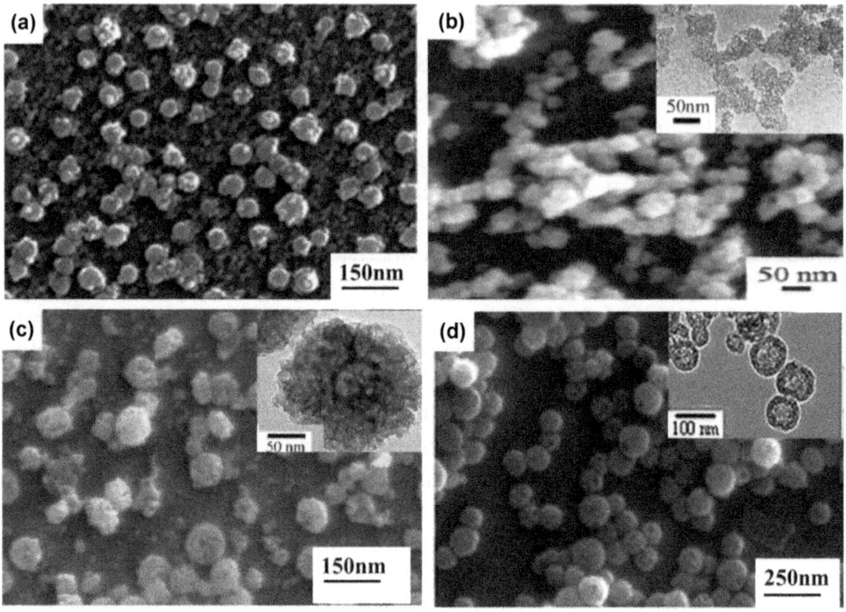

Figure 8.17 SEM images of (a) a LbL deposited $(PDDA/S\text{-}150)_1/(PDDA/S\text{-}30)_1$ coating on slide glass. Reproduced with permission from X. Liu and J. He, Superhydrophilic and antireflective properties of silica nanoparticle coatings fabricated via layer-by-layer assembly and postcalcination, *J. Phys. Chem. C*, 2009, **113**, 148–152, Copyright (2009) American Chemical Society.[19] (b) Coating of nonspherical mesoporous silica nanoparticles as prepared by deposition of one cycle of PDDA and nonspherical mesoporous silica nanoparticles (PDDA/MPSNP). Reproduced with permission from X. Li, X. Du and J. He J, Self-cleaning antireflective coatings assembled from peculiar mesoporous silica nanoparticles, *Langmuir*, 2010, **26**, 13528–12534, Copyright (2009) American Chemical Society.[56] (c) Coating of hierarchically mesoporous silica nanoparticles on slide glass by alternate LbL deposition of PDDA/MSNs of one cycle. Reproduced with permission from X. Du, X. Li X and J. He, Facile fabrication of hierarchically structured silica coatings from hierarchically mesoporous silica nanoparticles and their excellent superhydrophilicity and superhydrophobicity, *ACS Appl. Mater. Interfaces*, 2010, **2**, 2365–2372. Copyright (2009) American Chemical Society.[187] (d) Particulate coating of hollow mesoporous silica nanospheres prepared by deposition of one cycle of PDDA/HMSN followed by calcination. Reprinted from ref 45 with permission from John Wiley and Sons. Copyright © 2011 WILEY-VCH Verlag GmbH & Co. KGaA, Weinheim.

transmittance of 76–95% in the range of 400–800 nm and good superhydrophilicity after irradiation with UV light. They believed that such coated films will be important for improving the performance of solar cells and flat panel displays. Lately, SiO_2&TiO_2 double-shell hollow nanospheres (DSHNs) were designed and synthesized in He's group (Figure 8.18(d)).[198] They used the nanoparticles as building block for fabricating multifuncitonal AR thin

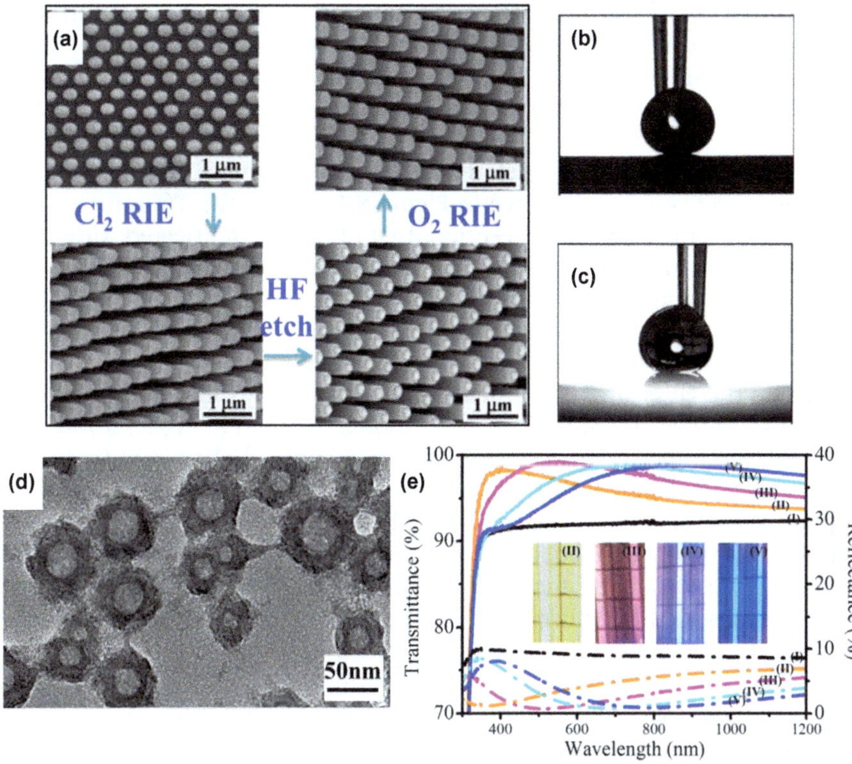

Figure 8.18 (a) Outline of the templating procedure for fabricating antireflective silicon pillar arrays by using a nonclose-packed colloidal monolayer as a template. Superhydrophobic coatings achieved on both templated silicon (b) and sol–gel glass pillar arrays (c) Reprinted from ref. 197 with permission from John Wiley and Sons. Copyright © 2008 WILEY-VCH Verlag GmbH & Co. KGaA, Weinheim. (d) TEM images of SiO_2&TiO_2 DSHNs, (e) UV-Vis-Near IR transmission spectra of (I) blank K-9 glass and SiO_2&TiO_2 DSHN thin film-coated K-9 glasses, which were fabricated by varied withdrawing speeds of (II) 50 mm min^{-1}, (III) 100 mm min^{-1}, (IV) 150 mm min^{-1} and (V) 200 mm min^{-1}, respectively. Reprinted from ref. 198 with permission from the Royal Society of Chemistry.

films. Substrates with DSHN thin films attained transmittance as high as 99.4% and average transmittance up to 98.5% in the visible region (Figure 8.18(e)). The nano-composite SiO_2–TiO_2 films exhibited intrinsic superhydrophilicity, anti-fogging, high photocatalytic activity, favorable robustness and function durability, which make them extremely attractive for applications in lenses, photovoltaic cells and windows of high-rise buildings.

Recently, antireflective and self-cleaning coatings on conductive substrates (such as indium-tin-oxide (ITO) glasses) have gradually attracted the attention of scientists. White organic light-emitting devices (OLEDs), which have many desirable characteristics such as being ultrathin, light weight, and having good flexibility, have been considered as a new generation of solid-state lighting sources. For the conventional bottom-emitting OLEDs fabricated on a flat glass substrate, however, only a small fraction (<20%) of the light generated in the device can escape and 80% is trapped in the glass substrate (glass mode) and the high index organic layer (waveguide modes). Wu *et al.*[200] came up with two reference bottom-emitting OLEDs without and with antireflective film, as shown in Figure 8.19(a) and (b). To improve light out coupling efficiency of conventional OLEDs, they fabricated an antireflective nanopillar array film onto the opposite side of the ITO-coated glass substrate by low-cost imprint lithography. Owing to the nanopillar structures and the inherent low-surface energy of PDMS, the film exhibited a high WCA (151°), as shown in Figure 8.19(d). After placing it in dirty water, a lot of dust resided on the glass surface, while the nanopillar array film remained clean. The transmittance and SEM characterization are shown in Figure 8.19(c), (d) and (f). The simple, cost-effective, and reproducible method may find great applications in illuminations and displays.

As we know, enhanced visible light is beneficial to improving the photoelectric conversion efficiency of solar cells. Similarly, antireflection near-infrared optics also has many important applications such as for night vision, sensors, dichroic mirrors, medical detection, body scans, and hot-spot detection.[201,202] Camargo *et al.*[203] prepared a two-layer film composed of multi-scale silica nano-particles and a poly-tetrafluoroethylene (PTFE)-like layer. An additional silica aerogel layer, combined with silica nanoparticle multi-scale clusters and PTFE-like layer, improves light transmittance further to 99.5% between 1600–2100 nm and a WCA of $158 \pm 2°$.

Recently, Park *et al.*[204] reported the fabrication of a multifunctional surfaces composed of tapered conical nanotextures by interference lithography. They designed an ingenious procedure to precisely geometrically control the conical shape and slenderness of the features as well as periodicity at the nanoscale. As a result, the desired aspect ratios and periodicities with few defects and large pattern area were achieved. The fabricated nanostructures demonstrated enhanced polarization-independent optical transmission (>98%) over a broad range of bandwidth and incident angles, and superhydrophilicity or robust superhydrophobicity after chemical vapor deposition treatment. However, in our opinion, though the fabricated structure shows good performance, the preparation conditions are too complex and demanding for practical applications. It is meaningful and necessary to further simplify the preparation process.

The performance of antireflective self-cleaning coatings has made good progress in the last few years, and is expected to be applied to many fields including optics, energy conversion and storage, and biomedical applications. However, very few products have been launched using such surfaces due to their susceptibility to mechanical contact during normal uses, such as finger contact, abrasive wear, and washing cycles.[137,205] Poor mechanical reliability and durability is still one of the major bottlenecks that prevents

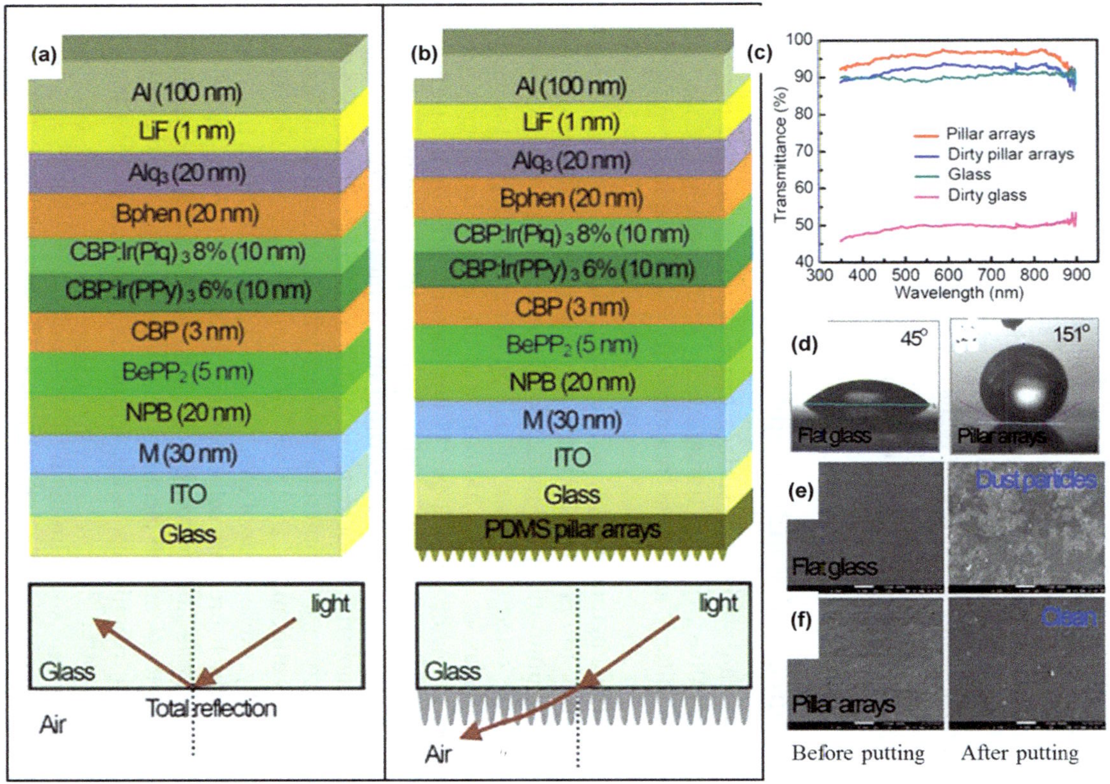

Figure 8.19 External quantum efficiency enhancement of white OLEDs (a) with antireflective film (b). (c) The optical transmittance before and after being immersed in dirty water. (d) The wetting property of flat glass and pillar arrays. (e) and (f) Comparison of the flat glass and pillar arrays characterized by the SEM images.[200] Copyright 2011 Optical Society of America.

them from being applied in commercial and industrial domains. Some efforts are apparently needed to overcome these drawbacks.

8.4 Applications

Multi-functional antireflective surfaces have attracted a lot of interest because the additional functions can improve the performance of optical devices. In the last years, antireflective and self-cleaning coatings have received tremendous attention. Antireflective self-cleaning coatings can be employed in a variety of applications: not only optical and electro-optical systems in telecommunications, glass lenses, eyeglasses, medicine, military products, but also lasers, mirrors, solar cells, IR diodes, multipurpose broad and narrow band-pass filters, architectural and automotive glasses and varied kinds of displays.[206]

8.4.1 Architectural Windows and Glasses

Glass coatings can modify the surface properties and optimize the visual transmittance (ARCs), thermal transmittance (low-E coatings), solar factor (solar control filters) and glass maintenance (self-cleaning coatings) of glasses.[207]

In the building engineering and construction industry, coated glasses act as an active filter to optimize the interactions between outdoor conditions (temperature, illumination) and indoor conditions. The main applications of ARC glasses are external glasses of tall buildings envelopes, transparent roofs, antiglare glasses, shop windows, museum and archaeological paths, reflector and projector glasses, *etc.*[207] For instance, the Amiran glass by Schott is a well-known example with very high light transmittance in the visible range. Simultaneously, concerning the maintenance cost of glasses, especially the innovative robot systems for tall buildings, self-cleaning ARCs glasses have become a commercial product with huge potential. In 2001, Pilkington Glass announced that they had developed the first self-cleaning windows. Usually, these kinds of windows are coated with a thin transparent layer of TiO_2, which presents self-cleaning properties through two distinct ways: photocatalysis and hydrophilicity. Saint-Gobain adopted the same principle in the Bioclean self-cleaning glass. Another important example of their application in buildings is the famous architecture of the China National Grand Theatre.[208]

However, there are some limits to the technologies available, such as the durability of treated glasses when exposed to the external environment.

8.4.2 Solar Collectors and Photovoltaic Modules

Solar cells are solid-state electrical devices that convert the energy of sunlight directly into electricity by the photovoltaic effect. Crystalline silicon is the most important material for solar cells. However, a common problem is the high RI of doped silicon and more than 30% of incident light is reflected back from the surface of crystalline silicon.[195] The reflection loss at the surface or interfaces is a disadvantage, which negatively affects the energy conversion efficiency of solar cells. For these reasons, scientists have devised a number of strategies for reducing the reflection of sunlight on the surface of solar cells, such as appending coatings with intermediate RI,[209] embedding metallic nanoparticles to induce surface plasmonic scattering,[210] micro-texturing[211] and absorption nanostructures with photonic crystals.[212] Bernhard discovered the moth cornea, which has uniformly arrayed conical protuberances in 1967. This structure was deemed to have the potential to prevent surface reflection for the whole solar radiation spectrum.[213] After that, various films of inorganic, polymeric, and inorganic/polymer hybrid materials have been fabricated by using diverse techniques. Some of them are assembled on solar cells to carry out detailed performance tests. Simultaneously, several compounds have been used as ARCs on crystalline silicon solar cells such as ITO, SnO_2, SiO_2, MgF_2, ZnS and ZnO.[214] The coatings can raise the efficiency of silicon-based PV devices by reducing the light energy lost due to reflection.

Nowadays, quarter-wavelength silicon nitride (SiN_x) films deposited by PECVD are the industrial standard for ARCs on crystalline silicon substrates.[197] This method offers high uniformity, high precision and high rates of deposition, but on the other hand, it is a high-cost chemical process that needs vacuum chambers. Other methods such as dip-coating, spray-coating, LbL, *etc.*, offer alternative approaches to low-cost and high-efficiency fabrication of ARCs. The controllability of thickness, RI and uniformity in these methods is still, however, a challenge and needs more deep research and development.

Glass is one of the most important optical substrates. Although the reflective loss is *ca.* 4% from each air/glass interface, which is not as serious as that of silicon, it could still degrade the performance of optical devices especially when multiple components are involved.[44,197] The use of glass with ARCs as a cover for solar thermal collectors has been proved to increase the efficiency of the system. For example, calculations have shown that the yearly energy output in a solar heating plant, where the temperature of the solar collector fluid is 100 °C, can be increased by about 20% by using ARCs on the solar glass cover. Generally, the material used as ARCs on glass is silicon dioxide.[215] However, a drawback limiting the use of ARCs is the adhesion of grime, which reduces transmittance constantly and significantly. For staying clean, outdoor packaging glasses usually mimic the lotus effect where dust and water cannot stick to the nano-structured surface. In other words, the ARCs are often endowed with superhydrophilic or superhydrophobic properties to prevent the accumulation of dust on modules which can reduce net efficiency by 32–40% over the first eight month period and up to 71.8% over a period of 15 years.[5,216,217] In addition, TiO_2 coatings are also used on solar cells for their photoinduced superhydrophilicity and photocatalytic property, which are very suitable in humid environments. The self-cleaning properties originate from a combination of the photocatalytic oxidative decomposition of organic contaminants and superhydrophilicity, which causes water droplets to spread on the TiO_2 surface, aiding the cleaning process. Whereas, it was also observed that most of the photocatalytic effect was lost within 5.5 years of outdoor exposure.[218] For dry environments, the glass itself should have a nano-structured surface so that the dust does not settle on the surface, and water is allowed to spread. Therefore, the combination of both self-cleaning and antireflective properties with strong durability is very desirable for applications in photovoltaic devices.

There are many other materials that can be used to prepare ARCs. In Chen's articles, the structures of solar selective surfaces are classified into seven different categories, *i.e.*, intrinsic selective materials, semiconductor–metal tandems, multilayer absorbers, textured surface absorbers, transition metal oxides, carbon nanoparticles embedded in ZnO and NiO and metal–dielectric composites.[219] Herein, ZnO single crystal nanorods or nanowires, which are cheap, stable, environmentally-friendly and nontoxic, have captured the attention of many researchers and show good prospects for optoelectronic applications because of their large exciton energy of 60 meV.[220]

8.4.3 Display Devices

Conventional display devices such as a cathode ray tube (CRT), liquid crystal display (LCD) and plasma display panel (PDP) are output devices for the presentation of information in visual or tactile form. The CRT creates images by accelerating and deflecting the electron beam onto the fluorescent screen, the LCD uses the light modulating properties of liquid crystals and the PDP utilizes small cells containing electrically charged ionized gases as fluorescent lamps. In these applications, reflection losses which affect the efficiency of transmitted energy should be avoided as much as possible.

Olifierczuk *et al.* once determined the reflection influences on the optical parameters for color visualization in the negative mode of a twisted nematic (TN) display.[8] A numerical program was worked out to obtain the contrast ratio and luminance in on-state for a reflective TN display working in real conditions. His results showed that the application of antireflective layers is necessary for obtaining a high value contrast ratio (CR) and could increase the value of CR about four times. Consequently, it is extremely important to eliminate harmful reflections in reflective TN displays.

In the latest displays, such as flat panel displays, the application of functional surface coatings with an antireflective property is a common solution. And the self-cleaning function may add new merits to the devices. Flat panel displays encompass a growing number of electronic visual display technologies. They are far lighter and thinner than traditional television sets and video displays that use cathode ray tubes (CRTs). They are widely applied in modern portable devices such as laptops, mobile phones, digital cameras and camcorders. In addition to LCD and PDP, examples of flat panel displays also include an organic light-emitting diode, electroluminescent display (ELD) and quantum dot display (QLED). In flat panel displays or detector applications, antireflective surfaces are usually employed to increase transmission and eliminate ghost images or veil glare caused by reflection from the optical surfaces, promoting the performance of the devices. In many applications (cell phones, cameras, *etc.*), the display must be magnified, but size and weight of the optical system must be minimized. Further, light is often polarization sensitive, which means that birefringence is undesirable. Hence, specialized coatings (including mirrors) are necessary.[221]

In some cases, wavelength-selective absorption films are attracting interest as a means of increasing the color purity of phosphors and improving their contrast. Ohishi[222] prepared a rhodamine B derivative with an ethoxysilano group by reacting rhodamine B sulfonyl chloride with aminopropyltriethoxysilane in dry pyridine. Then SiO_2 sol solution containing the new derivative was spin-coated onto the surface of a display device, such as CRT or PDP and then heated to form a thin film. This film exhibits good contrast enhancing properties because the organic dye can be incorporated into the film at high concentrations.

Moving further to possible applications, functional composite thin films with antireflective and self-cleaning properties are expected to be printed on a flexible substrate, such as a flexible conducting ITO substrate.[223] Therefore, the integration of superhydrophobicity and ARC into a surface is very desirable for various displays.

8.5 Conclusion and Outlook

In summary, recent developments in typical fabrication routes and applications of ARCs with a self-cleaning property have been described in this chapter. These surfaces have attracted significant attention within the scientific community over the last few decades because of their unique transparency-enhancing, self-cleaning properties and their potential for practical applications. The main contents of the chapter are summarized in Figure 8.20.

Although varied strategies have been developed to fabricate these functional surfaces, most of them are limited to laboratory research and are not suitable for industrial scale production. Several challenges still remain for practical applications. First, the ARCs are commonly susceptible to mechanical contact during normal use, such as finger contact, abrasive wear, washing cycles, and sand impact.[224] Mechanical wear on the surfaces could destroy the microscopic roughness features, or even remove the coatings from the substrates, finally causing a decline or complete loss of their optical property and wettability. Second, most of the reported preparation techniques involve multistep procedures or harsh conditions, or require special reagents and equipment. Many of those methods are expensive and only

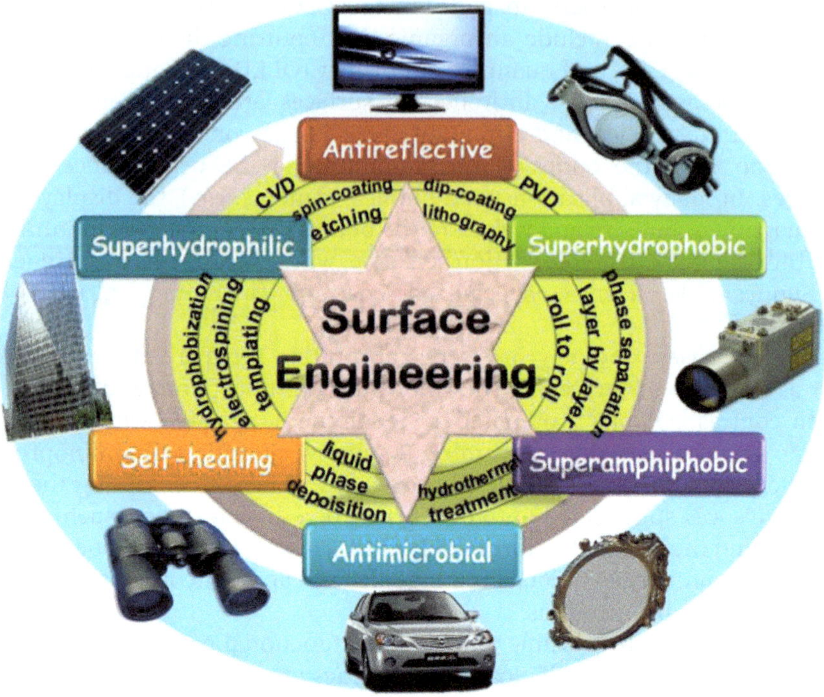

Figure 8.20 Main contents of the chapter.[228]
Reprinted from L. Yao and J. He, Recent progress in antireflection and self-cleaning technology – From surface engineering to fiunctional surfaces, *Prog. Mater. Sci.*, **61**, 94–143. Copyright 2014 with permission from Elsevier.

applicable to small flat surfaces or substrates of specific materials.[142,225] The limited durability and the high costs of coated surfaces are the two main barriers to the real applications of ARCs.

Most of the current work is still focused on increasing light transmission and wettability. In the near future, the following research directions should be given more consideration. (I) By combining theoretical simulation with modern analytical technologies and tools, further exploration and explanations should be conducted to reveal the mechanism of fabrication and the relationship between composition, structure, and property of surfaces.[226] (II) Fabricating ARC surfaces of high mechanical strength and abrasion resistance by cost-effective techniques is the key point in future work. It is significantly meaningful for large-area industrial production. (III) Environmental issues should be taken into account when preparing ARC surfaces. Not only fly ash but also water slurry or oil may fall on surfaces, and become contamination species.[227] Organic contaminants may also be adsorbed on the surfaces and change their optical and wetting properties. Therefore, multifunction surfaces with either enhanced durability or self-healing features are preferable for daily use. (IV) Interdisciplinary cooperation is necessary for researchers in the area of science and engineering. The increasing collaboration would be useful for identifying additional challenges, understanding fabrication mechanisms, and improving durability and abrasion resistance to adapt to practical applications. (V) Novel materials and coating techniques should be researched, which would expand the range of applications.

In conclusion, further fundamental investigations are still necessary to develop a simple, time- and cost-effective approach to achieve durable antireflective surfaces possessing sufficient self-cleaning and mechanical stability in future work. It will be beneficial for practical applications in large areas.

Acknowledgements

Financial support from the National Natural Science Foundation of China (Grant Nos. 21571182, 21271177), Beijing Science and Technology Plan (Z151100003315018), a Chinese Academy of Sciences Grant (CXJJ-14-M38), the National High Technology Research and Development Program ("863" Program) of China (Grant No. 2011AA050525), and the Knowledge Innovation Program of the Chinese Academy of Sciences (CAS) (Grant Nos. KGCX2-YW-370, KGCX2-EW-304-2) is greatly appreciated.

References

1. D. Stavenga, S. Foletti, G. Palasantzas and K. Arikawa, Light on the moth-eye corneal nipple array of butterflies, *Proc. R. Soc. B*, 2006, **273**, 661–667.
2. A. R. Parker and H. E. Townley, Biomimetics of photonic nanostructures, *Nat. Nanotechnol.*, 2007, **2**, 347–353.
3. T. Sun, L. Feng, X. Gao and L. Jiang, Bioinspired surfaces with special wettability, *Acc. Chem Res.*, 2005, **38**, 644–652.
4. Y. Li, J. Zhang and B. Yang, Antireflective surfaces based on biomimetic nanopillared arrays, *Nano Today*, 2010, **5**, 117–127.
5. L. Verma, M. Sakhuja, J. Son, A. Danner, H. Yang, H. Zeng *et al.*, Self-cleaning and antireflective packaging glass for solar modules, *Renewable Energy*, 2011, **36**, 2489–2493.
6. D. H. Wan, H. L. Chen, T. C. Tseng, C. Y. Fang, Y. S. Lai and F. Y. Yeh, Antireflective nanoparticle arrays enhance the efficiency of silicon solar cells, *Adv. Funct. Mater.*, 2010, **20**, 3064–3075.
7. J. H. Zhang and B. Yang, Patterning colloidal crystals and nanostructure arrays by soft lithography, *Adv. Funct. Mater.*, 2010, **20**, 3411–3424.
8. M. Olifierczuk and J. Zielinski, Significance of reflection reduction in a TN display for colour visualisation, *Opto-Electron. Rev.*, 2002, **10**, 35–38.
9. P. C. Tseng, M. A. Tsai, P. C. Yu and H. C. Kuo, Antireflection and light trapping of subwavelength surface structures formed by colloidal lithography on thin film solar cells, *Prog. Photovoltaics*, 2011, **20**, 135–142.

10. S. Chattopadhyay, Y. Huang, Y. Jen, A. Ganguly, K. Chen and L. Chen, Anti-reflecting and photonic nanostructures, *Mater. Sci. Eng., R*, 2010, **69**, 1–35.
11. J. Xiong, S. N. Das, B. Shin, J. P. Kar, J. H. Choi and J. M. Myoung, Biomimetic hierarchical ZnO structure with superhydrophobic and antireflective properties, *J. Colloid Interface Sci.*, 2010, **350**, 344–347.
12. Y. J. Wang, F. R. Hu, Y. Kanamori, T. Wu and K. Hane, Large area, freestanding GaN nanocolumn membrane with bottom subwavelength nanostructure, *Opt. Express.*, 2010, **18**, 5504–5511.
13. B. Yang, X. M. Zhang, J. H. Zhang, Z. Y. Ren, X. Li, X. Zhang *et al.*, Morphology and wettability control of silicon cone arrays using colloidal lithography, *Langmuir*, 2009, **25**, 7375–7382.
14. Z. Y. Liu, X. T. Zhang, T. Murakami and A. Fujishima, Sol-gel SiO_2/TiO_2 bilayer films with self-cleaning and antireflection properties, *Sol. Energy Mater. Sol. Cells*, 2008, **92**, 1434–1438.
15. C. Y. Kuo, Y. Y. Chen and S. Y. Lu, A facile route to create surface porous polymer films via phase separation for antireflection applications, *ACS Appl. Mater. Interfaces*, 2008, **1**, 72–75.
16. W. Zhu, X. Feng, L. Feng and L. Jiang, UV-Manipulated wettability between superhydrophobicity and superhydrophilicity on a transparent and conductive SnO_2 nanorod film, *Chem. Commun.*, 2006, 2753–2755.
17. L. Li, Y. Li, S. Gao and N. Koshizaki, Ordered Co_3O_4 hierarchical nanorod arrays: tunable superhydrophilicity without UV irradiation and transition to superhydrophobicity, *J. Mater. Chem.*, 2009, **19**, 8366–8371.
18. Y. J. Lee, D. S. Ruby, D. W. Peters, B. B. McKenzie and J. W. P. Hsu, ZnO nanostructures as efficient antireflection layers in solar cells, *Nano Lett.*, 2008, **8**, 1501–1505.
19. X. Liu and J. He, Superhydrophilic and antireflective properties of silica nanoparticle coatings fabricated via layer-by-layer assembly and post-calcination, *J. Phys. Chem. C*, 2009, **113**, 148–152.
20. N. Lu, Y. D. Wang, H. B. Xu, G. Shi, M. J. Xu, X. W. Lin *et al.*, Biomimetic corrugated silicon nanocone arrays for self-cleaning antireflection coatings, *Nano Res.*, 2010, **3**, 520–527.
21. K. N. Nakata, M. Sakai, T. Ochiai, T. Murakami, K. Takagi and A. Fujishima, Antireflection and self-cleaning properties of a moth-eye-Like surface coated with TiO_2 Particles, *Langmuir*, 2011, **27**, 3275–3278.
22. J. A. Howarter and J. P. Youngblood, Self-cleaning and next generation anti-fog surfaces and coatings, *Macromol. Rapid Commun.*, 2008, **29**, 455–466.
23. H. K. Raut, V. A. Ganesh, A. S. Nair and S. Ramakrishna, Anti-reflective coatings: a critical, in-depth review, *Energy Environ. Sci.*, 2011, **4**, 3779–3804.
24. H. A. Macleod, *Thin-film Optical Filters*, Taylor & Francis, 3rd edn, 2001.
25. A. Nakajima, A. Fujishima, K. Hashimoto and T. Watanabe, Preparation of transparent superhydrophobic boehmite and silica films

by sublimation of aluminum acetylacetonate, *Adv. Mater.*, 1999, **11**, 1365–1368.
26. R. N. Wenzel, Resisitance of solid Surfaces to wetting by water, *Ind. Eng. Chem.*, 1936, **28**, 988–994.
27. A. B. D. Cassie and S. Baxter, Wettability of porous surfaces, *Trans. Faraday Soc.*, 1944, **40**, 546.
28. E. Johnston, S. Bullock, J. Uilk, P. Gatenholm and K. J. Wynne, Networks from alpha, omega-dihydroxypoly(dimethylsiloxane) and (tridecafluoro-1,1,2,2-tetrahydrooctyl)triethoxysilane: Surface microstructures and surface characterization, *Macromolecules*, 1999, **32**, 8173–8182.
29. M. H. Sun, C. X. Luo, L. P. Xu, H. Ji, O. Y. Qi, D. P. Yu *et al.*, Artificial lotus leaf by nanocasting, *Langmuir*, 2005, **21**, 8978–8981.
30. X. Liu and J. He, Progress in antifogging technology-from surface engineering to functional surfaces, *Prog. Chem.*, 2010, **22**, 270 276.
31. S. N. Frank and A. J. Bard, Heterogeneous photocatalytic oxidation of cyanide ion in aqueous-solutions at TiO_2 powder, *J. Am. Chem. Soc.*, 1977, **99**, 303–304.
32. Y. Ohko, K. Hashimoto and A. Fujishima, Kinetics of photocatalytic reactions under extremely low-intensity UV illumination on titanium dioxide thin films, *J. Phys. Chem. A*, 1997, **101**, 8057–8062.
33. I. Sopyan, S. Murasawa, K. Hashimoto and A. Fujishima, Highly efficient TiO_2 film photocatalyst degradation of gaseous acetaldehyde, *Chem. Lett.*, 1994, 723–726.
34. J. Lee, H. Park and W. Choi, Selective photocatalytic oxidation of NH_3 to N_2 on plantation TiO_2 in water, *Environ. Sci. Technol.*, 2002, **36**, 5462–5468.
35. A. Fujishima and X. T. Zhang, Titanium dioxide photocatalysis: present situation and future approaches, *C. R. Chim.*, 2006, **9**, 750–760.
36. A. Fujishima, X. Zhang and D. A. Tryk, TiO_2 photocatalysis and related surface phenomena, *Surf. Sci. Rep.*, 2008, **63**, 515–582.
37. T. Minabe, D. A. Tryk, P. Sawunyama, Y. Kikuchi, K. Hashimoto and A. Fujishima, TiO_2-mediated photodegradation of liquid and solid organic compounds, *J. Photochem. Photobiol., A*, 2000, **137**, 53–62.
38. A. Heller, Chemistry and applications of photocatalytic oxidation of thin organic films, *Acc. Chem. Res.*, 1995, **28**, 503–508.
39. J. Schwitzgebel, J. G. Ekerdt, F. Sunada, S. E. Lindquist and A. Heller, Increasing the efficiency of the photocatalytic oxidation of organic films on aqueous solutions by reactively coating the TiO_2 photocatalyst with a chlorinated silicone, *J. Phys. Chem. B*, 1997, **101**, 2621–2624.
40. N. Sakai, A. Fujishima, T. Watanabe and K. Hashimoto, Quantitative evaluation of the photoinduced hydrophilic conversion properties of TiO_2 thin film surfaces by the reciprocal of contact angle, *J. Phys. Chem. B*, 2003, **107**, 1028–1035.
41. K. Hashimoto, H. Irie and A. Fujishima, TiO_2 photocatalysis: a historical overview and future prospects, *Jpn. J. Appl. Phys., Part 1*, 2005, **44**, 8269–8285.

42. R. K. Iler, Multilayers of colloidal particles, *J. Colloid Interface Sci.*, 1966, **21**, 569–594.
43. G. Decher and J. D. Hong, Buildup of unltrathin multilayer films by a self-assembly process. 1. consecutive adsorption of anionic and cationic bipolar amphiphiles on charged surfaces, *Makromol. Chem., Macromol. Symp.*, 1991, **46**, 321–327.
44. J. Hiller, J. D. Mendelsohn and M. F. Rubner, Reversibly erasable nanoporous anti-reflection coatings from polyelectrolyte multilayers, *Nat. Mater.*, 2002, **1**, 59–63.
45. X. Du and J. He, Facile fabrication of hollow mesoporous silica nanospheres for superhydrophilic and visible/near-IR antireflection coatings, *Chem. - Eur. J.*, 2011, **17**, 8165–8174.
46. P. Podsiadlo, L. Sui, Y. Elkasabi, P. Burgardt, J. Lee, A. Miryala *et al.*, Layer-by-layer assembled films of cellulose nanowires with antireflective properties, *Langmuir*, 2007, **23**, 7901–7906.
47. J. H. Cho, J. K. Hong, K. Char and F. Caruso, Nanoporous block copolymer micelle/micelle multilayer films with dual optical properties, *J. Am. Chem. Soc.*, 2006, **128**, 9935–9942.
48. K. Ariga, Y. Lvov and T. Kunitake, Assembling alternate dye-polyion molecular films by electrostatic layer-by-layer adsorption, *J. Am. Chem. Soc.*, 1997, **119**, 2224–2231.
49. Y. Li, X. Wang and J. Sun, Layer-by-layer assembly for rapid fabrication of thick polymeric films, *Chem. Soc. Rev.*, 2012, **41**, 5998–6009.
50. K. Ariga, J. P. Hill and Q. M. Ji, Layer-by-layer assembly as a versatile bottom-up nanofabrication technique for exploratory research and realistic application, *Phys. Chem. Chem. Phys.*, 2007, **9**, 2319–2340.
51. M. Osada and T. Sasaki, Two-dimensional dielectric nanosheets: novel nanoelectronics from nanocrystal building blocks, *Adv. Mater.*, 2012, **24**, 210–228.
52. J. Q. Sun, L. B. Zhang, Y. Li and J. C. Shen, Mechanically stable antireflection and antifogging coatings fabricated by the layer-by-layer deposition process and postcalcination, *Langmuir*, 2008, **24**, 10851–10857.
53. K. Ariga, Q. Ji, J. P. Hill, Y. Bando and M. Aono, Forming nanomaterials as layered functional structures toward materials nanoarchitectonics, *NPG Asia Mater.*, 2012, **4**, e17.
54. X. Liu and J. He, Hierarchically structured superhydrophilic coatings fabricated by self-assembling raspberry-like silica nanospheres, *J. Colloid Interface Sci.*, 2007, **314**, 341–345.
55. X. Du, X. Liu, H. Chen and J. He, Facile Fabrication of Raspberry-like Composite Nanoparticles and Their Application as Building Blocks for Constructing Superhydrophilic Coatings, *J. Phys. Chem. C*, 2009, **113**, 9063–9070.
56. X. Li, X. Du and J. He, Self-cleaning antireflective coatings assembled from peculiar mesoporous silica nanoparticles, *Langmuir.*, 2010, **26**, 13528–13534.

57. X. Du and J. He, Hierarchically mesoporous silica nanoparticles: extraction, amino-functionalization, and their multipurpose potentials, *Langmuir*, 2011, **27**, 2972–2979.
58. C. J. Brinker and A. J. Hurd, Fundamental of sol-gel dip-coating, *J. Phys. III*, 1991, **4**, 1231–1242.
59. L. Scriven, *Physics and Applications of Dip Coating and Spin Coating*, Cambridge University Press, 1988, pp, 717–729.
60. M. Faustini, L. Nicole, C. Boissière, P. Innocenzi, C. Sanchez and D. Grosso, Hydrophobic, antireflective, self-cleaning, and antifogging sol–gel coatings: an example of multifunctional nanostructured materials for photovoltaic cells, *Chem. Mater.*, 2010, **22**, 4406–4413.
61. Ö. Kesmez, H. Erdem Çamurlu, E. Burunkaya and E. Arpaç, Sol–gel preparation and characterization of anti-reflective and self-cleaning SiO_2–TiO_2 double-layer nanometric films, *Sol. Energy Mater. Sol. C*, 2009, **93**, 1833–1839.
62. M. Bautista and A. Morales, Silica antireflective films on glass produced by the sol–gel method, *Sol. Energy Mater. Sol. C*, 2003, **80**, 217–225.
63. G. Helsch, A. Mos, J. Deubener and M. Holand, Thermal resistance of nanoporous antireflective coatings on silica glass for solar tower receivers, *Sol. Energy Mater. Sol. C*, 2010, **94**, 2191–2196.
64. C. Lawrence, The mechanics of spin coating of polymer films, *Phys. Fluids*, 1988, **31**, 2786.
65. F. C. Krebs, Fabrication and processing of polymer solar cells: a review of printing and coating techniques, *Sol. Energy Mater. Sol. C*, 2009, **93**, 394–412.
66. D. B. Hall, P. Underhill and J. M. Torkelson, Spin coating of thin and ultrathin polymer films, *Polym. Eng. Sci.*, 1998, **38**, 2039–2045.
67. A. Steele, I. Bayer and E. Loth, Inherently superoleophobic nanocomposite coatings by spray atomization, *Nano Lett.*, 2009, **9**, 501–505.
68. B. J. Babu, A. Maldonado, S. Velumani and R. Asomoza, Electrical and optical properties of ultrasonically sprayed Al-doped zinc oxide thin films, *Mater. Sci. Eng., B*, 2010, **174**, 31–37.
69. T. Fender, Thermal spray high performance polymer coatings, *Mater. Technol.*, 1996, **11**, 16–20.
70. T. Schmidt, F. Gärtner, H. Assadi and H. Kreye, Development of a generalized parameter window for cold spray deposition, *Acta Mater.*, 2006, **54**, 729–742.
71. L. Gao, P. He, X. Li and J. He, Construction of nanostructured silica surfaces by spray-coating and their optical and wetting properties, *Imaging Sci. Photochem.*, 2012, **30**, 260–268.
72. X. H. M. X. H. Men, Z. Z. Zhang, J. Yang, X. T. Zhu, K. Wang and W. Jiang, Spray-coated superhydrophobic coatings with regenerability, *New J. Chem.*, 2011, **35**, 881–886.
73. C. H. Porcel, A. Izquierdo, V. Ball, G. Decher, J. C. Voegel and P. Schaaf, Ultrathin coatings and (poly(glutamic acid)/polyallylamine) films

deposited by continuous and simultaneous spraying, *Langmuir*, 2005, **21**, 800–802.
74. K. L. Choy, Chemical vapour deposition of coatings, *Prog. Mater. Sci.*, 2003, **48**, 57–170.
75. A. Biswas, I. S. Bayer, A. S. Biris, T. Wang, E. Dervishi and F. Faupel, Advances in top–down and bottom–up surface nanofabrication: Techniques, applications & future prospects, *Adv. Colloid Interface*, 2012, **170**, 2–27.
76. E. Randich and R. Pettit, Solar selective properties and high temperature stability of CVD ZrB_2, *Sol. Energy Mater.*, 1981, **5**, 425–435.
77. C. Yu, G. Sandhu and T. Doan, Antireflection coatings for advanced semiconductor device metallization using laser reflow and chemical mechanical planarization, *Appl. Phys. Lett.*, 1991, **59**, 1978–1980.
78. T. Mollart and K. Lewis, Transition metal oxide anti-reflection coatings for airborne diamond optics, *Diamond Relat. Mater.*, 2001, **10**, 536–541.
79. S. M. Yang, Y. C. Hsieh and C. A. Jeng, Optimal design of antireflection coating and experimental verification by plasma enhanced chemical vapor deposition in small displays, *J. Vac. Sci. Technol., A*, 2009, **27**, 336–341.
80. S. Mukhopadhyay, T. Jana and S. Ray, Development of low temperature silicon oxide thin films by photo-CVD for surface passivation, *J. Vac. Sci. Technol., A*, 2005, **23**, 417–422.
81. C. Van der Werf, H. Goldbach, J. Löffler, A. Scarfo, A. Kylner, B. Stannowski *et al.*, Silicon nitride at high deposition rate by Hot Wire Chemical Vapor Deposition as passivating and antireflection layer on multicrystalline silicon solar cells, *Thin Solid Films*, 2006, **501**, 51–54.
82. V. Verlaan, C. H. M. Van der Werf, Z. S. Flouweling, G. Romijn, A. W. Weeber, H. F. W. Dekkers *et al.*, Multi-crystalline si solar cells with very fast deposited (108 nm/min) passivating hot wire CVD silicon nitride as antireflection coating, *Prog. Photovoltaics*, 2007, **15**, 563–573.
83. M. Kuhr, S. Bauer, U. Rothhaar and D. Wolff, Coatings on plastics with the PICVD technology, *Thin Solid Films*, 2003, **442**, 107–116.
84. L. Remache, E. Fourmond, A. Mahdjoub, J. Dupuis and M. Lemiti, Design of porous silicon/PECVD SiOx antireflection coatings for silicon solar cells, *Mater. Sci. Eng., B*, 2011, **176**, 45–48.
85. X. Li and Y. C. Han, Tunable wavelength antireflective film by non-solvent-induced phase separation of amphiphilic block copolymer micelle solution, *J. Mater. Chem.*, 2011, **21**, 18024–18033.
86. S. Walheim, E. Schäffer, J. Mlynek and U. Steiner, Nanophase-separated polymer films as high-performance antireflection coatings, *Science*, 1999, **283**, 520–522.
87. M. Böltau, S. Walheim, J. Mlynek, G. Krausch and U. Steiner, Surface-induced structure formation of polymer blends on patterned substrates, *Nature*, 1998, **391**, 877–879.

88. N. Zhao, Q. Xie, L. Weng, S. Wang, X. Zhang and J. Xu, Superhydrophobic surface from vapor-induced phase separation of copolymer micellar solution, *Macromolecules*, 2005, **38**, 8996–8999.
89. B. C. Bunker, Molecular mechanisms for corrosion of silica and silicate-glasses, *J. Non-Cryst Solids*, 1994, **179**, 300–308.
90. J. M. Myoung, J. Xiong, S. N. Das, J. P. Kar and J. H. Choi, A multifunctional nanoporous layer created on glass through a simple alkali corrosion process, *J. Mater. Chem.*, 2010, **20**, 10246–10252.
91. X. Liu and J. He, One-step hydrothermal creation of hierarchical microstructures toward superhydrophilic and superhydrophobic surfaces, *Langmuir*, 2009, **25**, 11822–11826.
92. D. Qi, N. Lu, H. Xu, B. Yang, C. Huang, M. Xu *et al.*, Simple approach to wafer-scale self-cleaning antireflective silicon surfaces, *Langmuir*, 2009, **25**, 7769–7772.
93. F. Palumbo, R. Di Mundo, D. Cappelluti and R. d'Agostino, Superhydrophobic and superhydrophilic polycarbonate by tailoring chemistry and nano-texture with plasma processing, *Plasma Processes Polym.*, 2011, **8**, 118–126.
94. S. Pearton, High ion density dry etching of compound semiconductors, *Mater. Sci. Eng., B*, 1996, **40**, 101–118.
95. J. W. L. J. W. Leem and J. S. Yu, Broadband and wide-angle antireflection subwavelength structures of Si by inductively coupled plasma etching using dewetted nanopatterns of Au thin films as masks, *Thin Solid Films*, 2011, **519**, 3792–3797.
96. K. Gamo, Nanofabrication by FIB, *Microelectron. Eng.*, 1996, **32**, 159–171.
97. R. Ito and S. Okazaki, Pushing the limits of lithography, *Nature*, 2000, **406**, 1027–1031.
98. G. Wallraff and W. Hinsberg, Lithographic imaging techniques for the formation of nanoscopic features, *Chem. Rev.*, 1999, **99**, 1801–1821.
99. B. Yang, Y. F. Li, J. H. Zhang, S. J. Zhu, H. P. Dong, F. Jia *et al.*, Bioinspired silica surfaces with near-infrared improved transmittance and superhydrophobicity by colloidal lithography, *Langmuir*, 2010, **26**, 9842–9847.
100. S. J. Park, S. W. Lee, K. J. Lee, J. H. Lee, K. D. Kim, J. H. Jeong *et al.*, An antireflective nanostructure array fabricated by nanosilver colloidal lithography on a silicon substrate, *Nanoscale Res. Lett.*, 2010, **5**, 1570–1577.
101. J. Tommila, V. Polojarvi, A. Aho, A. Tukiainen, J. Viheriala, J. Salmi *et al.*, Nanostructured broadband antireflection coatings on AlInP fabricated by nanoimprint lithography, *Sol. Energy Mater. Sol. C*, 2010, **94**, 1845–1848.
102. A. del Campo and E. Arzt, Fabrication approaches for generating complex micro-and nanopatterns on polymeric surfaces, *Chem. Rev.*, 2008, **108**, 911–945.

103. C. Acikgoz, M. A. Hempenius, J. Huskens and G. J. Vancso, Polymers in conventional and alternative lithography for the fabrication of nanostructures, *Eur. Polym. J.*, 2011, **47**, 2033–2052.
104. S. Irie, M. Shigematsu, R. Sakamoto, K. Mizusawa, Y. Nakajima and T. Itani, A study of an organic bottom antireflective coating for 157-nm lithography, *J. Photopolym. Sci. Technol.*, 2003, **16**, 565–572.
105. Y. Lai, C. Lin, H. Wang, J. Huang, H. Zhuang and L. Sun, Superhydrophilic-superhydrophobic micropattern on TiO_2 nanotube films by photocatalytic lithography, *Electrochem. Commun.*, 2008, **10**, 387–391.
106. P. Sharma and K. Hanumantha Rao, Adhesion of Paenibacillus polymyxa on chalcopyrite and pyrite: surface thermodynamics and extended DLVO theory, *Colloids Surf., B*, 2003, **29**, 21–38.
107. Z. Wu, D. Lee, M. F. Rubner and R. E. Cohen, Structural color in porous, superhydrophilic, and self-cleaning SiO_2/TiO_2 bragg stacks, *Small*, 2007, **3**, 1445–1451.
108. Y. Lai, Y. Tang, J. Gong, D. Gong, L. Chi, C. Lin *et al.*, Transparent superhydrophobic/superhydrophilic TiO_2-based coatings for self-cleaning and anti-fogging, *J. Mater. Chem.*, 2012, **22**, 7420–7426.
109. V. A. Ganesh, A. S. Nair, H. K. Raut, T. M. Walsh and S. Ramakrishna, Photocatalytic superhydrophilic TiO_2 coating on glass by electrospinning, *RSC Adv.*, 2012, **2**, 2067–2072.
110. Y. Lai, Z. Lin, J. Huang, L. Sun, Z. Chen and C. Lin, Controllable construction of ZnO/TiO_2 patterning nanostructures by superhydrophilic/superhydrophobic templates, *New J. Chem.*, 2010, **34**, 44–51.
111. H. S. Lim, D. Kwak, D. Y. Lee, S. G. Lee and K. Cho, UV-driven reversible switching of a roselike vanadium oxide film between superhydrophobicity and superhydrophilicity, *J. Am. Chem. Soc.*, 2007, **129**, 4128–4129.
112. M. Miyauchi, M. Shibuya, Z. G. Zhao and Z. Liu, Surface wetting behavior of a WO_3 electrode under light-irradiated or potential-controlled conditions, *J. Phys. Chem. C*, 2009, **113**, 10642–10646.
113. R. Blossey, Self-cleaning surfaces-virtual realities, *Nat. Mater.*, 2003, **2**, 301–306.
114. B. T. McDonald and T. Cui, Superhydrophilic surface modification of copper surfaces by layer-by-layer self-assembly and liquid phase deposition of TiO_2 thin film, *J. Colloid Interface Sci.*, 2011, **354**, 1–6.
115. K. R. Denison and C. Boxall, Photoinduced "stick-slip" on superhydrophilic semiconductor surfaces, *Langmuir*, 2007, **23**, 4358–4366.
116. Y. Liu, L. Qian, C. Guo, X. Jia, J. Wang and W. Tang, Natural superhydrophilic TiO_2/SiO_2 composite thin films deposited by radio frequency magnetron sputtering, *J. Alloy Compd.*, 2009, **479**, 532–535.
117. D. Tang, K. Cheng, W. Weng, C. Song, P. Du, G. Shen *et al.*, TiO_2 nanorod films grown on Si wafers by a nanodot-assisted hydrothermal growth, *Thin Solid Films*, 2011, **519**, 7644–7649.

118. M. Kobayashi, Y. Terayama, H. Yamaguchi, M. Terada, D. Murakami, K. Ishihara *et al.*, Wettability and antifouling behavior on the surfaces of superhydrophilic polymer brushes, *Langmuir*, 2012, **28**, 7212–7222.
119. J. Zang, C. M. Li, S. J. Bao, X. Cui, Q. Bao and C. Q. Sun, Template-free electrochemical synthesis of superhydrophilic polypyrrole nanofiber network, *Macromolecules*, 2008, **41**, 7053–7057.
120. X. N. Xie, S. X. Lim, Y. Wang, X. Gao, K. K. Lee, C. H. Sow *et al.*, A nanosegregant approach to superwettable and water – attracting surfaces, *Macromol. Chem. Phys.*, 2010, **211**, 2187–2192.
121. Y. Gao, Y. Masuda and K. Koumoto, Light-excited superhydrophilicity of amorphous TiO_2 thin films deposited in an aqueous peroxotitanate solution, *Langmuir*, 2004, **20**, 3188–3194.
122. H. Pizem, C. N. Sukenik, U. Sampathkumaran, A. K. McIlwain and M. R. De Guire, Effects of substrate surface functionality on solution-deposited titania films, *Chem. Mater.*, 2002, **14**, 2476 2485.
123. V. Rico, P. Romero, J. Hueso, J. Espinós and A. González-Elipe, Wetting angles and photocatalytic activities of illuminated TiO_2 thin films, *Catal. Today*, 2009, **143**, 347–354.
124. C. S. Lee, J. Kim, G. H. Gu, D. H. Jo, C. G. Park, W. Choi *et al.*, Photocatalytic activities of TiO_2 thin films prepared on Galvanized Iron substrate by plasma-enhanced atomic layer deposition, *Thin Solid Films*, 2010, **518**, 4757–4761.
125. C. S. Lee, J. Kim, J. Son, W. Choi and H. Kim, Photocatalytic functional coatings of TiO_2 thin films on polymer substrate by plasma enhanced atomic layer deposition, *Appl. Catal., B*, 2009, **91**, 628–633.
126. V. Pore, T. Kivelä, M. Ritala and M. Leskelä, Atomic layer deposition of photocatalytic TiO_2 thin films from TiF_4 and H_2O, *Dalton Trans.*, 2008, 6467–6474.
127. Y. Tang, Y. Liu, U. Sampathkumaran, M. Z. Hu, R. Wang and M. R. De Guire, Particle growth and particle–surface interactions during low-temperature deposition of ceramic thin films, *Solid State Ionics*, 2002, **151**, 69–78.
128. D. Gutiérrez-Tauste, X. Domènech, M. A. Hernández-Fenollosa and J. A. Ayllón, Alternative fluoride scavengers to produce TiO_2 films by the liquid phase deposition (LPD) technique, *J. Mater. Chem.*, 2006, **16**, 2249–2255.
129. Y. Aoi, H. Kambayashi, E. Kamijo and S. Deki, Synthesis of mesoscopic hollow spheres and inner surface functionalized hollow spheres of titanium dioxide by the liquid phase deposition method, *J. Mater. Res.*, 2003, **18**, 2832–2836.
130. H. Strohm, M. Sgraja, J. Bertling and P. Löbmann, Preparation of TiO_2-polymer hybrid microcapsules, *J. Mater. Sci.*, 2003, **38**, 1605–1609.
131. Y. Tsuge, J. Kim, Y. Sone, O. Kuwaki and S. Shiratori, Fabrication of transparent TiO_2 film with high adhesion by using self-assembly methods: application to super-hydrophilic film, *Thin Solid Films*, 2008, **516**, 2463–2468.

132. S. Deki, Y. Aoi, J. Okibe, H. Yanagimoto, A. Kajinami and M. Mizuhata, Preparation and characterization of iron oxyhydroxide and ironoxide thin films by liquid-phase deposition, *J. Mater. Chem.*, 1997, **7**, 1769–1772.
133. X. Chen and S. S. Mao, Titanium dioxide nanomaterials: synthesis, properties, modifications, and applications, *Chem. Rev.*, 2007, **38**, 2891–2959.
134. L. Sirghi, T. Aoki and Y. Hatanaka, Hydrophilicity of TiO_2 thin films obtained by radio frequency magnetron sputtering deposition, *Thin Solid Films*, 2002, **422**, 55–61.
135. M. Yamagishi, S. Kuriki, P. Song and Y. Shigesato, Thin film TiO_2 photocatalyst deposited by reactive magnetron sputtering, *Thin Solid Films*, 2003, **442**, 227–231.
136. E. Hosono, H. Matsuda, I. Honma, M. Ichihara and H. Zhou, Synthesis of a perpendicular TiO_2 nanosheet film with the superhydrophilic property without UV irradiation, *Langmuir*, 2007, **23**, 7447–7450.
137. X. Zhang, F. Shi, J. Niu, Y. Jiang and Z. Wang, Superhydrophobic surfaces: from structural control to functional application, *J. Mater. Chem.*, 2008, **18**, 621–633.
138. B. Bhushan, Y. C. Jung and K. Koch, Self-cleaning efficiency of artificial superhydrophobic surfaces, *Langmuir*, 2009, **25**, 3240–3248.
139. X. M. Li, D. Reinhoudt and M. Crego-Calama, What do we need for a superhydrophobic surface? A review on the recent progress in the preparation of superhydrophobic surfaces, *Chem. Soc. Rev.*, 2007, **36**, 1350–1368.
140. Z. Guo, W. Liu and B. L. Su, Superhydrophobic surfaces: from natural to biomimetic to functional, *J. Colloid Interface Sci.*, 2011, **353**, 335–355.
141. C. Neinhuis and W. Barthlott, Characterization and distribution of water-repellent, self-cleaning plant surfaces, *Ann. Bot.*, 1997, **79**, 667–677.
142. C. H. Xue, S. T. Jia, J. Zhang and J. Z. Ma, Large-area fabrication of superhydrophobic surfaces for practical applications: an overview, *Sci. Technol. Adv. Mater.*, 2010, 11.
143. T. Onda, S. Shibuichi, N. Satoh and K. Tsujii, Super-water-repellent fractal surfaces, *Langmuir*, 1996, **12**, 2125–2127.
144. M. Yoon, Y. Kim and J. Cho, Multifunctional colloids with optical, magnetic and superhydrophobic properties derived from nucleophilic substitution-induced layer-by-layer assembly in organic media, *ACS Nano*, 2011, **5**, 5417–5426.
145. Y. Li, F. Liu and J. Sun, A facile layer-by-layer deposition process for the fabrication of highly transparent superhydrophobic coatings, *Chem. Commun.*, 2009, 2730–2732.
146. L. Xu and J. He, Fabrication of highly transparent superhydrophobic coatings from hollow silica nanoparticles, *Langmuir*, 2012, **28**, 7512–7518.

147. J. Feng, M. T. Tuominen and J. P. Rothstein, Hierarchical superhydrophobic surfaces fabricated by dual-scale electron-beam-lithography with well-ordered secondary nanostructures, *Adv. Funct. Mater.*, 2011, **21**, 3715–3722.
148. T. Ishizaki, J. Hieda, N. Saito and O. Takai, Corrosion resistance and chemical stability of super-hydrophobic film deposited on magnesium alloy AZ31 by microwave plasma-enhanced chemical vapor deposition, *Electrochim. Acta*, 2010, **55**, 7094–7101.
149. K. K. S. Lau, J. Bico, K. B. K. Teo, M. Chhowalla, G. A. J. Amaratunga, W. I. Milne *et al.*, Superhydrophobic carbon nanotube forests, *Nano Lett.*, 2003, **3**, 1701–1705.
150. Y. Miyauchi, B. Ding and S. Shiratori, Fabrication of a silver-ragwort-leaf-like super-hydrophobic micro/nanoporous fibrous mat surface by electrospinning, *Nanotechnology*, 2006, **17**, 5151–5156.
151. H. Tang, H. Wang and J. He, Superhydrophobic titania membranes of different adhesive forces fabricated by electrospinning, *J. Phys. Chem. C*, 2009, **113**, 14220–14224.
152. K. Acatay, E. Simsek, C. Ow-Yang and Y. Z. Menceloglu, Tunable, superhydrophobically stable polymeric surfaces by electrospinning, *Angew Chem., Int. Ed.*, 2004, **43**, 5210–5213.
153. H. Shang, Y. Wang, S. Limmer, T. Chou, K. Takahashi and G. Cao, Optically transparent superhydrophobic silica-based films, *Thin Solid Films*, 2005, **472**, 37–43.
154. N. J. Shirtcliffe, G. McHale, M. I. Newton, C. C. Perry and P. Roach, Porous materials show superhydrophobic to superhydrophilic switching, *Chem. Commun.*, 2005, 3135–3137.
155. B. Cortese and H. Morgan, Controlling the wettability of hierarchically structured thermoplastics, *Langmuir*, 2012, **28**, 896–904.
156. J. P. Fernández-Blázquez, D. Fell, E. Bonaccurso and A. Campo, Superhydrophilic and superhydrophobic nanostructured surfaces via plasma treatment, *J. Colloid Interface Sci.*, 2011, **357**, 234–238.
157. M. Nicolas, Acids and alkali resistant sticky superhydrophobic surfaces by one-pot electropolymerization of perfluoroalkyl alkyl pyrrole, *J. Colloid Interface Sci.*, 2010, **343**, 608–614.
158. X. Yang, J. Zhuang, X. Li, D. Chen, G. Ouyang, Z. Mao *et al.*, Hierarchically nanostructured rutile arrays: acid vapor oxidation growth and tunable morphologies, *ACS Nano*, 2009, **3**, 1212–1218.
159. C. T. Hsieh, S. Y. Yang and J. Y. Lin, Electrochemical deposition and superhydrophobic behavior of ZnO nanorod arrays, *Thin Solid Films*, 2010, **518**, 4884–4889.
160. T. Darmanin, E. Taffin de Givenchy, S. Amigoni and F. Guittard, Hydrocarbon versus fluorocarbon in the electrodeposition of superhydrophobic polymer films, *Langmuir*, 2010, **26**, 17596–17602.
161. M. Ma, R. M. Hill, J. L. Lowery, S. V. Fridrikh and G. C. Rutledge, Electrospun poly (styrene-block-dimethylsiloxane) block copolymer fibers exhibiting superhydrophobicity, *Langmuir*, 2005, **21**, 5549–5554.

162. D. H. Reneker and A. L. Yarin, Electrospinning jets and polymer nanofibers, *Polymer*, 2008, **49**, 2387–2425.
163. L. Jiang, Y. Zhao and J. Zhai, A lotus-leaf-like superhydrophobic surface: a porous microsphere/nanofiber composite film prepared by electrohydrodynamics, *Angew. Chem., Int. Ed.*, 2004, **116**, 4438–4441.
164. S. Agarwal, S. Horst and M. Bognitzki, Electrospinning of fluorinated polymers: formation of superhydrophobic surfaces, *Macromol. Mater. Eng.*, 2006, **291**, 592–601.
165. D. Han and A. J. Steckl, Superhydrophobic and oleophobic fibers by coaxial electrospinning, *Langmuir*, 2009, **25**, 9454–9462.
166. F. Shi, Z. Liu, G. L. Wu, M. Zhang, H. Chen, Z. Wang *et al.*, Surface imprinting in layer-by-layer nanostructured films, *Adv. Funct. Mater.*, 2007, **17**, 1821–1827.
167. C. Jeffree, The cuticle, epicuticular waxes and trichomes of plants, with reference to their structure, functions and evolution. *Insects and the Plant Surface Edward Arnold*, London, 1986, p. 23–64.
168. Y. C. Jung and B. Bhushan, Mechanically durable carbon nanotube-composite hierarchical structures with superhydrophobicity, self-cleaning, and low-drag, *ACS Nano*, 2009, **3**, 4155–4163.
169. X. Deng, L. Mammen, H. J. Butt and D. Vollmer, Candle soot as a template for a transparent robust superamphiphobic coating, *Science*, 2012, **335**, 67–70.
170. Y. Lee, Y. Yoo, J. Kim, S. Widhiarini, B. Park, H. C. Park *et al.*, Mimicking a superhydrophobic insect wing by argon and oxygen ion beam treatment on polytetrafluoroethylene film, *J. Bionic Eng.*, 2009, **6**, 365–370.
171. X. Zhang, J. Zhang, Z. Ren, X. Li, D. Zhu, T. Wang *et al.*, Morphology and wettability control of silicon cone arrays using colloidal lithography, *Langmuir*, 2009, **25**, 7375–7382.
172. W. K. Cho and I. S. Choi, Fabrication of hairy polymeric films inspired by geckos: wetting and high adhesion properties, *Adv. Funct. Mater.*, 2008, **18**, 1089–1096.
173. X. Sheng and J. Zhang, Superhydrophobic behaviors of polymeric surfaces with aligned nanofibers, *Langmuir*, 2009, **25**, 6916–6922.
174. H. M. Bok, T. Y. Shin and S. Park, Designer binary nanostructures toward water slipping superhydrophobic surfaces, *Chem. Mater.*, 2008, **20**, 2247–2251.
175. H. Hoefnagels, D. Wu, G. de With and W. Ming, Biomimetic superhydrophobic and highly oleophobic cotton textiles, *Langmuir*, 2007, **23**, 13158–13163.
176. B. Qian and Z. Shen, Fabrication of superhydrophobic surfaces by dislocation-selective chemical etching on aluminum, copper, and zinc substrates, *Langmuir*, 2005, **21**, 9007–9009.
177. C. H. Xue, S. T. Jia, J. Zhang, L. Q. Tian, H. Z. Chen and M. Wang, Preparation of superhydrophobic surfaces on cotton textiles, *Sci. Technol. Adv. Mater.*, 2008, **9**, 035008.

178. C. H. Xue, S. T. Jia, H. Z. Chen and M. Wang, Superhydrophobic cotton fabrics prepared by sol–gel coating of TiO_2 and surface hydrophobization, *Sci. Technol. Adv. Mater.*, 2008, **9**, 035001.
179. C. H. Xue, S. T. Jia, J. Zhang and L. Q. Tian, Superhydrophobic surfaces on cotton textiles by complex coating of silica nanoparticles and hydrophobization, *Thin Solid Films*, 2009, **517**, 4593–4598.
180. K. Ramaratnam, V. Tsyalkovsky, V. Klep and I. Luzinov, Ultrahydrophobic textile surface via decorating fibers with monolayer of reactive nanoparticles and non-fluorinated polymer, *Chem. Commun.*, 2007, 4510–4512.
181. X. Wang and J. Shen, A review of contamination-resistant antireflective sol–gel coatings, *J. Sol–Gel Sci. Technol.*, 2011, 1–7.
182. U. Schulz, Review of modern techniques to generate antireflective properties on thermoplastic polymers, *Appl. Opt.*, 2006, **45**, 1608–1618.
183. J. Genzer and K. Efimenko, Recent developments in superhydrophobic surfaces and their relevance to marine fouling: a review, *Biofouling*, 2006, **22**, 339–360.
184. V. A. Ganesh, H. K. Raut, A. S. Nair and S. Ramakrishna, A review on self-cleaning coatings, *J. Mater. Chem.*, 2011, **21**, 16304–16322.
185. Y. F. Si and Z. G. Guo, Superhydrophobic nanocoatings: from materials to fabrications and to applications, *Nanoscale*, 2015, **7**, 5922–5946.
186. *Handbook of Photovoltaic Science and Engineering EAL*, S. Hegedus, John Wiley & Sons, Chichester, 2003.
187. X. Du, X. Li and J. He, Facile fabrication of hierarchically structured silica coatings from hierarchically mesoporous silica nanoparticles and their excellent superhydrophilicity and superhydrophobicity, *ACS Appl. Mater. Interfaces*, 2010, **2**, 2365–2372.
188. X. Li and J. He, In situ assembly of raspberry-and mulberry-like silica nanospheres toward antireflective and antifogging coatings, *ACS Appl. Mater. Interfaces*, 2012, **4**, 2204–2211.
189. L. Xu and J. He, Anti-fogging and-reflection coatings fabricated by integrating solid and mesoporous silica nanoparticles without any post-treatments, *ACS Appl. Mater. Interfaces*, 2012, **4**, 3293–3299.
190. L. Xu, X. Li and J. He, Mechanically durable superhydrophilic antireflective nanoparticles coatings via layer-by-layer assembly followed by post-treatments, *Acta Chim. Sin.*, 2011, **69**, 2648–2652.
191. Z. Geng, J. He and L. Xu, Fabrication of superhydrophilic and antireflective silica coatings on poly (methyl methacrylate) substrates, *Mater. Res. Bull.*, 2012, **45**, 1562–1567.
192. L. Yao and J. He, Multifunctional Surfaces with Outstanding Mechanical Stability on Glass Substrates by Simple H_2SiF_6-Based Vapor Etching, *Langmuir*, 2013, **29**, 3089–3096.
193. N. C. Linn, C. H. Sun, P. Jiang and B. Jiang, Self-assembled biomimetic antireflection coatings, *Appl. Phys. Lett.*, 2007, **91**, 101108.
194. C. H. Sun, A. Gonzalez, N. C. Linn, P. Jiang and B. Jiang, Templated biomimetic multifunctional coatings, *Appl. Phys. Lett.*, 2008, **92**, 051107.

195. C. H. Sun, P. Jiang and B. Jiang, Broadband moth-eye antireflection coatings on silicon, *Appl. Phys. Lett.*, 2008, **92**, 061112.
196. C. H. Sun, W. L. Min, N. C. Linn, P. Jiang and B. Jiang, Templated fabrication of large area subwavelength antireflection gratings on silicon, *Appl. Phys. Lett.*, 2007, **91**, 231105.
197. W. L. Min, B. Jiang and P. Jiang, Bioinspired self-cleaning antireflection coatings, *Adv. Mater.*, 2008, **20**, 3914–3918.
198. L. Yao, J. H. He, Z. Geng and T. T. Ren, Fabrication of mechanically robust, self-cleaning and optically high-performance hybrid thin films by SiO2&TiO2 double-shelled hollow nanospheres, *Nanoscale*, 2015, **7**, 13125–13134.
199. X. T. Zhang, O. Sato, M. Taguchi, Y. Einaga, T. Murakami and A. Fujishima, Self-cleaning particle coating with antireflection properties, *Chem. Mater.*, 2005, **17**, 696–700.
200. D. Wu, Y. B. Zhao, S. Z. Wu, Y. F. Liu, H. Zhang, S. Zhao *et al.*, Simultaneous efficiency enhancement and self-cleaning effect of white organic light-emitting devices by flexible antireflective films, *Opt. Lett.*, 2011, **36**, 2635–2637.
201. M. S. Park and J. K. Kim, Broad-band antireflection coating at near-infrared wavelengths by a breath figure, *Langmuir*, 2005, **21**, 11404–11408.
202. L. Zhang, Y. Li, J. Sun and J. Shen, Layer-by-layer fabrication of broad-band superhydrophobic antireflection coatings in near-infrared region, *J. Colloid Interface Sci.*, 2008, **319**, 302–308.
203. K. C. Camargo, A. F. Michels, F. S. Rodembusch and F. Horowitz, Multi-scale structured, superhydrophobic and wide-angle, antireflective coating in the near-infrared region, *Chem. Commun.*, 2012, **48**, 4992–4994.
204. K. C. Park, H. J. Choi, C. H. Chang, R. E. Cohen, G. H. McKinley and G. Barbastathis, Nanotextured silica surfaces with robust superhydrophobicity and omnidirectional broadband supertransmissivity, *ACS Nano*, 2012, **6**, 3789–3799.
205. T. Verho, C. Bower, P. Andrew, S. Franssila, O. Ikkala and R. H. A. Ras, Mechanically durable superhydrophobic surfaces, *Adv. Mater.*, 2011, **23**, 673–678.
206. O. Duyar and H. Z. Durusoy, Design and preparation of antireflection and reflection optical coatings, *Turk. J. Phys.*, 2004, **28**, 139–144.
207. A. Cannavale, F. Fiorito, M. Manca, G. Tortorici, R. Cingolani and G. Gigli, Multifunctional bioinspired sol-gel coatings for architectural glasses, *Build. Environ.*, 2010, **45**, 1233–1243.
208. J. Chen and C. S. Poon, Photocatalytic construction and building materials: From fundamentals to applications, *Build. Environ.*, 2009, **44**, 1899–1906.
209. P. Doshi, G. E. Jellison and A. Rohatgi, Characterization and optimization of absorbing plasma-enhanced chemical vapor deposited antireflection coatings for silicon photovoltaics, *Appl. Opt.*, 1997, **36**, 7826–7837.

210. B. K. Shin, T. I. Lee, J. Xiong, C. Hwang, G. Noh, J. H. Cho *et al.*, Bottom-up grown ZnO nanorods for an antireflective moth-eye structure on $CuInGaSe_2$ solar cells, *Sol. Energy Mater. Sol. C*, 2011, **95**, 2650–2654.
211. J. Müller, B. Rech, J. Springer and M. Vanecek, TCO and light trapping in silicon thin film solar cells, *Sol. Energy*, 2004, **77**, 917–930.
212. O. El Daif, E. Drouard, G. Gomard, A. Kaminski, A. Fave, M. Lemiti *et al.*, Absorbing one-dimensional planar photonic crystal for amorphous silicon solar cell, *Opt. Express.*, 2010, **18**, A293–A299.
213. C. Bernhard, G. Kolmodin and B. Meyerson, On the prenatal development of function and structure in the somesthetic cortex of the sheep, *Prog. Brain Res.*, 1967, **26**, 60–77.
214. M. Oliveira Jr., D. Silva, A. Côrtes, M. Namani and F. Marques, Diamond like carbon used as antireflective coating on crystalline silicon solar cells, *Diamond Relat. Mater.*, 2009, **18**, 1028–1030.
215. S. Furbo and L. Jivan Shah, Thermal advantages for solar heating systems with a glass cover with antireflection surfaces, *Sol. Energy*, 2003, **74**, 513–523.
216. A. Salim, F. Huraib and N. Eugenio, *PV Power-study of System Options and Optimization*, 1988, p. 688–692.
217. J. C. Ward and G. O. G. Lof, Long-term (18 years) performance of a residential solar heating system, *Sol. Energy*, 1976, **18**, 301–308.
218. K. Motohashi and T. Inukai, Self-cleaning performance evaluation of commercial photocatalyst coating materials through 5 years outdoor exposure. *RILEM International Symposium On Photocatalysis, Environment and Construction Materials Italy*, 2007, pp. 307–313.
219. J. Chen, H. Ye, L. Aé, Y. Tang, D. Kieven, T. Rissom *et al.*, Tapered aluminum-doped vertical zinc oxide nanorod arrays as light coupling layer for solar energy applications, *Sol. Energy Mater. Sol. C*, 2011, **95**, 1437–1440.
220. C. Lee, T. Lee, S. Lyu, Y. Zhang, H. Ruh and H. Lee, Field emission from well-aligned zinc oxide nanowires grown at low temperature, *Appl. Phys. Lett.*, 2002, **81**, 3648.
221. M. Foley, Technical advances in microstructured plastic optics for display applications. *SID Symposium Digest of Technical Papers*, Wiley Online Library, 1999, p. 1106.
222. T. Ohishi, An ethoxy-silano rhodamine B derivative and its application to surface coatings on display devices, *J. Non-Cryst Solids*, 2003, **332**, 80–86.
223. J. Park, L. D. Fouché and P. T. Hammond, Multicomponent patterning of layer-by-layer assembled polyelectrolyte/nanoparticle composite Thin films with controlled alignment, *Adv. Mater.*, 2005, **17**, 2575–2579.
224. X. Zhu, Z. Zhang, X. Men, J. Yang, K. Wang, X. Xu *et al.*, Robust superhydrophobic surfaces with mechanical durability and easy repairability, *J. Mater. Chem.*, 2011, **21**, 15793–15797.

225. P. A. Levkin, F. Svec and J. M. J. Fréchet, Porous polymer coatings: a versatile approach to superhydrophobic surfaces, *Adv. Funct. Mater.*, 2009, **19**, 1993–1998.
226. K. Liu, X. Yao and L. Jiang, Recent developments in bio-inspired special wettability, *Chem. Soc. Rev.*, 2010, **39**, 3240.
227. A. Tuteja, W. Choi, M. Ma, J. M. Mabry, S. A. Mazzella, G. C. Rutledge *et al.*, Designing superoleophobic surfaces, *Science*, 2007, **318**, 1618–1622.
228. L. Yao and J. He, Recent progress in antireflection and self-cleaning technology – From surface engineering to functional surfaces, *Prog. Mater. Sci.*, 2014, **61**, 94–143.

CHAPTER 9

Advances in Oil/Water Separation of Biomimetic Superhydrophobic Coatings

ZHIGUANG GUO*[a,b] AND FUCHAO YANG[a,c]

[a] State Key Laboratory of Solid Lubrication, Lanzhou Institute of Chemical Physics, Chinese Academy of Sciences, Lanzhou 730000, People's Republic of China; [b] Hubei Collaborative Innovation Centre for Advanced Organic Chemical Materials and Ministry of Education Key Laboratory for the Green Preparation and Application of Functional Materials, Hubei University, Wuhan 430062, People's Republic of China; [c] University of Chinese Academy of Sciences, Beijing 100049, People's Republic of China
*Email: zguo@licp.cas.cn

9.1 Introduction

As a general consensus, environmental protection has long been a particularly important issue. Unfortunately, oil or organic chemical leaks during industrial accidents or oil tankers sinking happen frequently.[1] For example, oil spill accidents have continued to occur, from the 1967 Torrey Canyon oil spill to the latest 2011 Bohai Bay oil spill.[2] The explosion of the "Deep water Horizon" oil rig,[3] the most serious pollution incident of the last decade, is catastrophic for the environment and the effect is long term and lethal, especially for marine and aquatic ecosystems. It has also produced an abominable influence on people's normal life.[4] To address the oil spill, artificial separation, rather than burning, of the oily water is a more satisfactory way. Burning the as-spilled oil would waste energy and create

RSC Smart Materials No. 21
Self-cleaning Coatings: Structure, Fabrication and Application
Edited by Junhui He

Published by the Royal Society of Chemistry, www.rsc.org

more environmental pollution while the separation of as-spilled oil can be re-collected and reused in various industries. Aside from these, wastewater from industries such as petrochemicals, food, textiles, leather, and metal processing, has not been treated completely and discharged, resulting in lasting detrimental ecological effects. In order to deal with these issues, researchers have paid more and more attention to investigating separating oil/water mixtures.[5]

Until now, methods applied to separating oil/water mixtures have been gradually put forward.[6] Previously, some conventional techniques were employed to deal with oily wastewater, *i.e.*, centrifuges, oil skimmers, coalescers, settling tanks, gravity separations, ultrasonic separations, coagulation, an electric field, and flotation technologies.[7] During these separation processes, low separation efficiency, high energy consumption, and the secondary pollutants are limited to their wide applications. Thus, it is highly desirable to look for more efficient and economical methods to solve these problems.[5]

Essentially speaking, the oil/water separation is the wettability behavior that occurs at the interface of the solid, air, water and oil phase. Materials with superhydrophobic surfaces, one of the extreme states of surface wettability, have been inspired, intensively explored and accelerated by the discovery of a lotus leaf's self-cleaning phenomenon in nature.[8] Various porous or pulverous superhydrophobic materials have been investigated for the absorption and removal of various kinds of oils and organic solvents from oily wastewater. If the oil/water mixture is in an emulsified state, materials usually used for immiscible oil/water separation may be invalid because of the smaller droplet size. An elaborate membrane filtration technology was developed and can often be used to separate various emulsions.

In this chapter, we will begin with the understanding and design of the superhydrophobic surface, its definition, approaches to superhydrophobic surfaces and endowing some special materials with superhydrophobic property. Next, we will discuss various oil–water separations based on various superhydrophobic or superwettable materials. As for the separation of emulsified oil/water mixtures, the materials for the separation of water-in-oil, oil-in-water, and both water-in-oil and oil-in-water emulsions will be sequentially introduced. Followed by these, the principles to optimal design of materials bearing on separating oil–water mixtures are presented. Finally, the summary and outlook concerning future development of oil–water separations will be included.

9.2 Understanding and Design of the Superhydrophobic Surface

9.2.1 Understanding the Superhydrophobic Surface

It is widely accepted that wettability is the tendency of a liquid to spread on a solid substrate and is characterized by the water contact angle (WCA).

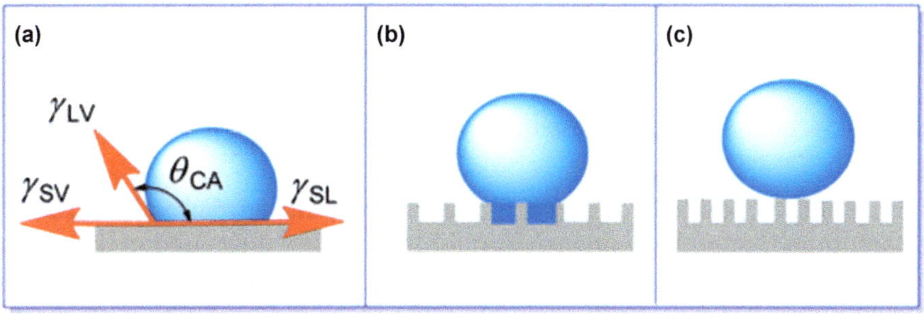

Figure 9.1 Schematic illustration of a droplet placed onto a flat substrate (a) and rough substrates (b) and (c). The droplet on a rough surface is either in a Wenzel state (b) or a Cassie–Baxter state (c).
Reproduced from ref. 11 with permission from the Royal Society of Chemistry.

The WCA is defined geometrically as the angle formed by a liquid droplet at the three phase boundary where a liquid, gas and solid intersect each other as shown in Figure 9.1(a).[9] The value of WCA on a surface can be used to determine the degree of affinity between liquid and solid, denoting a surface with contact angle (CA) below 90° as hydrophilic. If the CA value on a surface is larger than 90°, then it can be denoted as a hydrophobic surface.[10] The sliding angle (SA), is a threshold tilting value of the angle between the surface and horizon line, above which a liquid droplet starts to roll/slide upon elevating an end of the surface.[11] The SA of a water droplet on a solid surface is also an important parameter for predicting its wetting behavior in practical applications.[12] Generally, a superhydrophobic surface is defined as having a WCA greater than 150° but a SA less than 10° for a 5 μL droplet. Such surfaces commonly exist in the biological world and many plant leaves (lotus leaf,[9b] ramee rear surface[13]) and specific surfaces of animals (water striders,[14] eyes of mosquito[15]) in nature exhibit their superhydrophobic surface. As for a superhydrophobic surface, the interplay of surface microstructure and chemical composition causes water droplets to remain spherical on this kind of surface or slide away easily.

Next, we will discuss the effect of roughness factor and free surface energy on the wettability of solid surface. For an ideal smooth solid surface, Young's equation can be expressed as follows:

$$\cos\theta_0 = \frac{\gamma_{SA} - \gamma_{SL}}{\gamma_{LA}} \tag{9.1}$$

where γ_{SA} and γ_{LA} are the surface energies (surface tensions) of the solid and liquid against air, and γ_{SL} is the interface energy (interface tension) between solid and liquid. In a practical situation, the roughness of a solid surface can greatly magnify the wetting properties of this solid even without taking into account any other potential influence factors. The "roughness factor" can be denoted by r and is given as:

$$r = \text{roughness factor} = \frac{\text{actual surface}}{\text{geometric surface}} \tag{9.2}$$

It should be pointed out that the area of the actual surface will be always greater than that of the geometric surface (the surface measured in the plane of the interface). Wenzel proposed an equation that gives a relation between the equilibrium CA and the apparent angle formed on a rough surface, referred to as the Wenzel equation:

$$\cos\theta_W = r\cos\theta_0 \tag{9.3}$$

where θ_0 is the equilibrium CA, θ_W is the apparent CA on a rough surface and the Wenzel state is schematically shown in Figure 9.1(b). This equation directly tells us that the roughness can amplify the wetting property of the solid surface. That is to say, if a water droplet is initially wetting on a smooth surface, the roughness would lead to greater hydrophilicity; if a water droplet is originally non-wetting on the smooth surface, then the introduction of the roughness can make the surface of the solid substrate even more non-wetting, resulting in an enhancement of hydrophobicity.

The Wenzel equation has revealed the apparent CA on a rough surface, but it would be invalid while the solid surface is porous or composed of diffident composition. Cassie and Baxter further amended the equations proposed by introducing the surface coefficient, referred to as the Cassie–Baxter equation given below:

$$\cos\theta_{CB} = f_{SL}(1 + \cos\theta_W) - 1 \tag{9.4}$$

where f_{SL} represents the solid–liquid fraction under the contact area and the Cassie–Baxter state is schematically shown in Figure 9.1(c). When $f_{SL} = 1$, the Cassie–Baxter equation turns into the Wenzel equation.

On the other hand, water droplets initially wetting or non-wetting an ideal smooth surface depends on its surface properties. Surface atoms or molecules of liquids or solids have higher energy than similar atoms and molecules in the interior, which results in surface tension or surface free energy (surface energy) so as to reach a stable state with a relatively lower energy. So the surface energy or surface tension can be used to evaluate the surface properties. In general, increasing the surface roughness was not directly based on a low-surface-energy substrate and then modifying the as-prepared rough surface with low-surface-energy materials is obligatory. Special low-surface-energy materials will be illustrated later in this chapter. This elucidation regarding roughness and surface properties is helpful for us to make surfaces with a superhydrophobic property.

9.2.2 Approaches to a Superhydrophobic Surface

Based on the preceding section, we bear in mind that the surface roughness and surface energy (chemical composition) are both crucial factors for the formation of superhydrophobic surfaces. Numerous fabrication methods have been developed for the preparation of rough surfaces, each having their own strengths and weaknesses.[16] They are often divided into two basic approaches: the top-down method and the bottom-up method.[17]

As for the top-down method, etching, lithography and anodization are widely used. Etching—removal of material from original surfaces—is a facile process to produce a rough structure and can be separated into wet etching and dry etching. Based on wet etching, Guo *et al.* have produced rough surfaces on aluminium and its alloy with binary structures at micro- and nanometer scales by subtle oxidation[18] and Zhang *et al.* prepared micro-flowers and nanorod array hierarchical structures on a copper substrate by immersing it into a mixed solution of NaOH and ammonium persulfate ($(NH_4)_2S_2O_8$).[19] Laser processing is an example of dry etching to construct a special surface topography. Yong *et al.* prepared spike-like structures on a flat Si surface *via* femtosecond laser micromachining by a line-by-line and serial scanning process, as shown in Figure 9.2.[20] Utilizing the technique of lithography, surfaces can be patterned with different shapes and different sizes. Jo *et al.* firstly prepared a ZnO nanoparticles dispersion resin and then patterned a circular cone-shaped micro-pattern with the as-prepared resin by ultraviolet nanoimprint lithography.[21] Anodization, often used as a material protection technology, is a kind of metal surface treatment in which the surfaces of a metal and its alloy form an oxide film through an impressed anodic current in an electrolyte solution. Barthwal *et al.* fabricated a rough surface based on a Ti plate *via* anodization.[22] The TiO_2 nanotube arrays are tightly formed on the entire surface of the Ti substrate with a satisfactory appearance.

As for the bottom-up methods, electrodeposition, the hydrothermal method and the sol–gel process are selected as representatives. Electrodeposition is aimed at covering the substrate with a layer of metal. The surface morphology can be controlled by changing monomer structures and the electrochemical parameters. We have fabricated rough copper mesh films *via* a facile and environmentally friendly method by electroplating Cu nanoparticles on as-cleaned copper mesh films.[23] The hydrothermal method, the products of which are high purity uniformly disperse, is another useful technique to create roughness. We have prepared a rough surface based on hierarchical rutile TiO_2 flowers through a simple one-step hydrothermal method.[24] The flower-like structure was made up of many petals on the surface obtained and each petal consisted of many ultrathin nano-needles as shown in Figure 9.3 and these rough coatings show super-amphiphilicity in air. As for the sol–gel process, chemical solution or sol is utilized as a precursor. When a large amount of solvent remains in the network during the network formation process, a gel is formed after a series

Figure 9.2 Typical SEM images of a structure irradiated by femtosecond laser. (a) 45° tilted view SEM image. (b) Higher resolution 45° tilted view SEM image. (c) Top view SEM image. (d) High magnification SEM image of a single spike decorated with nanoscale protrusions.
Reproduced from ref. 20 with permission from the Royal Society of Chemistry.

of hydrolyses of the precursor.[25] The sol–gel method can be applied to many kinds of substrates, such as metals, silicon wafers, glass, and textiles, fabricating superhydrophobic coatings.[26] We have developed a feasible method to fabricate a spray of sol–gel nanocoatings which can realize transparent superhydrophobic coatings on most solid surfaces through a mild process without any pre-treatment.[27]

We found that the top-down method achieves a surface topography with high control, and the bottom-up method usually leads to a random structure.[28] However, the top-down method is difficult to put into effect.

These methods are applied to construct a rough surface. Usually, surface modification with low surface energy materials is required for fabricating a superhydrophobic surface.[29] Until now, various kinds of low-surface-energy coatings have been developed as shown in Figure 9.4, such as, alkanethiols,[30] organic silanes,[31] fatty acids,[32] aromatic azide,[33] and fluorocarbon.[34] Alkanethiols are a kind of low-surface-energy materials bearing a hydrophobic alkyl chain and a thiol group as a surface anchor. Organic silane was used to modify surfaces bearing hydroxyl groups (hydrophobicity of its alkyl chain), such as silicon wafers, glass slides, silicon nanoparticles

Figure 9.3 FE-SEM images of hierarchical rutile TiO_2 flowers: (a) top view of flower clusters, (b) top view of an individual flower, and (c) the close-up of nanopetals. (d) The close-up of a selected area enclosed by a black rectangle in inset; the inset shows TEM image of a single nanopetal.

and zeolite, to create low-surface-energy coatings.[31] Fatty acid is another important low-surface-energy material and surface wetting ability with different contact angles can be obtained by modifying rough surfaces with different lengths of alkyl chain. By use of a surface-reactive molecule of 4-azido-*N*-dodecylbenzamide with a molecule bearing an alkyl chain as a hydrophobic tail and an azide group as the reactive surface anchor, Zhang *et al.* developed a simple and convenient method to provide stable low-surface-energy coatings on organic surfaces.[33] The chemical composition determines the low surface free energy, and a lower surface energy leads to higher hydrophobicity.[35] Generally, these low-surface-energy coatings were usually anchored on the rough substrate by self-assembled or spin-coating methods. Self-assembled monolayers (SAMs) are molecular assemblies that are formed spontaneously by the immersion of an appropriate substrate into a solution of an active surfactant in an organic solvent.[29]

9.2.3 Endowing Special Materials with a Superhydrophobic Property

The special materials here have the potential to be used for oil–water separations. These functional materials mainly contain sponge and foam-based

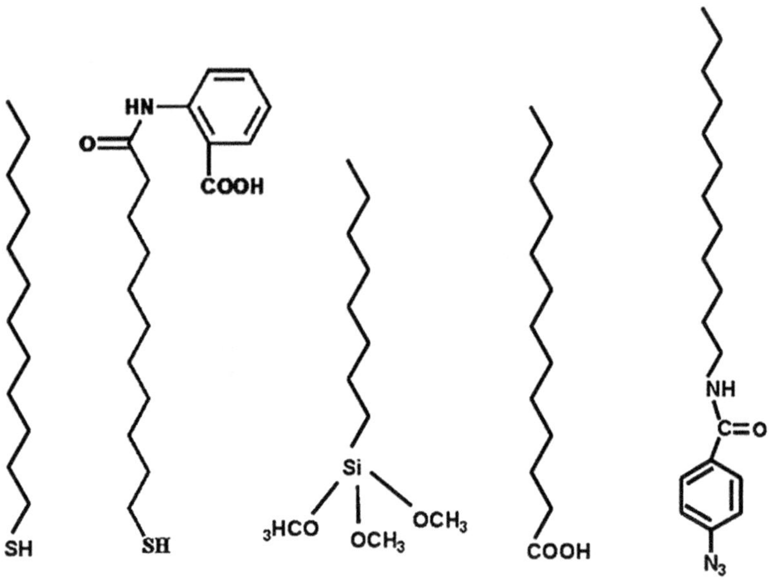

Figure 9.4 Examples of low-surface-energy modifications: long alkyl chain thiols, long alkyl chain thiols bearing benzoic acid, alkyl or fluorinated organic silanes, long alkyl chain fatty acids, and alkyl chain modified aromatic azides (from left to right).
Reproduced from ref. 29 with permission from the Royal Society of Chemistry.

materials, fabric-based materials, metallic mesh-based materials, particles and powdered materials.

9.2.3.1 Sponge and Foam-based Materials

Sponges and foams, with their inherently plentiful holes and hydrophilic properties, are cheaply available materials. Sponges and foams show an interconnected three-dimensional skeleton supported by chaotic fiber assemblies, thus endowing the sponges and foams with a huge space for oil absorption and storage. Generally, they can absorb various liquids (including water and oils or organics) with poor selectivity, which make them unrealistic for removing oils/organics from the water phase. Importantly, high selective absorption can be realized on a sponge after a construction of a befitting surface topography and modification with low surface energy substances. Pan *et al.* used a solution-immersion process to prepare superhydrophobic polyurethane (PU) sponges bearing a Ag coating.[36] They claimed that the as-prepared sponges quickly and selectively absorbed various kinds of oils over 13 times the sponges' weight while completely repelling water through a combination of porous, superhydrophobic, and superoleophilic properties. We have used a facile and easily scalable fabrication technique based on commercially available PU foams functionalized with polypyrrole (PPy) by *in situ* polymerization. This sort of PPy-coated sponge comprising nano-scale rough coatings and micro-structured porous substrates is shown in Figure 9.5(a).[37] Such a superhydrophobic PPy-coated sponge is quite stable in several harsh conditions, including heating at 200 °C for 1 h, and freezing at −20 °C for 1 h. As shown in Figure 9.5(b), when the sponge was dropped on a *n*-hexane/water surface, it floated easily on the water and set out to absorb the red *n*-hexane quickly. By simply moving them around oil-polluted waters using tweezers, they can absorb the floating oil from the polluted regions, thereby purifying the contaminated water. In Figure 9.5(c), the underwater dichloroethane (higher density than water) can also be removed and recycled efficaciously using this superhydrophobic sponge. Endowed with a superhydrophobic property, sponge and foam-based materials are one of the promising routes for developing oil/water separation materials.

9.2.3.2 Fabric-based Materials

After endowing fabric with a superhydrophobic property, it is also considered a good candidate for oil/water separation because the ready-made fibers in the integral fabric provide microscale roughness and their native porosity ensures the free passage of liquids. It should be noted that the initial roughness of the fabric is not sufficient and establishing a hierarchical rough structured surface on the fabric is needed. Many research groups have utilized a dip coating method, one of the simplest approaches, to achieve this goal. Xue *et al.* have reported that by coating fibers with titania sol to generate a dual-size surface roughness, followed by hydrophobization with stearic acid, 1*H*,1*H*,2*H*,2*H*-perfluorodecyltrichlorosilane or their combination, hydrophilic cotton fabrics were made superhydrophobic.[38] We have proposed a general methodology for robust superhydrophobic fabrics *via* the *in situ* growth of both transition-metal oxides and metallic nanocrystals, including the simple neutralization reaction and oxidation–reduction reaction. The porous surfaces coated with Group VIII and IB nanocrystals (such as Fe, Co, Ni, Cu, and Ag) can not only present multiscale surface roughness, but also readily coordinate with thiols, leading to special wettability.[39] Frankly, fabric has its weakness, such as a thin, paper-like two-dimensional structure, which seriously limits its oil capture ability.

9.2.3.3 Metallic Mesh-based Materials

As important and irreplaceable engineering materials, metals are widely used in our daily life.[40] Metallic mesh-based materials possess relative uniform and different sizes of pores with different meshes (200, 400 or 800 meshes and so on) per square inch. There are many types of metallic mesh-based materials, such as stainless steel mesh, copper mesh, titanium mesh, *etc.* To enhance the superhydrophobic behavior of a porous metallic

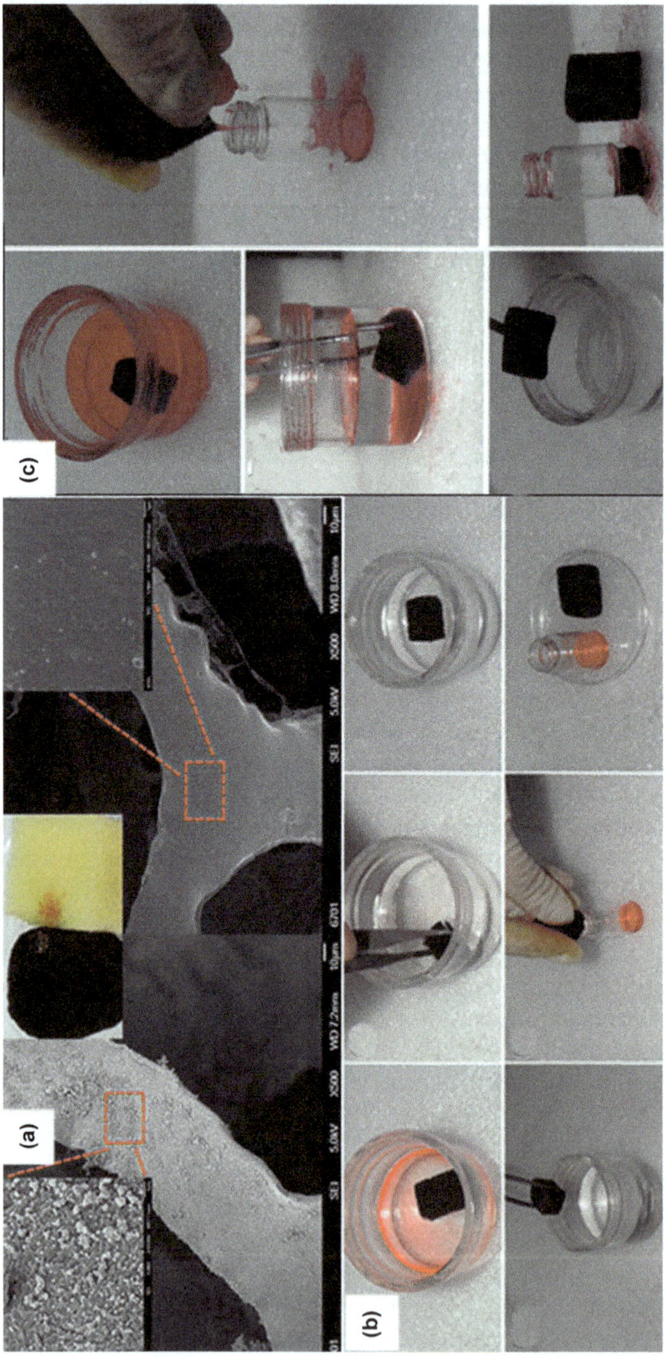

Figure 9.5 (a) SEM images of the superhydrophobic PPy-coated and pristine PU sponge. (b) The absorption and recycling process of *n*-hexane. (c) The absorption and recycling process of the underwater dichloroethane. Reproduced from ref. 37 with permission from the Royal Society of Chemistry.

mesh film, a filtering mesh with selective permeability to water and oil was prepared, resulting in oil/water separating properties. A stainless steel mesh was the first material for fabricating superhydrophobic surfaces *via* a spray-and-dry method by Jiang *et al.* in 2004.[41] The pre-mixed aqueous emulsion containing Teflon, adhesive (polyvinyl acetate), dispersant (polyvinyl alcohol), and surfactant (sodium dodecyl benzene sulfonate) was sprayed evenly on the stainless mesh with compressed air and then it underwent a high-temperature drying process (350 °C) to decompose the adhesive, dispersant and surfactant. The prepared mesh was so water-repellent that a water droplet was unstable on it and rolled off easily We have reported that mixed modifiers of methyl-terminated thiol and carboxyl-terminated thiol were successfully assembled on stainless steel meshes utilizing polydopamine as the adhesion layer and the strong thiol ligand with Ag.[42] Importantly, the surfaces modified by mixed thiol show different responsive behavior to non-basic and basic water droplets. Plus, the selectiveness of high water–oil repellence reveals the unique and smart tendencies of the as-prepared functional stainless steel mesh.

Copper mesh is another frequently used substrate for oil/water separation. Inspired by a ramee leaf, rod-like unitary structured films were fabricated on copper meshes *via* a facile and simple chemical vapor corrosion method by us.[43] Following the surface modification of 1-decanethiol ($C_{10}H_{22}S$), the as-prepared films exhibiting superhydrophobicity and superoleophilicity can be applied to separate an oil–water mixture. In a comparative perspective, these superhydrophobic surfaces with unitary microstructures have better surface mechanical properties than those with binary micro- and nano-structures. Furthermore, we have fabricated a superhydrophobic copper mesh film, which can orderly separate the oil/water mixtures.[44] During the separation process, oil selectively permeated from water first, but then the water also spontaneously passed through later. For all we know, there has been no report about separating oil/water mixtures sequentially before. Compared to previous reports, it effectively simplifies the operation with no need for excluding water after oil separation. Besides, instead of making efforts to design materials responding to external stimuli (such as pH, temperature, UV, and magnetic field), it is much more simple to control the oil and water collection step by step *via* regulation of the oil/water mixture. What is more, the raw material used here is cheap, commercially available, and easily fabricated, which is of great importance to design an ideal separation material for industrial applications.

9.2.3.4 Particles and Powdered Materials

In the preceding subsections, the superhydrophobic surfaces used in water–oil separation were mainly fabricated on porous substrates, such as meshes, fabrics, membranes, and so on. These oil/water separation surfaces allow oil to penetrate through them whereas they resist water due to negative capillary effects. Although they can efficiently separate a mixture of oil and water, they cannot selectively remove oil *in situ* from the oil–water interface. Recently, some novel hydrophobic and oleophilic particles without porous skeletons like sponges have been developed to achieve selective oil separation *in situ* from an oil/water mixture. Mahapatra *et al.* developed a simple, cost-effective and "green" route for the synthesis of highly hydrophobic and oleophilic $CaCO_3$ particles in the absence of any solvent and at ambient temperature conditions.[45] Oleic acid as an additive promotes polymorph selectivity from calcite to vaterite with "flaky-floret" like morphology with concomitant changes in its surface wettability from hydrophilic to hydrophobic. The materials synthesized have been successfully employed for selectively mopping up a variety of oils such as sunflower oil, engine oil and diesel from the surface of water. This has been established in its powder form as well as in the format of a "spill-pouch". In addition, the material has been used to fabricate functionalized polyurethane foam, which mops up oil spillages from water and renders the oil easy to recover. Although these particles and powders can be used in oil spill clean-ups, they are difficult to transfer and recycle. Naturally, magnetic particles/powders were developed to overcome the recycling problem because particles/powders with magnetism can be easily collected using an external magnetic field. We have reported a facile method to synthesize Fe_3O_4@polydopamine (Fe_3O_4@PDA) composite microspheres as shown in Figure 9.6.[1] These functional Fe_3O_4@PDA core–shell composite microspheres with magnetic response and special wetting properties were successfully assembled utilizing the modification of 1*H*,1*H*,2*H*,2*H*-perfluorodecyltrichlorosilane. In this paper, we demonstrated

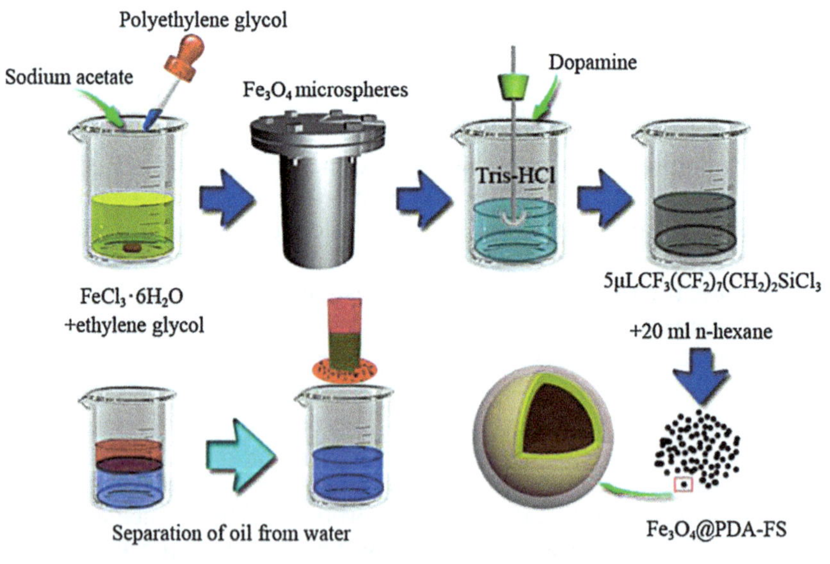

Figure 9.6 Schematic representation of the preparation and application of superhydrophobic Fe_3O_4@PDA-FS microspheres.

that the as-obtained product with designed functionality has favorable superhydrophobic and superparamagnetic properties, showing significant advantages in both magnetic control behavior and oil/water separation.

9.3 Various Oil–Water Separations Call for Various Superwettable Materials

Even superhydrophobic materials have been widely applied for oil–water separations, they cannot cover all this research field. As mentioned previously, we will expand to other extreme states of surface wettability, superhydrophilic/underwater superoleophobic surfaces. We should point out that the surfaces with superhydrophilic property often exhibit underwater superoleophobicity naturally and the difference between superhydrophobicity and superhydrophilicity is only whether the rough substrate is modified by low surface-energy coatings or not. Here, we regard the superwettable materials with extreme states of surface wettability including superhydrophobic/superoleophilic materials and superhydrophilic/underwater superoleophobic materials.

9.3.1 The Background of Oil–Water Mixture Formation

Mixtures of oil and water can be classified in terms of the diameter (d) of the dispersed phase: free oil (oil droplet size >150 μm), dispersed oil (20 μm < oil droplet size < 150 μm), and emulsified oil (oil droplet size <20 μm).[46] Here, we denote the mixtures of oil and water with oil droplet size >150 μm as layered oil-and-water mixtures (or immiscible oil–water mixtures), and denote the mixtures of oil and water with oil droplet size <20 μm as emulsions. Utilizing the special wetting behavior of functional surfaces, two different types of these materials—superhydrophobic/superoleophilic surfaces and superhydrophilic/underwater superoleophobic surfaces—have been successfully designed, fabricated, and employed in the separation of layered oil–water mixtures, as given in Section 9.3.2.

Emulsions tend to have a cloudy appearance because the multiple phase-interfaces scatter light as they pass through the emulsion. An oil–water emulsion is a complex form of oil–water coexistence, especially in the presence of surfactants (or dispersants). Common emulsions are inherently unstable and, thus, do not tend to form spontaneously. Even to form a temporary emulsion, energy input through stirring, homogenizing, shaking, or exposure to power ultrasound is needed. Over time, emulsions tend to revert to the stable states of the phases comprising tiny parts of emulsion. To obtain a stable state of emulsions, an emulsifier (or surfactant), a substance with both hydrophobic and hydrophilic groups, which stabilizes an emulsion by increasing its kinetic stability, are necessary.

Despite this, oily wastewater, whether industrial effluent or sanitary wastewater, exists in emulsion form. So the separation of oil–water emulsions has a double significance in ecology and economics, and special attention is paid to the latest development in advanced materials for oil–water emulsion separation below.

9.3.2 Types of Superwettable Surfaces Applied for Immiscible Oil–Water Separations

9.3.2.1 *Superhydrophobic/Superoleophilic Surfaces Applied for Oil–Water Separations*

Due to the lower surface tension of oil, it is usually hard for the superhydrophobic surfaces to repel oil droplets and in many cases, superhydrophobic surfaces exhibit a superoleophilic property. Superhydrophobic–superoleophilic materials are typical oil-removing materials. The superhydrophobic and superoleophilic properties will make an oil phase spread easily while the water phase will be repelled, thus separating oils from an oil–water mixture. These types of ideal oil removing materials are generally considered to have a high oil absorption capacity as well as a low water pick up, low density, they are environmentally friendly, and have good recyclability for a wide range of oils/organics.

We have presented that stable superhydrophobic and superoleophilic nanocoatings on textiles and sponges, successfully achieved from thiol-ligand nanocrystals, can effectively be used for application in oil/water separation.[47] The nanocrystals include VIII and IB metals and oxide nanoparticles, such as Fe, Co, Ni, Cu and Ag. Textiles with superhydrophobic VIII and IB nanocoatings with *n*-octadecyl thiol as a modifier have a similar surface energy to oil droplets, and these modified textiles will favor superoleophilicity. The difference between water and oil on these surfaces with special wettability results in one intrinsic application in oil/water separation.

We have developed a kind of stable superhydrophobic/superoleophilic soft porous material for oil/water separation *via* the oxidative chemical polymerization of aniline on the surface of fabrics as shown in Figure 9.7(a).[48] Interestingly, the as-prepared fabrics showed stable and robust superhydrophobic properties towards many corrosive solutions (acidic, basic, salt liquids), hot water, and mechanical abrasion. In addition, it was proven that this method can also be applied to other porous materials with different pore diameters and chemical composition, such as stainless steel meshes with different pore diameters and sponges. As shown in Figure 9.7(b), when a mixture of water and oil (dyed with Solvent Red 24 for clear observation) was poured from a separating funnel onto a mesh, the water flowed to the outside of the bottle into the beaker, while oil quickly permeated though the mesh and dropped into the bottle. The results indicated that the oil/water mixture could be effectively separated by the as-prepared superhydrophobic PANI-coated mesh. Importantly, oil/water separation has successfully been achieved *via* the superhydrophobic/superoleophilic PANI-coated fabric, as illustrated in Figure 9.7(c). When the hexane and water mixture was poured onto the superhydrophobic fabric,

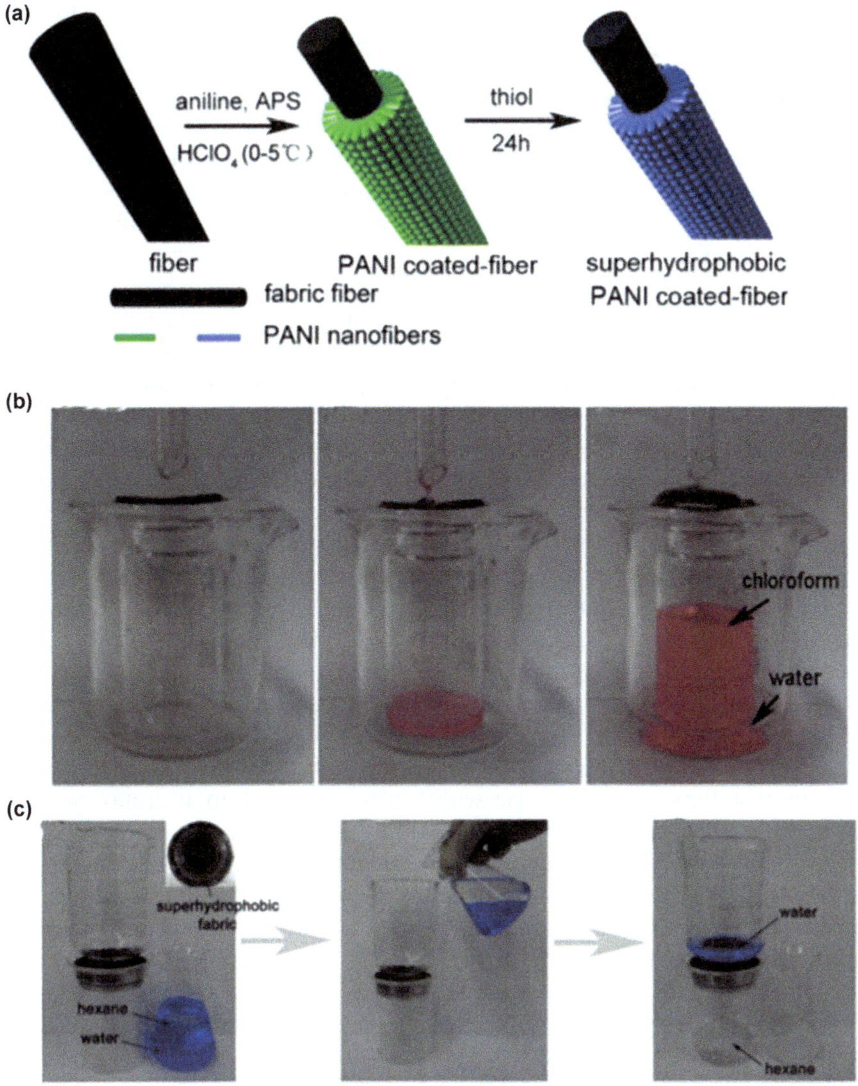

Figure 9.7 (a) Schematic drawing of the synthesis procedure of the superhydrophobic PANI-coated fabric. (b) Oil/water separation process of the superhydrophobic mesh coated with PANI nanofibers; the chloroform was dyed with Solvent Red 24 for clear observation. (c) Oil/water separation process of hexane and water using the superhydrophobic PANI-based fabric; the water was dyed by methylene blue for clear observation. Reproduced from ref. 48 with permission from the Royal Society of Chemistry.

hexane quickly spread and permeated through the fabric and then rapidly dropped into the beaker underneath, whereas the water (dyed by methylene blue for clear observation) resided on the fabric.

9.3.2.2 Superhydrophilic/Underwater Superoleophobic Surfaces Applied for Oil–Water Separations

Gradually, researchers have discovered some drawbacks of superhydrophobic/superoleophilic materials, *i.e.*, they are easily fouled or blocked by oil due to high viscosity and these materials are usually unsuitable for oils that are less dense than water. Fish scales, which are known to be well protected from contamination by oil pollution in the sea, have stimulated particular interest. Inspired by fish scales, a promising oil–water separation material with hydrophilic and underwater superoleophobic properties has been developed. Superhydrophobic/superoleophilic materials are applied to allow the oil phase to penetrate the surface (or absorb the oil phase) but repelling water while superhydrophilic and underwater superoleophobic surfaces can let the water pass through freely, however repelling oil totally. In 2014, our group achieved some good results regarding this aspect. Dong *et al.* demonstrated a novel method to fabricate a hydrophilic graphene oxide (GO) coating onto stainless steel meshes.[49] Compared to neat meshes, GO-coated meshes become more hydrophilic in air and superoleophobic under water. Taking advantage of this completely opposite wettability, it can separate various oils from water under gravity. Yu *et al.* reported an outstanding superhydrophilic and underwater superoleophobic film that can separate water from oil prepared by growing pure silica zeolite crystals on a stainless steel mesh.[50] Based on the excellent superhydrophilicity and underwater superoleophobicity of their zeolite surface, high separation efficiency of various oils can be achieved. Flux and intrusion pressure are tunable by simply changing the pore size, dependent on the crystallization time of the zeolite crystals. Based on an important type of zeolite (Silicalite-1: structure type MFI) coating on the stainless steel, we have prepared special wettable films that are superamphiphilic in air and superoleophobic underwater as shown in Figure 9.8.[51] This prepared film can effectively collect the oil from an oil/water mixture underwater driven by gravity several times, showing good durability and high separation efficiency, which is very helpful in the promising application of energy-efficient membranes for reducing the environmental impact of oil spills.

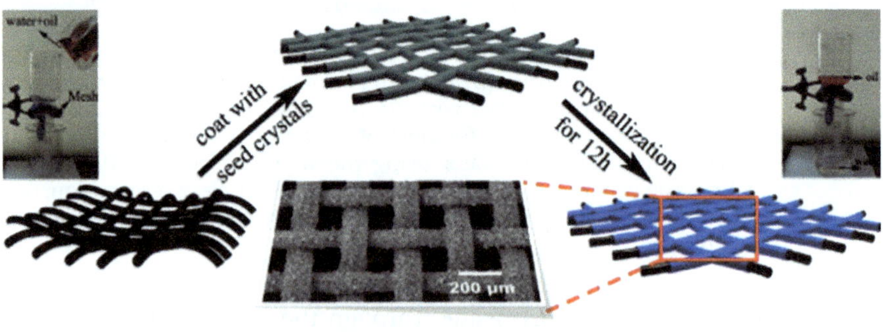

Figure 9.8 Superhydrophilic and underwater superoleophobic MFI zeolite-coated film for oil/water separation.

9.3.3 Superwettable Surface Applied for Emulsified Oil–Water Separations

In realistic situations, oil–water mixtures are not always well layered and a large amount of oil–water mixtures needed to be processed exists in the form of an emulsion. The materials mentioned above are commonly not suitable for emulsified oil/water separation, especially for surfactant-stabilized emulsions. The pore sizes of these materials are usually tens of micrometers. However, the oil dispersion phase in water or water dispersion phase in oil are generally considered as automatic layers because oil and water are immiscible in nature and their densities are different. Generally, an emulsion is defined as a drop in the diameter of the dispersed phase (oil or water) lower than 20 μm. Here, we defined an oil dispersion phase in a water continuous phase (size of oil droplet less than 20 μm) as an "oil-in-water emulsion" and defined a water dispersion phase (size of water droplet less than 20 μm) in an oil continuous phase as "water-in-oil emulsion". Whether an emulsion of oil and water turns into an "oil-in-water emulsion" or a "water-in-oil emulsion" depends on the volume fraction of both and the type of emulsifier (surfactant) present. Some progress has been made in this field and the emulsion separation materials can also be divided into the same types as the layered oil/water separation materials: superhydrophobic/superoleophilic materials and superhydrophilic/underwater superoleophobic materials.

9.3.3.1 Superwettable Surface Applied for Emulsified Oil-in-water Separations

Chen *et al.* report a novel kind of superhydrophilic hybrid membrane, prepared by depositing a $CaCO_3$-based mineral coating on PAA-grafted polypropylene microfiltration membranes, for effective oil-in-water emulsion separation.[52] In their work, diesel was utilized as the target oil for preparing oil-in-water emulsions. Experimental results show that the permeate flux can reach 2000 L m^{-2} h^{-1}. This value is much higher than conventional ultrafiltration membranes (300 L m^{-2} h^{-1}), while the external pressure used (0.05 MPa) is lower than that of the latter (>0.1 MPa).

In Wang's work, they synthesized a hierarchically porous monolith with macro- and meso-pores *via* a sol–gel and phase separation process.[53] Due to the surface modification by organic silanes, the wettability of the silica material was effectively controlled. A series of hydrophobic porous silica monoliths (HPSM) were obtained. Using a "reverse membrane emulsification" process, the HPSM not only cleared oil away from water, but also broke the micro-emulsion efficiently, even when emulsion stabilizer was in the system, as shown in Figure 9.9. They speculated that HPSM exhibited excellent demulsification ability *via* adsorbing the emulsifier from the emulsion, leading to the complete breakdown of surfactant stabilized emulsions (Figure 9.9(c)). The demulsification ratio is as high as 99.95% and the HPSM were reusable.

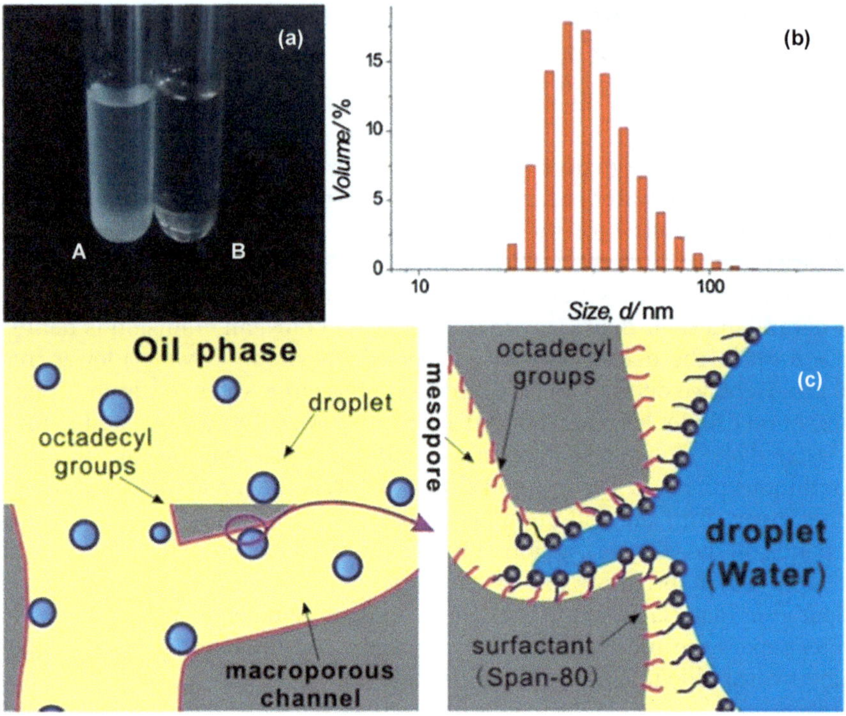

Figure 9.9 (a) Image of the filtrated solutions which were partly, A, and completely, B, demulsified. (b) The diameter distribution of droplets in the micro-emulsion and (c) the mechanism of the demulsification by HPSM-18. Reproduced from ref. 53 with permission from the Royal Society of Chemistry.

9.3.3.2 *Superwettable Surface Applied for Emulsified Water-in-oil Separations*

Amar *et al.* have prepared one kind of fluorosilane-grafted ceramic membranes that can be used to separate waterin-oil emulsions, through a thermally-driven separation process.[54] After grafting 1*H*,1*H*,2*H*,2*H*-perfluorodecyltriethoxysilane on the ceramic membrane, their hydrophilic character (CA = 25°) changed into hydrophobic (CA = 160°). High salt and oil retention higher than 99% were obtained using a modified ceramic membrane. Huang *et al.* developed a novel fluorinated benzoxazine, namely 3-(3-(trifluoromethyl)phenyl)-2*H*-benzoxazine-6-carbaldehyde (BAF-CHO) as a starting monomer, which was synthesized *via* a one-step Mannich reaction.[55] The as-synthesized *in situ* polymerized superhydrophobic and superoleophilic nanofibrous membranes can effectively separate water-in-oil microemulsions, as schematically shown in Figure 9.10, which exhibit an extremely high flux of 892 L m^{-2} h^{-1} solely driven by gravity, as well as good antifouling properties, thermal stability and durability, suggesting their use as promising materials

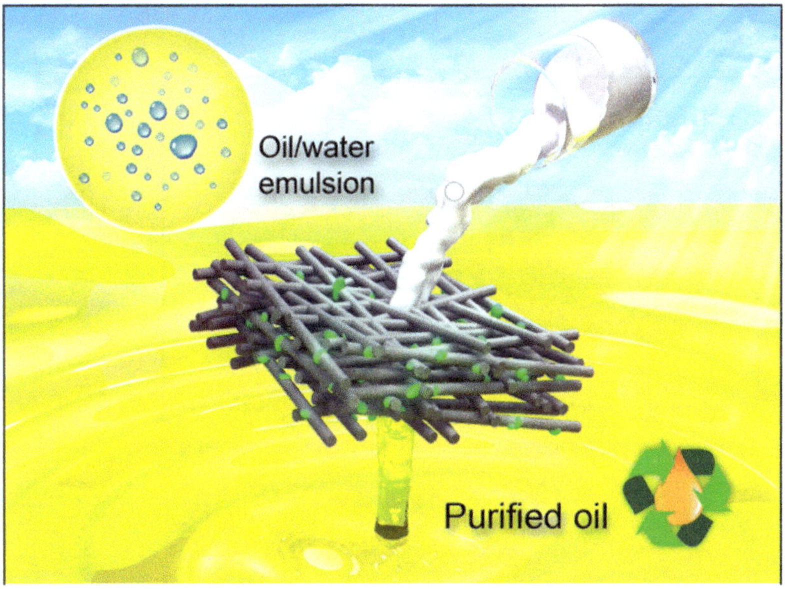

Figure 9.10 Schematic of the emulsified water–oil separations from Huang's work. Reproduced from ref. 55 with permission from the Royal Society of Chemistry.

for practical emulsified wastewater treatment. (Even though the phrase "oil/water emulsion" is used in the picture, the microemulsion is an water-in-oil emulsion as the partially enlarged drawing in the image indicates.)

Zhang and Chen *et al.* prepared superhydrophobic polymer/carbon nanotube hybrid membranes *via* covalent attachment of a hydrophobic polymer.[56] The membrane obtained shows excellent separation properties for surfactant-stabilized water-in-oil emulsions with separation efficiency as high as 99.94% and high flux (5000 L m^{-2} h^{-1} bar^{-1}). As shown in Figure 9.11, a dispersed water/toluene emulsion was poured into the filtration cell and was suction filtered under a pressure of 10 kPa. Emulsion droplets de-emulsified once touching the PS-g-CNTs membrane, toluene immediately permeated through the membrane and water was retained above the membrane. In Figure 9.11(b), before separating, there are numerous water droplets with a size of about 70 nm in the feed solution. In Figure 9.11(c), after separating, there are no droplets that can be observed in the collected filtrate in the whole view. All this evidence demonstrates the excellent separating properties of the PS-g-CNTs membrane. It is worth noting that the membrane exhibited good antifouling properties, outstanding recyclability, robust stability and durability, showing attractive potential for practical oil/water separation.

In realistic situations, oily water produced and emitted by human activities or industries are extremely complex situations regarding the coexistence of oil-in-water emulsions and water-in-oil emulsions. It is hard to judge

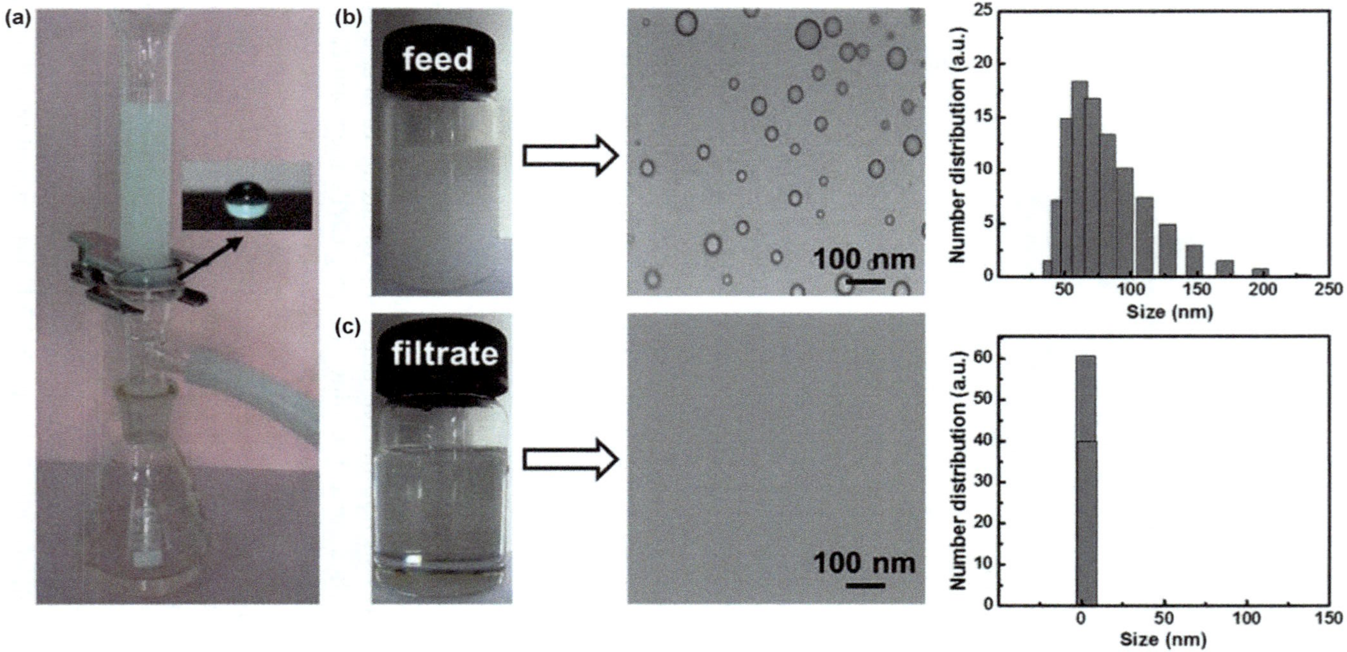

Figure 9.11 (a) Photograph of separating a water/toluene emulsion where toluene selectively permeates through the PS-CNTs membrane. Photographs and DLS data of water/toluene emulsion before (b) and after (c) filtration.
Reproduced from ref. 56 with permission from the Royal Society of Chemistry.

their kind and, in most cases, there is probably a mixture of these two types. Killing two birds with one stone is interesting and satisfying. Therefore, it is of great importance to seek better functional materials in response to these circumstances. Gu *et al.* achieved Janus (with different properties on the same object) polymer/CNTs hybrid membranes *via* self-initiated photo-grafting and photopolymerization (SIPGP).[57] This novel membrane can effectively separate both surfactant stabilized oil-in-water and water-in-oil emulsions because of the anisotropic wettability of the membranes. The WCA of the membrane increases to 153° after modification with hydrophobic poly(styrene) (PS); on the other hand, grafting hydrophilic poly(*N*,*N*-dimethylaminoethyl methacrylate) (PDMAEMA) from the lower surface (opposite side of the PS brushes) leads to a WCA decreasing to 17°. These remarkable membranes make them promising candidates for various practical applications such as controllable oil/water separation and selective oil absorption.

9.4 The Principles to Optimal Design of Oil–Water Separations Materials

In essence, oil/water separation is the wettability behavior that occurs at the interface of the solid, air, water and oil phase. When a liquid droplet was presented on a solid surface in air, we investigated a series of situations. Theoretically, we have optimally designed superhydrophobic surfaces using a paraboloid and semicircular protrusion microtexture.[58] The effects of all the geometrical parameters for such a paraboloid microtexture on free energy (FE) and free energy barrier (FEB) as well as equilibrium contact angle (ECA) and contact angle hysteresis (CAH) of a superhydrophobic surface have been systematically investigated in detail. We found that the droplet position for a metastable state is closely related to the intrinsic CA of the surface. Furthermore, the paraboloid base steepness plays a significant important role in ECA and CAH, and a critical steepness is necessary for the transition from non-composite to composite states. In addition, it has been revealed experimentally that some superhydrophobic surfaces in nature, such as the rice leaf, show strong anisotropic wetting behavior. In our published work, based on a thermodynamic approach, the effects of the profile shape of a parallel grooved microstructure on FE with its barrier (FEB) and ECA with its hysteresis (CAH) for various orientations of different parallel micro texture surfaces have been systematically investigated in detail.[59] The results indicated that the anisotropy of wetting properties strongly depended on the specific topographical features and wetting state.

To easily understand wetting behavior underwater, we firstly and comparatively give the multi-situations in air (Figure 9.12(a)–(c)). When wetting behavior occurs underwater, we take the CA of a smooth surface into account firstly (Figure 9.12(d)). As suggested by Jung and Bhushan, the wetting equation at a solid–water–oil interface can be conducted by combining the Young's equation (eqn (9.1)) of a solid–air–water interface and a

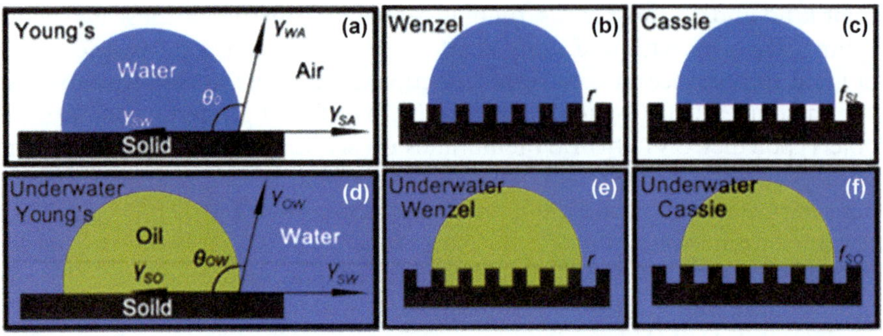

Figure 9.12 Comparative illustration of the in-air and underwater wetting mechanism of water/oil droplets residing on solid surfaces at different wetting states. (a) Water droplet on a smooth surface. Water droplet on rough surfaces with (b) a Wenzel (homogeneous) wetting state and (c) a Cassie (heterogeneous) wetting state. (d) Underwater oil droplets on a smooth surface. Underwater oil droplets on rough surfaces with (e) a Wenzel (homogeneous) wetting state and (f) a Cassie (heterogeneous) wetting state.
Reproduced from ref. 61 with permission from the Royal Society of Chemistry.

solid–air–oil interface.[60] The apparent OCA (θ_{OW}) in an aqueous environment can be given as:

$$\cos\theta_{OW} = \frac{\gamma_{OA}\cos\theta_O - \gamma_{WA}\cos\theta_W}{\gamma_{OW}} \tag{9.5}$$

where θ_W and θ_O are the WCA and OCA in air. γ_{OA}, γ_{WA} and γ_{OW} are interface tensions of oil–air, water–air and oil–water interfaces, respectively. θ_W and θ_O represent the WCA and OCA in air. As predicted by eqn (9.5), for a hydrophilic surface, the surface is simultaneously oleophilic due to the lower surface tension of oil (γ_{OA}) in air than that of water (γ_{WA}) ($\theta_O < \theta_W < 90°$). Therefore, the values of $\cos\theta_O$ and $\cos\theta_W$ are all positive. Since the surface tensions of oil/organic liquids are much lower than that of water ($\gamma_{OA} \ll \gamma_{WA}$), the value of $\gamma_{OA}\cdot\cos\theta_O - \gamma_{WA}\cdot\cos\theta_W$ is commonly negative and thus it can be concluded that most hydrophilic surfaces in air show an oleophobic property underwater at the solid–water–oil interface.

Similar to the Wenzel and Cassie equations in air, the underwater Wenzel equation (Figure 9.12(e)) and Cassie equation (Figure 9.12(f)) can be obtained by introducing the surface roughness and contact phase fractions (solid–oil and solid–water interfaces).

$$\text{Wenzel equation: } \cos\theta_W = r\cos\theta_{OW} \tag{9.6}$$

$$\begin{aligned}\text{Cassie equation: } \cos\theta_{CB} &= r_f f_{SO}\cos\theta_{OW} + f_{SO} - 1 \\ &= r_f\cos\theta_{OW} - f_{SW}(r_f\cos\theta_{OW} + 1)\end{aligned} \tag{9.7}$$

where f_{SO} is the solid–oil fraction under the contact area. It is obvious that the underwater wettability of oil is also determined by the surface roughness, solid–oil fraction (solid–water fraction), as well as the Young's CA of oil in an aqueous environment.

The above analysis is of great concern to optimally separate layered oil/water mixtures. As for oil/water emulsion separation, more factors should be taken into account and a size-sieving mechanism is the foundation model. Except that the wettability of the membrane must be controlled in a certain range, the pore sizes of the filtering membrane are the key factors. To make the continuous phase pass through the membrane quickly, the pore sizes must be large enough. To avoid droplets of the dispersed phase squeezing through the pores, the size of the pores must be small enough as shown in Figure 9.13.

To quantitatively estimate the speed of continuous phase passing through the membrane, a classical fluid dynamic theory is the Hagen–Poiseuille equation:[63]

$$J = \frac{\varepsilon \pi r_p^2 \Delta p}{8 \mu L} \tag{9.8}$$

where the filtration rate J is described as a function of the surface porosity ε, the membrane's pore radius r_p, the pressure drop Δp, the viscosity of the liquid μ, and the total distance L of the liquid running through the membrane.[62] This equation predicts that the flux is directly proportional to the square of the pore size of the membrane and inversely proportional to the thickness of the membrane. So, we should work on trying to prepare thinner membranes with an optimal pore size.

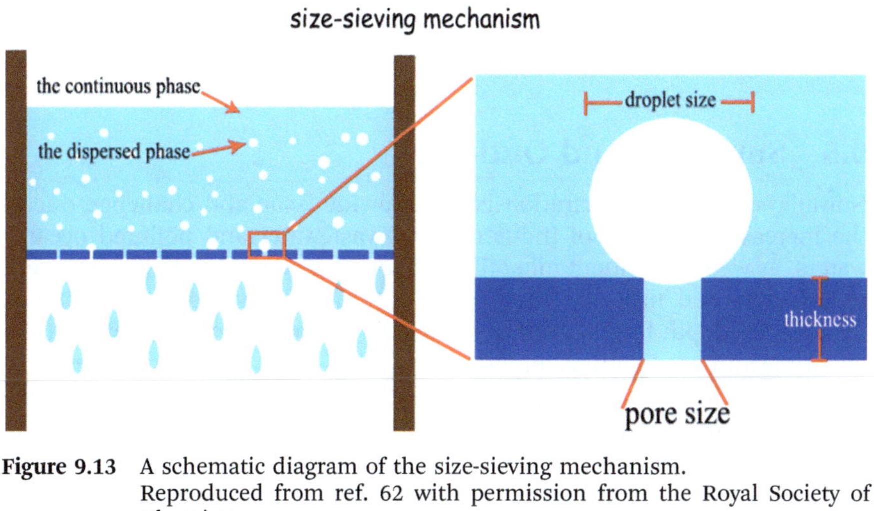

Figure 9.13 A schematic diagram of the size-sieving mechanism. Reproduced from ref. 62 with permission from the Royal Society of Chemistry.

However, if the filtering membrane is too thin, then the problem of mechanical strength would stand out. Here, we introduce two important performance assessments for the filter membrane: trans-membrane pressure (P_T) and breakthrough pressure (P_B). P_T is the external pressure on the separation system and the flux increases linearly with increasing the P_T. However, the maximum P_T that a certain membrane can bear is limited by its mechanical strength. The P_B is the maximum pressure at which the filtering membranes still possess permselectivity. Also, the external P_T has the risk of exceeding the oil P_B. In this case, both the continuous phase and emulsions are forced to permeate through the membranes, decreasing the separation efficiency. Theoretically, the relationship between these parameters can be expressed as follows:[64]

$$P_B = \frac{2\gamma_{wo} \cos \theta_{adv,ws(o)}}{r_p} \tag{9.9}$$

$$Q = \frac{-kA}{4\mu} \frac{\Delta P}{L} \sim r^4 \tag{9.10}$$

$$Q_{max} = \frac{\rho_p A \gamma_{wo} \pi \cos \theta_{adv,ws(o)}}{4\mu} \frac{r_p^3}{L} \tag{9.11}$$

Here Q_{max} is the maximum flow rate while still rejecting emulsions, ρ_p is the number pore density, A is the area of the membrane, γ_{wo} is the interfacial tension between water and oil, and $\cos \theta_{adv,ws(o)}$ is the advancing contact angle of a water droplet on the membrane surface in the presence of oil. These equations explicitly reveal how flow rate is increasingly limited for decreasing pore sizes and how the surface wettability, membrane diameter, permeability and pressure will affect the maximum flow. For the optimal design of superwettable materials applied to oil/water mixtures, we can benefit a lot from these principles.

9.5 Summary and Outlook

Nowadays, oil–water separation is a worldwide issue and challenge due to the increasing amount of industrial oily wastewater and polluted oceanic waters, as well as frequent oil spill accidents. Oil-contaminated water needs to be addressed urgently. Traditional techniques such as oil skimmers, centrifuges, depth filters, coalescers, and flotation technologies often take a long time and need tedious manual operation. These separation processes of an oil/water mixture are empirical to a great extent, thus making the separation incomplete with either oil remaining in water or water remaining in oil. So more feasible methods should be improved to separate oil/water mixtures. Also, a higher flow rate, better selectivity, and excellent separation efficiency are required to meet the rapid demands.

Superhydrophobic or other special wettability materials show enormous potential in the treatment of oil spill accidents and industrial oily water due to their high selective absorbing/filtering features. We begin with the understanding and design of superhydrophobic surfaces. In this chapter, we have introduced their origin and definition, approaches to superhydrophobic surfaces and endowing porous materials or particles with superhydrophobic properties that have potential ability for separating oil/water mixtures. Next, we reviewed recent progress in superhydrophobic/superoleophilic surfaces and superhydrophilic/underwater superoleophobic surfaces applied for separating layered oil/water mixtures. These materials used for the separation of immiscible oil–water mixtures are usually not effective for emulsified oil–water mixtures, especially not for surfactant-stabilized micro emulsions with a dispersed droplet size below 20 μm. Recently reported membrane technologies based on superwettable materials with potential for efficient separation of emulsified oil–water mixtures were highlighted by sorting them into two categories: separation of oil-in-water or water-in-oil emulsions.

Although the investigations in this field have made tremendous strides, there are still a lot of challenges, and some of the problems still need to be solved, such as separation efficiency, flow rate and service life. In theory, we have pointed out the underlying mechanism to improve the performance of oil/water separation.

To achieve the industrialization of oil/water separation materials at an early stage, future work will mainly concentrate on aspects as follows from our viewpoint. First and foremost, the design and synthesis of stable and durable superhydrophobic materials are the key issues. To obtain an extreme wetting state, most of the superhydrophobic surfaces are elaborated with micro and nano-scale fine structures, which can be easily damaged by external influences including mechanical stress and chemical contamination, restricting the material's practical applications. Secondly, the simple and mass-production techniques of oil/water separation materials for large-area oil spills are still required. A large portion of the synthesis methods cannot be carried out on a large scale, such as the hydrothermal method, electrochemical deposition and the *in situ* growth method. Thirdly, it is challenging to realize effective and high-throughput separation of a wide range of oil/water emulsions with small droplet sizes from the micrometer to the nanometer range, because the separation speed and the membrane pore sizes are considered to be contradictory. Fourthly, separation of high-viscous oil/water mixtures is glamorous and should be taken into consideration. Last but not the least, smart interfacial materials for different oil/water separation purposes are also needed to be put forward. These smart interfacial materials are external stimulus-responsive, ranging from single to dual or multiple stimulus-responsive.

Acknowledgements

This work is supported by the National Nature Science Foundation of China (NO11172301 and 21203217), the "Funds for Distinguished Young

Scientists" of Hubei Province (2012FFA002), the "Western Light Talent Culture" Project, the Co-joint Project of Chinese Academy of Sciences and the "Top Hundred Talents" Program of Chinese Academy of Sciences and the National 973 Project (2013CB632300). Many thanks to Mr Ben Wang and Yifan Si for thoughtful discussions and kind help.

References

1. F. Yang, Y. Dong and Z. Guo, *Colloids Surf., A*, 2014, **463**, 101–109.
2. (a) J. F. Piatt and P. Anderson, *Response of Common Murres to the Exxon Valdez Oil Spill and Long-term Changes in the Gulf of Alaska Marine Ecosystem*, ed. S. D. Rice, R. B. Spies, D. A. Wolfe and B. A. Wright, 1996, vol. 18, pp. 720–737; (b) C. H. Peterson, S. D. Rice, J. W. Short, D. Esler, J. L. Bodkin, B. E. Ballachey and D. B. Irons, *Science*, 2003, **302**, 2082–2086.
3. P. Lin and L. Guo, *Mar. Chem.*, 2015, **174**, 13–25.
4. Z. Chu, Y. Feng and S. Seeger, *Angew. Chem., Int. Ed.*, 2015, **54**, 2328–2338.
5. G. Kwon, A. K. Kota, Y. Li, A. Sohani, J. M. Mabry and A. Tuteja, *Adv. Mater.*, 2012, **24**, 3666–3671.
6. Z. Xue, S. Wang, L. Lin, L. Chen, M. Liu, L. Feng and L. Jiang, *Adv. Mater.*, 2011, **23**, 4270–4273.
7. J. Rubio, M. L. Souza and R. W. Smith, *Miner. Eng.*, 2002, **15**, 139–155.
8. S. Wang, K. Liu, X. Yao and L. Jiang, *Chem. Rev.*, 2015, **115**, 8230–8293.
9. (a) B. Wang, Y. Zhang, L. Shi, J. Li and Z. Guo, *J. Mater. Chem.*, 2012, **22**, 20112; (b) F. Yang and Z. Guo, *J. Bionic Eng.*, 2015, **12**, 88–20197.
10. V. A. Ganesh, H. K. Raut, A. S. Nair and S. Ramakrishna, *J. Mater. Chem.*, 2011, **21**, 16304–16322.
11. Z. Chu and S. Seeger, *Chem. Soc. Rev.*, 2014, **43**, 2784–2798.
12. T. Darmanin and F. Guittard, *J. Mater. Chem. A*, 2014, **2**, 16319–16359.
13. Z. Guo and W. Liu, *Plant Sci.*, 2007, **172**, 1103–1112.
14. X. Gao and L. Jiang, *Nature*, 2004, **432**, 36.
15. S. Yu, Z. Guo and W. Liu, *Chem. Commun.*, 2014, **51**, 1775–1794.
16. C.-H. Xue and J.-Z. Ma, *J. Mater. Chem. A*, 2013, **1**, 4146–4161.
17. T. Jiang, Z. Guo and W. Liu, *J. Mater. Chem. A*, 2014, **3**, 1811–1827.
18. Z. Guo, F. Zhou, J. Hao and W. Liu, *J. Am. Chem. Soc.*, 2005, **127**, 15670–15671.
19. X. Zhu, Z. Zhang, X. Xu, X. Men, J. Yang, X. Zhou and Q. Xue, *J. Colloid Interface Sci.*, 2012, **367**, 443–449.
20. J. Yong, F. Chen, Q. Yang, D. Zhang, U. Farooq, G. Du and X. Hou, *J. Mater. Chem. A*, 2014, **2**, 8790–8795.
21. H.-B. Jo, J. Choi, K.-J. Byeon, H.-J. Choi and H. Lee, *Microelectron. Eng.*, 2014, **116**, 51–57.
22. S. Barthwal, Y. S. Kim and S.-H. Lim, *J. Colloid Interface Sci.*, 2013, **400**, 123–129.
23. B. Wang and Z. Guo, *Appl. Phys. Lett.*, 2013, **103**, 063704.

24. H. Wang and Z. Guo, *Appl. Phys. Lett.*, 2014, **104**, 183703.
25. Y. Si and Z. Guo, *Nanoscale*, 2015, **7**, 5922–5946.
26. Y. Li, G. Duan and W. Cai, *J. Colloid Interface Sci.*, 2007, **314**, 615–620.
27. B. Dong, D. Fan, S. Wang and Z. Guo, *Chem. Lett.*, 2015, **44**, 1245–1247.
28. H. Bellanger, T. Darmanin, E. Taffin de Givenchy and F. Guittard, *Chem. Rev.*, 2014, **114**, 2694–2716.
29. X. Zhang, F. Shi, J. Niu, Y. Jiang and Z. Wang, *J. Mater. Chem.*, 2008, **18**, 621–633.
30. B. Wang and Z. Guo, *Chem. Commun.*, 2013, **49**, 9416–9418.
31. Y. Si, H. Zhu, L. Chen, T. Jiang and Z. Guo, *Chem. Commun.*, 2015, **51**, 16794–16797.
32. J. Zhang, Y. Liu, Z. Wei and J. Zhang, *Appl. Surf. Sci.*, 2013, **265**, 363–368.
33. F. Shi, J. Niu, Z. Liu, Z. Wang, M. Smet, W. Dehaen, Y. Qiu and X. Zhang, *Langmuir*, 2007, **23**, 1253–1257.
34. X. Zhu, Z. Zhang, X. Xu, X. Men, J. Yang, X. Zhou and Q. Xue, *Langmuir*, 2011, **27**, 14508–14513.
35. B. Xin and J. Hao, *Chem. Soc. Rev.*, 2010, **39**, 769–782.
36. Q. Zhu, Q. Pan and F. Liu, *J. Phys. Chem. C*, 2011, **115**, 17464–17470.
37. S. Yu and Z. G. Guo, *RSC Adv.*, 2015, **5**, 107880–107888.
38. C.-H. Xue, S.-T. Jia, H.-Z. Chen and M. Wang, *Technol. Adv. Mater.*, 2008, **9**, 035001.
39. B. Wang, J. Li, G. Wang, W. Liang, Y. Zhang, L. Shi, Z. Guo and W. Liu, *ACS Appl. Mater. Interfaces*, 2013, **5**, 1827–1839.
40. J. Guo, S. Yu, J. Li and Z. Guo, *Chem. Commun.*, 2015, **51**, 6493–6495.
41. L. Feng, Z. Zhang, Z. Mai, Y. Ma, B. Liu, L. Jiang and D. Zhu, *Angew. Chem., Int. Ed.*, 2004, **43**, 2012–2014.
42. F. Yang and Z. Guo, *RSC Adv.*, 2015, **5**, 13635–13642.
43. H. Zhu and Z. Guo, *Chem. Lett.*, 2014, **43**, 1645–1647.
44. H. Zhu and Z. Guo, *Chem. Lett.*, 2015, **44**, 1431–1433.
45. A. Sarkar and S. Mahapatra, *J. Mater. Chem. A*, 2014, **2**, 3808–3818.
46. Y. Zhu, F. Zhang, D. Wang, X. F. Pei, W. Zhang and J. Jin, *J. Mater. Chem. A*, 2013, **1**, 5758–5765.
47. J. Li, L. Shi, Y. Chen, Y. Zhang, Z. Guo, B.-L. Su and W. Liu, *J. Mater. Chem.*, 2012, **22**, 9774.
48. W. Liang and Z. Guo, *RSC Adv.*, 2013, **3**, 16469–16474.
49. Y. Dong, J. Li, L. Shi, X. Wang, Z. Guo and W. Liu, *Chem. Commun.*, 2014, **50**, 5586–5589.
50. Q. Wen, J. Di, L. Jiang, J. Yu and R. Xu, *Chem. Sci.*, 2013, **4**, 591–595.
51. J. Zeng and Z. Guo, *Colloids Surf., A*, 2014, **444**, 283–288.
52. P. C. Chen and Z. K. Xu, *Sci. Rep.*, 2013, **3**, 2776.
53. Y. Wang, S. Tao and Y. An, *J. Mater. Chem. A*, 2013, **1**, 1701–1708.
54. M. Khemakhem, S. Khemakhem and R. Ben Amar, *Colloids Surf., A*, 2013, **436**, 402–407.
55. M. Huang, Y. Si, X. Tang, Z. Zhu, B. Ding, L. Liu, G. Zheng, W. Luo and J. Yu, *J. Mater. Chem. A*, 2013, **1**, 14071–14074.

56. J. Gu, P. Xiao, J. Chen, F. Liu, Y. Huang, G. Li, J. Zhang and T. Chen, *J. Mater. Chem. A*, 2014, **2**, 15268–15272.
57. J. Gu, P. Xiao, J. Chen, J. Zhang, Y. Huang and T. Chen, *ACS Appl. Mater. Interfaces*, 2014, **6**, 16204–16209.
58. (a) L. Tie, Z. Guo and W. Li, *J. Colloid Interface Sci.*, 2014, **436C**, 19–28; (b) L. Tie, Z. Guo and W. Li, *RSC Adv.*, 2014, **5**, 8446–8454.
59. L. Tie, Z. Guo and W. Liu, *J. Colloid Interface Sci.*, 2015, **453**, 142–150.
60. Y. C. Jung and B. Bhushan, *Langmuir*, 2009, **25**, 14165–14173.
61. B. Wang, W. Liang, Z. Guo and W. Liu, *Chem. Soc. Rev.*, 2015, **44**, 336–361.
62. Y. Si and Z. Guo, *Chem. Lett.*, 2015, **44**, 874–883.
63. Z. Shi, W. Zhang, F. Zhang, X. Liu, D. Wang, J. Jin and L. Jiang, *Adv. Mater.*, 2013, **25**, 2422–2427.
64. B. R. Solomon, N. Hyder and K. K. Varanasi, *Sci. Rep.*, 2014, **4**, 5504.

CHAPTER 10

Superhydrophobic/ Superhydrophilic Property in Functionally Cooperated Smart Device

FENG SHI* AND MENG XIAO

State Key Laboratory of Chemical Resource Engineering & Key Laboratory of Carbon Fibre and Functional Polymer, Ministry of Education, Beijing University of Chemical Technology, Beijing, China
*Email: shi@mail.buct.edu.cn

10.1 Introduction

Smart materials are materials that can regulate their properties responding to external stimuli, and there has been a boom in their development since the 1990s. Smart materials usually possess three essential elements: sensibility, actuation and controllability. Smart materials are often defined according to their corresponding stimulus, such as photo-responsive materials (responding to light exposure), pH-responsive materials (responding to external pH change), thermo-responsive materials (responding to temperature change), electro-, stress-, humidity-, chemical- and redox-responsive materials (responding to electricity, stress, humidity, chemical substances and redox reactions, respectively) *etc.*[1–23] Smart materials can sense stimulation and self-actuate to perform some reaction (Figure 10.1); upon removal of the stimulation, the materials can recover to their initial status.

RSC Smart Materials No. 21
Self-cleaning Coatings: Structure, Fabrication and Application
Edited by Junhui He

Published by the Royal Society of Chemistry, www.rsc.org

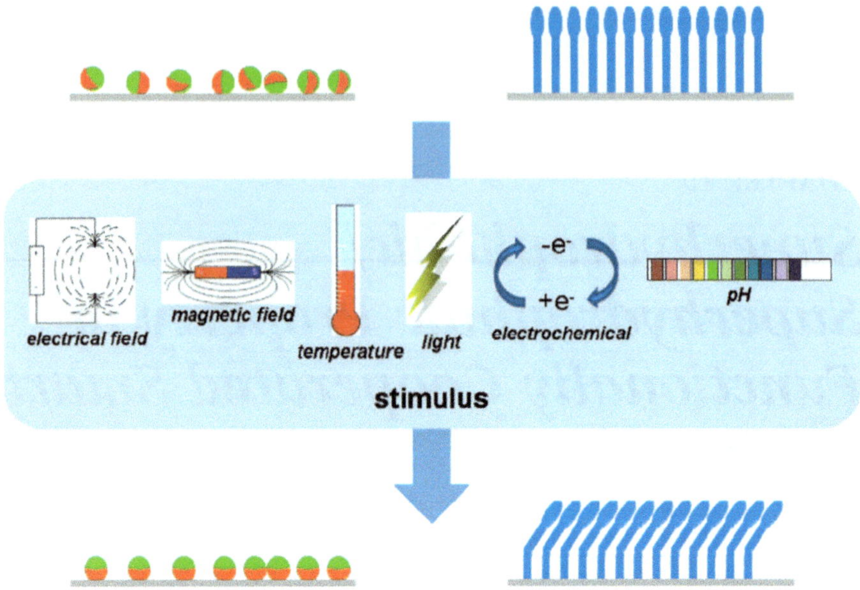

Figure 10.1 Smart materials responding to external stimuli. Reproduced with permission from M. Yoshida and J. Lahann, *ACS Nano*, 2008, **2**, 1101. Copyright (2008) American Chemical Society.[23]

Research on smart materials has extended the application range of materials from singular surface modification to multi-disciplinary applications, such as bionics,[24–26] the medical field[27,28] and nanotechnology.[29] As an important branch, smart materials with a reversible transformation between superhydrophobicity and superhydrophilicity play a vital role in many research fields, such as oil/water separation, biological field and bionics of natural entities.[30,31]

The wetting behaviour of a liquid on a solid surface is a crucial surface property, which plays important roles in industry, agriculture, and daily life. Wettability is also a fundamental property in biological systems. Since the lotus[32–34] effect was found in the 1970s, bionics has attracted a large amount of attention. Besides lotus leaves, a large number of flora and fauna have demonstrated special wettability, such as rice,[35] peanuts[36] and *Nepenthes*;[37–39] flowers of the rose;[40] wings of the butterfly,[41] dragonfly,[42–44] and cicada;[45] legs of the water strider[46] and gecko;[47] hairs of the dandelion[48] and weasel,[49] and so on[50] (Figure 10.2). Most of them show superhydrophobicity or superhydrophilicity, which is a result of their environment and living habitat. Nature gives us much inspiration to learn. Biomimetic studies on the aforementioned special surfaces indicate that surfaces with special wettability are a combined effect of morphology structure and chemical compositions. For example, papillae are found on the surface of lotus leaves, and a wax layer with low free energy covers the papilla to give a superhydrophobic surface with a self-cleaning property.[33]

Figure 10.2 Multiscale structures in nature: leaves of (a) lotus, (b) rice and (c) peanut; (d) rose flower; (e) wing of a butterfly; leg of (f) a water spider and (g) gecko.
(a) W. Barthlott, Purity of the scared lotus, or escape from contamination in biological surfaces, *Planta*, 1997, **202**, with permission of Springer,[33] (b) Reproduced from ref. 35 with permission from John Wiley and Sons. Copyright © 2002 WILEY-VCH Verlag GmbH & Co. KGaA, Weinheim. (c) Reproduced from ref. 36 with permission from John Wiley and Sons. Copyright © 2013 WILEY-VCH Verlag GmbH & Co. KGaA, Weinheim. (d) Reproduced with permission from L. Feng, Y. N. Zhang, J. M. Xi, Y. Zhu, N. Wang, F. Xia and L. Jiang, *Langmuir*, 2008, **24**, 4114.[40] Copyright (2008) American Chemical Society. (f) Reprinted by permission from Macmillan Publishers Ltd: *Nature* (ref. 46), Copyright (2004). (g) From M. Sitti and R. S. Fearing, *J. Adhes. Sci. Technol.*, 2003, **18**, 1055 reprinted by permission of the publisher (Taylor and Francis Ltd, http://www.tandfonline.com).[47]

Simulated by the above natural surfaces with special wettability, scientists have fabricated a large number of new materials.[51–55] Inspired by the legs of a water strider, Zhang's group fabricated a gold thread by combining a layer-by-layer assembling technique, electrochemical deposition of gold aggregates and modification of a low-surface-energy compound (Figure 10.3A).[56] Inspired by the silk of the *Uloborus walckenaerius*, Zheng *et al.* designed an artificial fibre that exhibited a directional water-collecting ability (Figure 10.3B).[57] Inspired by *Nepenthes* pitcher plants, Aizenberg and co-workers developed an approach to create a porous surface with liquid- and ice-repellence and self-healing abilities (Figure 10.3C).[58] Simply mimicking

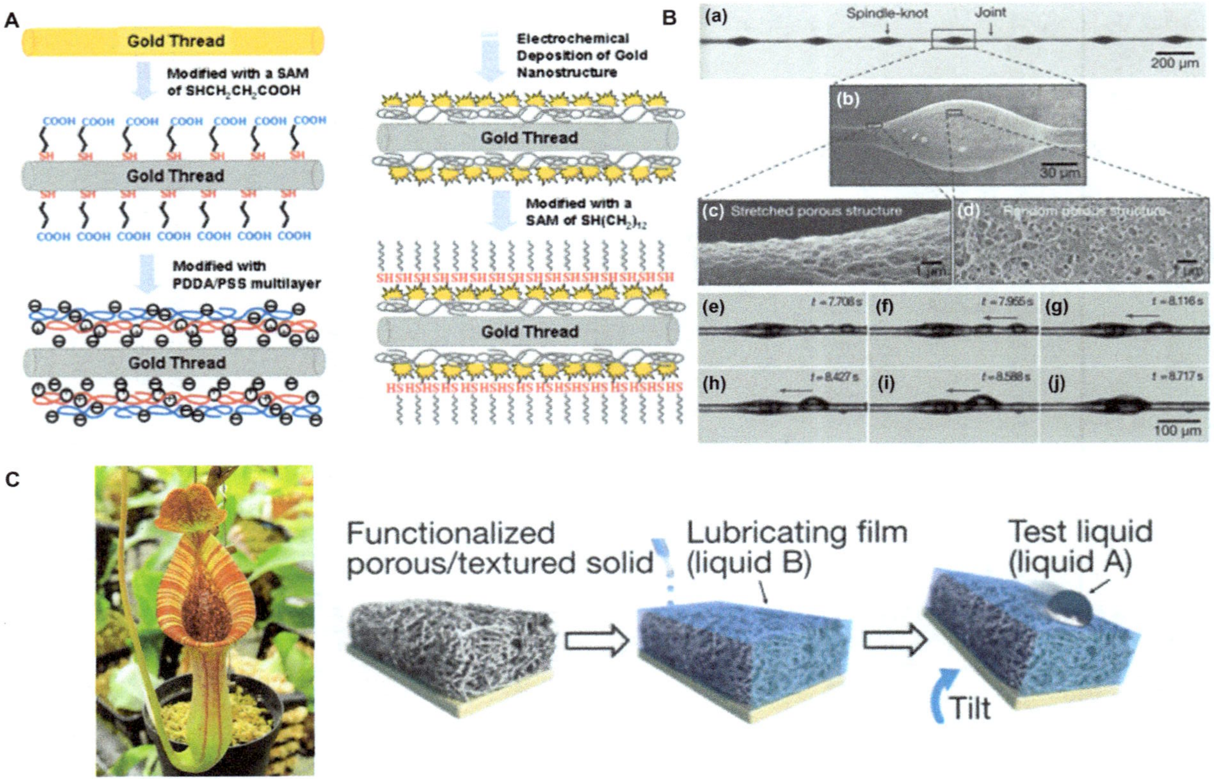

Figure 10.3 Materials inspired by (A) legs of a water spider (reproduced from ref. 56 with permission from John Wiley and Sons. Copyright © 2007 WILEY-VCH Verlag GmbH & Co. KGaA, Weinheim), (B) silk from *Uloborus walckenaerius* (reprinted by permission from Macmillan Publishers Ltd: *Nature* (ref. 57), Copyright 2010) and (C) *Nepenthes* pitcher (reprinted by permission from Macmillan Publishers Ltd: *Nature* (ref. 58), Copyright 2011).

bionts living in nature has provided a large amount of inspiration in the development of material science; however, the simplicity of the surface limits the applications of the artificial materials. Therefore, controlling the specific wettability of a solid surface is of great importance in academic sciences and material technologies. The fact that simple smart surfaces/materials cannot satisfy humans' need in real applications can be compensated through functionally cooperative systems.

The concept of a functionally cooperative system originated from the establishment of the National Nanotechnology Initiative (NNI) by the National Science and Technology Council in 2003. Since the establishment of the NNI, scientists have a new understanding of material science. In the initiative, the development of nanotechnology until the year 2015 was thought to face several steps (Figure 10.4):[59] the first was the development of passive nanostructures that show no intellectuality, such as coatings, nanoparticles, nanostructured metals, polymers and ceramics; the second was the development of active nanostructures, which are similar to smart nanomaterials, such as targeted drugs and adaptive structures; the third was the development of nanosystems, which may be an assembly of several active nanostructures. In this way, to achieve a given complicated goal, several smart materials should be integrated into one system, work in an ordered manner, and realize a given complex intention. The description above is what a functionally cooperated smart device means.

Shi *et al.* first introduced the concept of the functionally cooperative system for the construction of smart devices.[60] The concept was utilized to identify the orderly manner in a diving-surfacing process. Shi *et al.* fabricated a device consisting of a pH-responsive cube and a hydrogen peroxide-responsive catalyst. When the device was placed in an acidic solution,

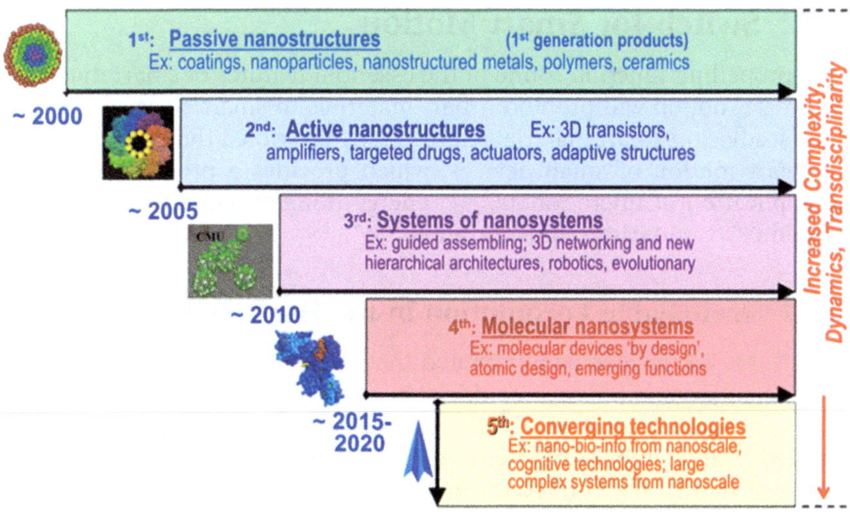

Figure 10.4 Timeline for nanotechnology commercialization (2000–2020).

it floated on the water surface; when the solution was adjusted to an alkaline solution, the surface of the cube became superhydrophilic, and the solution penetrated into the device and increased the specific gravity of the device, leading to diving of the device. Subsequently, the hydrogen peroxide added under acid conditions was decomposed to oxygen gas, which accumulated in the device and decreased the specific gravity of the device, inducing the surfacing process. This controlled locomotion is a combined result of the reversible superhydrophobic/superhydrophilic property and the catalyzation of platinum on the decomposition of hydrogen peroxide; these two functional materials were cooperated to realize the goal of a controlled diving–surfacing locomotion. Therefore, the functionally cooperated smart device is an integrated system to perform a series of behaviors responding to external stimuli, which demands two or more functional surfaces or materials.

Functionally cooperated smart devices have been developed in many fields, such as switches for smart motion, oil/water separation, ion channels, selective permeability and bio-applications. These developments will be discussed according to different classifications of the functionally cooperated smart devices based on their various applications. The implications of the functional cooperation between different organizations in the smart devices will be discussed as well. This chapter is organized into six sections. The first section gives a brief introduction to smart devices with a switchable superhydrophobic/superhydrophilic property. The second, third, fourth, fifth and six sections provide a comprehensive overview of the smart device in five different research areas: switch for smart motion, oil/water separation, ion channel, selective permeability and bio-applications. Lastly, an outlook for the future of these smart materials is revealed.

10.2 Switch for Smart Motion

A self-propelling function is one of the essential features of smart materials, and smart motion will promote smart materials to smart devices. In recent years, studies on smart materials have greatly promoted the development of the smart motion of small devices, which provides a promising route to the application of micro-robots,[61–63] energy transformation,[64–68] biological detection[69–71] and cargo transportation.[72–74]

10.2.1 Switchable Locomotion in the Horizontal Direction

Since G. M. Whitesides demonstrated that the decomposition of hydrogen peroxide to oxygen bubbles could be used to efficiently propel the locomotion of small objects on the centimetre scale,[75] bubble propulsion has attracted a large amount of attention in the field of smart locomotion. The most commonly used bubbles are oxygen bubbles decomposed from hydrogen peroxide, hydrogen bubbles generated from the reaction between an active metal and an acid and carbon dioxide bubbles from the reaction between calcium carbonate and an acid. Moreover, since a reversible superhydrophobic/superhydrophilic smart surface has been demonstrated to be able to repel and adsorb water, respectively, it is thought to show potential in controlling the contact between actuators and fuels.

By introducing platinum for the decomposition of hydrogen peroxide onto one of the ends of a gold thread, Zhang's group reported a pH-responsive smart motion. The gold thread was covered by gold nanoparticles through electrochemical deposition and then modified with a pH-responsive monolayer of 2-(11-mercaptoundecanamido)benzoic acid. In an acidic solution, the pH-responsive surface of the gold thread was superhydrophobic and suffered almost no friction, thus performing a linear locomotion in a water channel (Figure 10.5(a) and (b)); while, in an alkaline solution, the pH-responsive surface was superhydrophilic and was wetted by water, the friction suffered prevented the gold thread moving in the horizontal direction (Figure 10.5(c) and (d)). Therefore, the smart device showed a pH-responsive smart locomotion.[76]

By introducing decomposition of hydrogen peroxide into a moving system, Shi *et al.* realized a smart ON–OFF–ON locomotion of a device. The device consisted of three parts: a superhydrophobic boat with drag-reducing property; a smart nickel mesh that could reversibly be transformed between a superhydrophobic and superhydrophilic property; and an actuator that was covered by platinum aggregates for the amplified catalysis. When the device was placed in a hydrogen peroxide solution under acid conditions, the device floated and remained still on the water surface (Figure 10.6(a)); after the addition of basic solution, the smart nickel mesh transformed from superhydrophobic to superhydrophilic, the water level rose and the platinum right above the mesh touched the solution and began to catalyze the decomposition of hydrogen peroxide into oxygen bubbles, propelling the device forward (Figure 10.6(b) and (c)). Then, an acidic solution was added to transform the nickel mesh from superhydrophilic to hydrophobic, preventing the catalyzation of platinum on hydrogen peroxide, thus the device stopped in the middle of the channel (Figure 10.6(d)). When a basic solution was added for a second time, the platinum catalyzed the decomposition of hydrogen peroxide for a second time and propelled the device forward (Figure 10.6(e) and (f)). The smart motion could be repeated for a long time.[77]

Besides bubble propulsion, the Marangoni effect is also one of the most commonly used driving forces. When locomotion of small objects is performed on a water/air interface in the horizontal direction, the Marangoni effect,[78,79] which often occurs on the interface, is an important driving force. In nature, beetles of the genus *Stenus* were found to propel themselves forward by releasing a kind of low-surface-energy compound onto the surface of water in the 1900s. By mimicking the insects, Shi's group integrated a superhydrophobic boat to replace the insect's body, a reservoir to act as the pygidial gland and a reversible superhydrophobic–superhydrophilic smart material to mimic its gland end into one device (Figure 10.7). When the

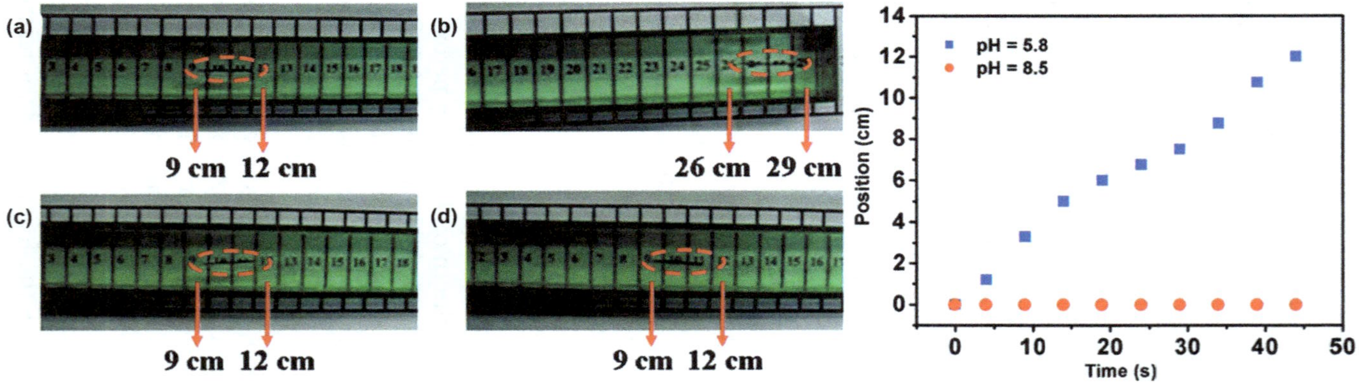

Figure 10.5 pH-Responsive smart locomotion propelled by oxygen bubbles. Reproduced with permission from X. X. Chen, J. Gao, B. Song, M. Smet and X. Zhang, *Langmuir*, 2010, **26**, 104.[76] Copyright (2010) American Chemical Society.

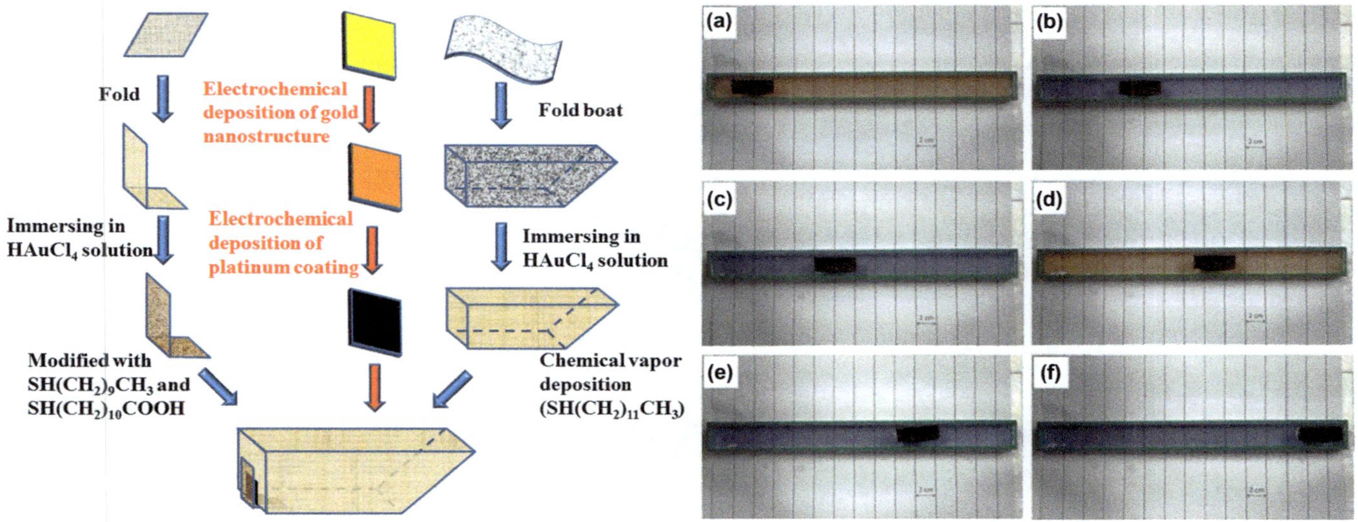

Figure 10.6 pH-Responsive ON–OFF–ON motion propelled by bubble propulsion generated from hydrogen peroxide. Reproduced from ref. 77 with permission from John Wiley and Sons. Copyright © 2013 WILEY-VCH Verlag GmbH & Co. KGaA, Weinheim.

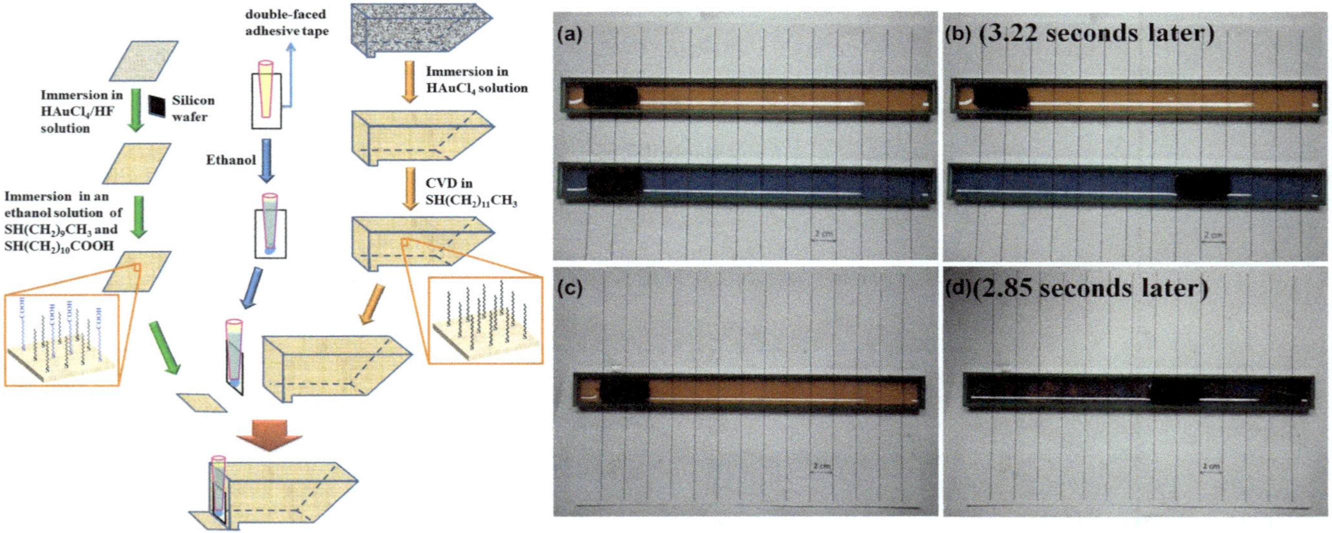

Figure 10.7 Horizontal locomotion with a pH-responsive initiation propelled by the Marangoni effect to mimic the beetles of genus *Stenus*.
Reproduced from ref. 80 with permission from John Wiley and Sons. Copyright © 2013 WILEY-VCH Verlag GmbH & Co. KGaA, Weinheim.

device was placed on the surface of an acidic solution, it floated and remained still. After the addition of a basic solution, the smart surface transformed from superhydrophobic to superhydrophilic, leading to a rise in the water level; the hanging ethanol droplet below the reservoir touched the water and diffused on the water's surface, which induced the surface tension gradient and consequent locomotion of the device.[80]

The photoresist has been demonstrated to be a UV-responsive material. Under UV illumination, the photoresist decomposes and dissolves in a basic solution. Integrating such a smart material with a superhydrophobic object into one device, Shi's group realized a smart light-induced ON–OFF–ON motion. The device consisted of a superhydrophobic boat with drag-reducing property and an actuator made up of a photoresist and a surfactant mixture through spinning on a silicon wafer (Figure 10.8A). In the dark, the device floated on the basic solution's surface and stayed still. When the actuator was exposed to UV light, the photoresist decomposed and dissolved in the basic solution, and the mixed surfactant was released and diffused on the water's surface, inducing a surface tension gradient, thus propelling the device forward (Figure 10.8B). After the device stopped because of the disappearance of the surface tension gradient, a second illumination of UV light was operated on the actuator, and the device re-started to move forward for a second time. Moreover, the direction of the moving object could be tuned by introducing another actuator to the system (Figure 10.8C).[81]

10.2.2 Switchable Locomotion in the Vertical Direction

Owing to the lower density of gas compared to that of water, bubbles are usually used to realize diving–surfacing locomotion. The key to realize vertical locomotion is to control the buoyancy of the device to balance with its gravity. Generation of bubbles in the device is the most important element to enhance the buoyancy to realize the surfacing process; while on the water's surface, the internal gas is released to the atmosphere, decreasing buoyancy for a diving process.

It has been demonstrated that poly(*N*-isopropylacrylamide) (PNIPAAm) is a thermo-responsive polymer with a lower critical solution temperature (LCST) of 32–33 °C in water. By modifying PNIPAAm onto the rough surface of a device, the device obtained presented a thermo-responsive property. When the device was placed into a beaker with water at a temperature of 41 °C, the device floated on the water's surface (Figure 10.9B(a)); when the temperature was adjusted to 28 °C and hydrogen peroxide solution was added, the device dived to the solution's bottom (Figure 10.9B(b)–(d)). By regulating the temperature between 41 and 28 °C, a diving–surfacing cycled motion could be performed. This thermo-responsive smart motion could be integrated as follows (Figure 10.9A): when the temperature is fixed at 41 °C, the PNIPAAm chains are in a shrinking state and intramolecular hydrogen bonds are formed, exhibiting superhydrophobicity with the amplification of

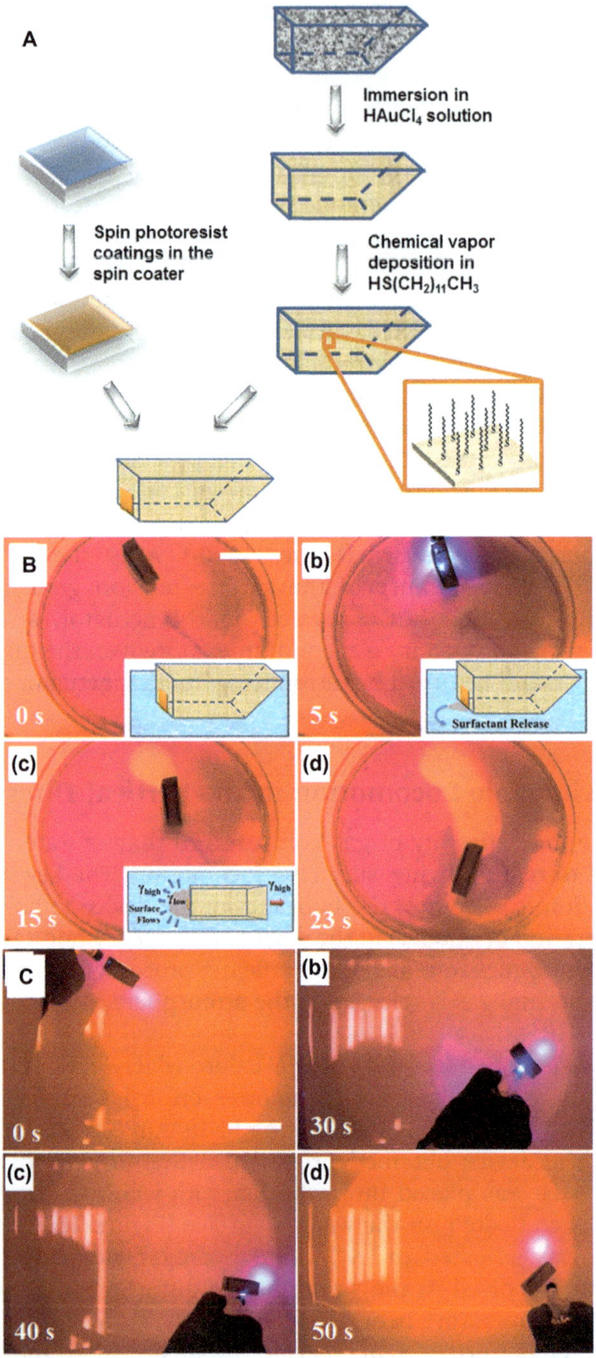

Figure 10.8 Photo-responsive horizontal locomotion propelled by the Marangoni effect.
Reprinted by permission from Macmillan Publishers Ltd: *Nature* (ref. 81), Copyright 2014.

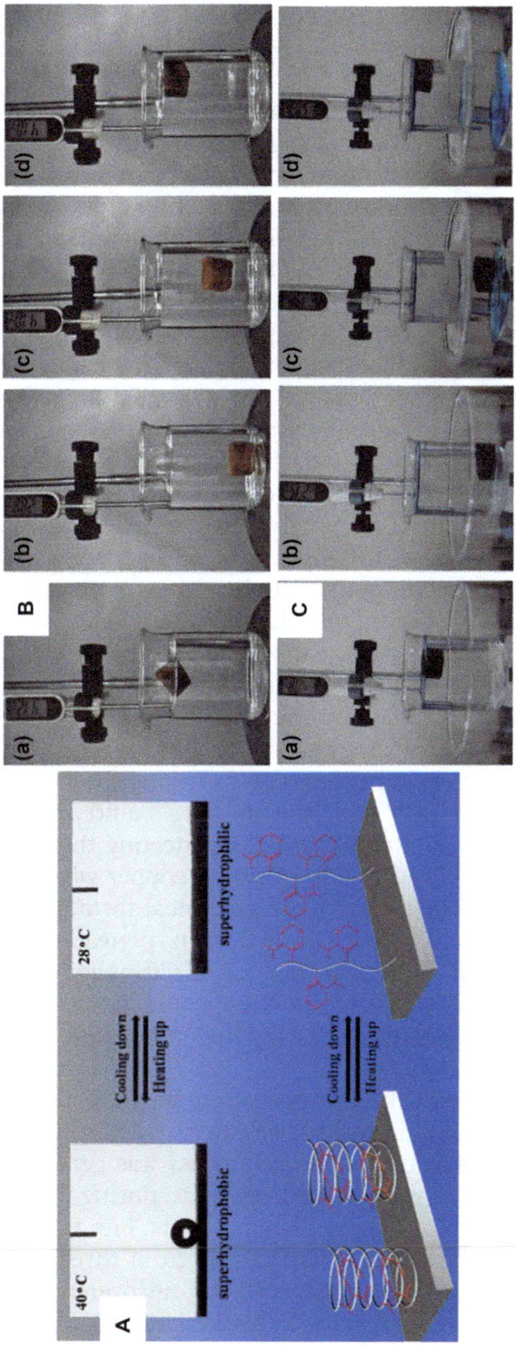

Figure 10.9 Thermo-responsive diving–surfacing cycled locomotion propelled by bubble propulsion generated from hydrogen peroxide and its application in smart transportation.
Reproduced from ref. 82 with permission from John Wiley and Sons. Copyright © 2013 WILEY-VCH Verlag GmbH & Co. KGaA, Weinheim.

a rough surface; when at 28 °C, the PNIPAAm chains swell and form intermolecular hydrogen bonds with water molecules, performing a superhydrophilic property. Moreover, the as-prepared device could be used to transport small molecules directionally (Figure 10.9C).[82]

Besides thermo-responsive materials, pH-responsive materials with a superhydrophobic–superhydrophilic transformation provide another valid switch for smart locomotion. Combining decomposition of hydrogen peroxide under the catalysis of platinum and a smart surface with a reversible superhydrophobic–superhydrophilic property, Shi's group realized smart diving–surfacing cycled motion through regulating the pH of the solution. A cube-shaped device made up of nickel foam was covered by a pH-responsive coating with a micro-/nano-hierarchical structure, resulting in a superhydrophobic property with pH-responsiveness. Placed in an acidic solution, the device floated because of its superhydrophobic property under acid conditions; when the pH value of the solution was adjusted to an alkaline one, the device's surface gradually became superhydrophilic and the solution penetrated into the device, resulting in an increase of the specific gravity of the device and consequent diving process. After the addition of an acidic solution and hydrogen peroxide, the internal platinum segment catalyzed the decomposition of hydrogen peroxide into oxygen bubbles, which accumulated in the upper space of the device, repelling the internal solution out of the device and thus leading to the reduction of the specific gravity of the device; meanwhile, the surface of the device gradually transformed from superhydrophilic to hydrophobic. After the device floated on the water's surface for a while, the internal gas was released to the external atmosphere and resulted in the re-diving process of the device. Then, repeating the addition of basic solution and acidic solution/hydrogen peroxide, the authors could realize a cycled diving–surfacing motion (Figure 10.10A).[60] Connecting the device with an electrochemical workstation through a flexible copper wire, they realized an energy conversion process from the mechanical form, which was generated from the chemical energy of hydrogen peroxide, to electricity (Figure 10.10B) through Faraday's law of electromagnetic induction. Moreover, the electromotive force obtained could be adjusted through regulating the concentration of fuel, *i.e.* hydrogen peroxide, and the shape of the device.[83]

During the decomposition of hydrogen peroxide, reactive oxygen that seriously damaged the stimuli-responsive surface and thus reduced the sustainability of the electricity generating process was generated. To solve this problem, Shi's group introduced hydrogen gas, during the generation of which there were no harmful intermediate products, into the smart diving–surfacing locomotion. The smart device consisted of three parts: two pH-responsive uncovered boxes, one quartz two-end uncovered cell acting as a window and a magnesium segment to react with acid and produce hydrogen gas. When the device was placed into an acidic solution, it floated on the water's surface because of the superhydrophobicity of the pH-responsive

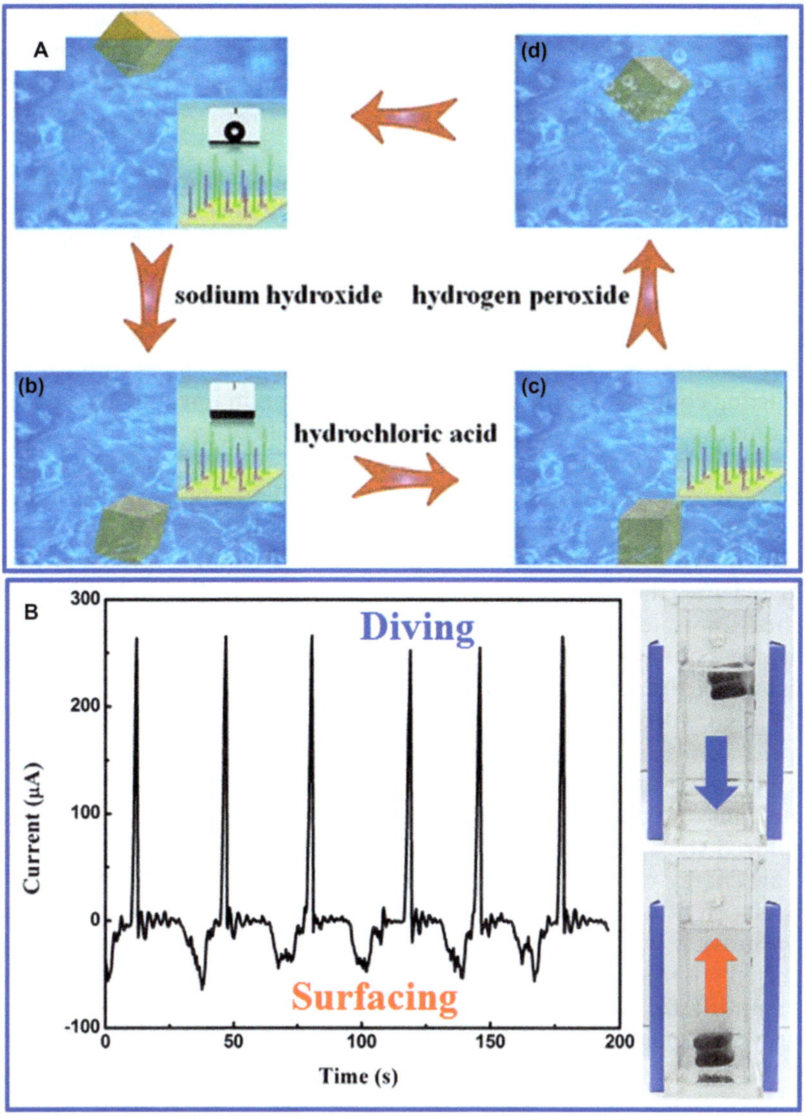

Figure 10.10 pH-Responsive diving–surfacing cycled locomotion propelled by bubble propulsion generated from hydrogen peroxide and its application in the electricity generation process.
(A) Reproduced from ref. 60 with permission from John Wiley and Sons. Copyright © 2010 WILEY-VCH Verlag GmbH & Co. KGaA, Weinheim and (B) reprinted from M. Song *et al.*, Generating induced current through diving-surface motion of a stimulus-responsive smart device, *Nano Energy*, **20**, 233–243, Copyright 2016 with permission from Elsevier.[83]

parts (Figure 10.11A(a)); after the solution was adjusted to basic, the pH-responsive parts transformed from superhydrophobic to superhydrophilic, increasing the specific gravity of the device, thus the device dived to the solution's bottom (Figure 10.11A(b) and (c)). Then acidic solution was added to adjust the solution back to acidic, the acid added reacted with the internal magnesium segment and produced hydrogen bubbles, which accumulated in the upper part of the device; when the specific gravity of the device was lower than that of water when approximately two-thirds of the inner space were occupied by hydrogen gas, the device re-floated to the water surface (Figure 10.11A(d)–(f)). Connecting the smart device with an electrochemical workstation through a flexible copper wire and putting the locomotion system between a pair of magnets, the diving–surfacing induced the conducting copper wire to cut the magnetic induction line and generated an electromotive force (Figure 10.11B(a)). However, because the upper surface of the device was wetted by water, the inner gas could not easily be released to the external atmosphere, resulting in the failure of a rapid re-diving locomotion. To solve this problem, the authors replaced one of the two pH-responsive parts with a hydrophobic part. Therefore, when the device re-floated after the addition of acidic solution, once the hydrophobic surface touched the water's surface, the internal gas easily penetrated the hydrophobic wall and was released to the external atmosphere, leading to a rapid re-diving locomotion and consequently, rapid positive–negative current peaks (Figure 10.11B(b)). Moreover, hydrogen gas showed little damage to the pH-responsive surface.[84]

Apart from hydrogen gas, carbon dioxide has been demonstrated to be an inert gas that does not destroy the stimuli-responsive surface. Shi's group introduced the reaction between calcium carbonate and hydrochloric acid into an energy conversion system, and realized the electricity generation process with little impact on the surface of device. As shown in Figure 10.12, the pH-responsiveness of the device originated from the modification of mixed thiol molecules ($HS(CH_2)_9CH_3$ and $HS(CH_2)_{10}COOH$). With the amplification of the rough surface, the device could perform a superhydrophobic–superhydrophilic reversible property under different pHs. When the device was placed into a beaker filled with an acidic solution, the device floated on the water's surface because of its superhydrophobic property. When the solution was adjusted to basic, the superhydrophobic surface transformed to a superhydrophilic one, and the solution penetrated through the surface. Then the solution was adjusted back to acidic after the addition of hydrochloric acid, the internal acid solution reacted with calcium carbonate to produce carbon dioxide, resulting in the reduction of the specific gravity of the device and consequent surfacing process. Upon touching the water's surface, the device immediately re-dived to the solution's bottom because the internal gas was released to the external atmosphere. When the device was connected with an electrochemical workstation through a flexible copper wire, the mechanical energy obtained was converted into electricity with a high conversion efficiency.[85]

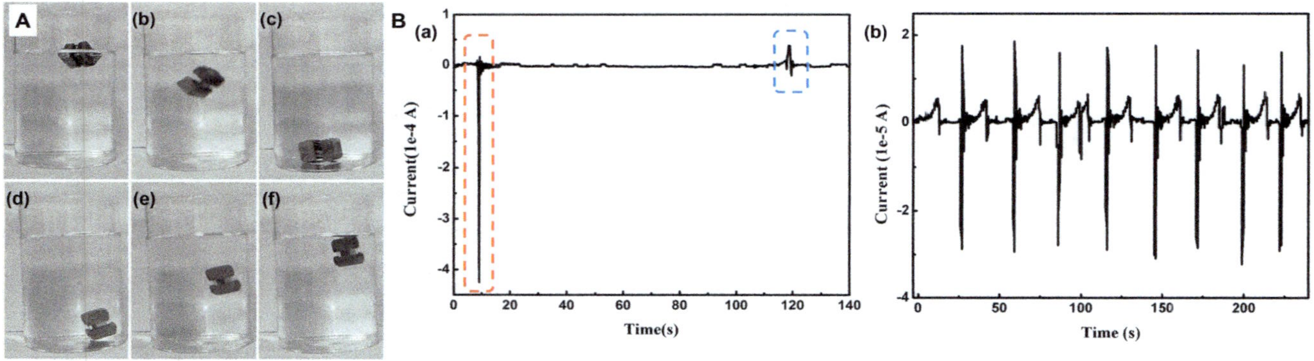

Figure 10.11 pH-Responsive diving–surfacing cycled locomotion propelled by bubble propulsion generated from hydrogen peroxide and its application in the electricity generation process.
Reproduced from ref. 84 with permission from John Wiley and Sons. Copyright © 2014 WILEY-VCH Verlag GmbH & Co. KGaA, Weinheim.

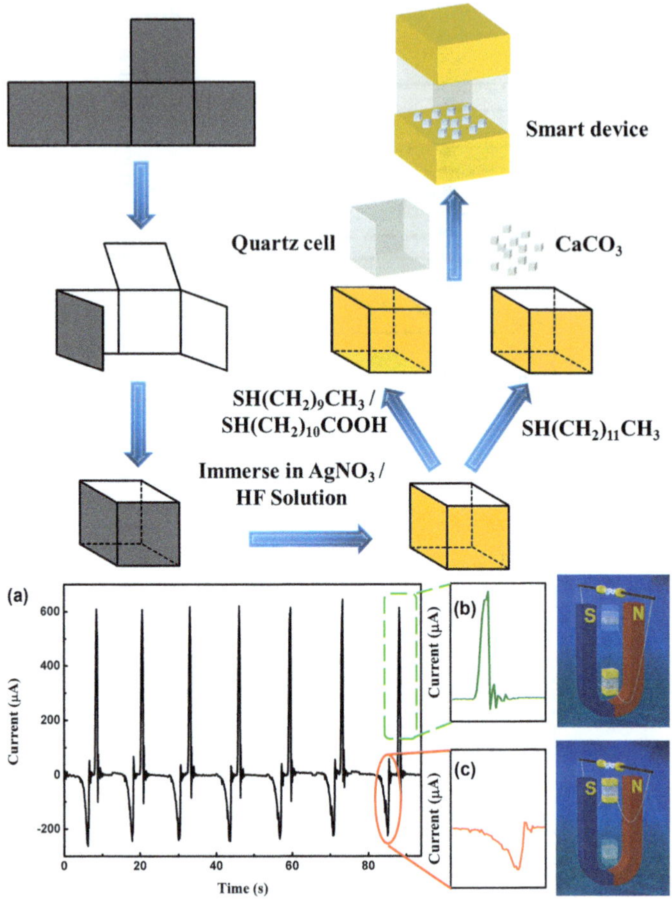

Figure 10.12 Diving–surfacing cycled locomotion with a pH-responsive initiation propelled by bubble propulsion generated from the reaction between calcium carbonate and an acid, and its application in energy conversion. Reproduced from ref. 85 with permission from John Wiley and Sons. Copyright © 2015 WILEY-VCH Verlag GmbH & Co. KGaA, Weinheim.

10.3 Oil/Water Separation

Because of the similar surface energy between the compound modified on the rough surface and oil, and the surface roughness enhancing the property, superhydrophobic/superhydrophilic surfaces (Figure 10.13(a) and (b), respectively) usually display superoleophilicity/superoleophobicity (Figure 10.13(c) and (d), respectively). One essential application for such a surface integrating both watery super-wettability and the opposite oily super-wettability is to separate oil/water mixtures. The key element for an oil/water separating material is either hydrophilic or oleophilic so that it allows one phase of liquid to penetrate through while repelling the other one; moreover, the introduction of high roughness enhances such an effect.

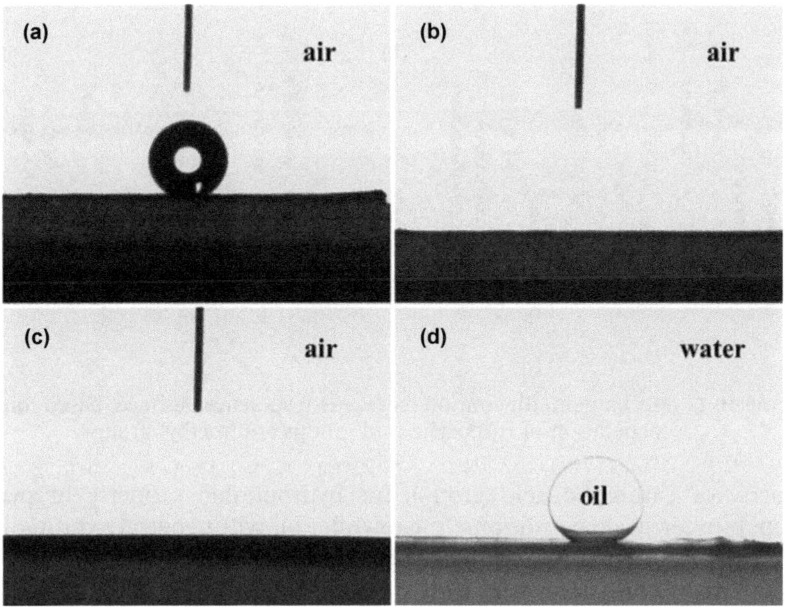

Figure 10.13 Contact angles of a water drop on (a) a superhydrophobic and (b) a superhydrophilic surface and an oil drop on (c) a superoleophilic and (d) a superoleophobic surface.

Based on the cooperation of surface morphology and chemical composition, super-wettability has shown tremendous potential in oil/water separation. Until now, two strategies have been developed to separate oil/water mixtures: the first is to directly utilize superhydrophobic materials to make the oil phase spread easily, adsorb and penetrate on the material while repelling the water phase; the second is to utilize a smart material that can reversibly switch between superhydrophobicity (superoleophilicity) and superhydrophilicity (superoleophobicity) to intelligently separate the oil/water mixture. The introduction of smart materials effectively magnifies the application range of oil/water separating materials, which are discussed in detail in the following section.

10.3.1 pH-Responsive Oil/Water Separation

Smart materials that have the ability to control oily wettability in an aqueous media have shown great potential for underwater applications in oil/water separation. The most commonly adjusted element is the pH value of the solution, which is attributed to the protonation of functional groups in a low pH and deprotonation in a high pH, such as carboxylic acid and pyridyl. Therefore, through adjusting the pH of the solution, we can obtain surfaces with the hydrophobic property of carboxylic acid groups (or hydrophilic property of pyridyl groups) under acidic conditions and the hydrophilic

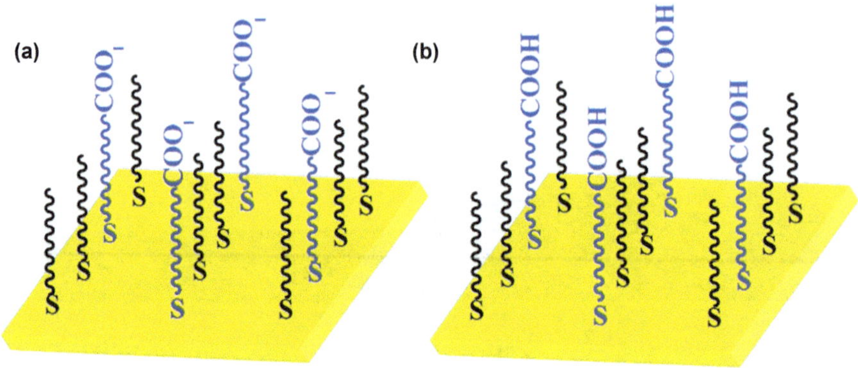

Scheme 10.1 Mechanism illustration of a pH-responsive surface based on the cooperation of carboxylic acid groups and methyl groups.

property of carboxylic acid groups (or hydrophobic property of pyridyl groups) under basic conditions; meanwhile, oil will penetrate through the material with a hydrophobic/oleophilic property, and be repelled by the hydrophilic/oleophobic surface. As a result, the oil/water mixture will be separated successfully.

The carboxylic acid groups are often used along with methyl groups with a certain ratio. Under basic conditions, carboxylic acid groups are deprotonated to carboxylate radicals (Scheme 10.1(a)); on the contrary, the carboxylic acid groups are protonated to carboxyl (Scheme 10.1(b)). Because the deprotonated groups show higher surface energy than the protonated groups, the combined effect of carboxylic acid groups and methyl groups results in a relatively hydrophilic property under acidic conditions and hydrophobic property under basic conditions, and a controlled wettability under a controlled pH.

Shi's group introduced such a pH-responsive material with a reversible superhydrophobic/superhydrophilic property, which was attributed to the cooperation of carbonate acid groups and methyl groups, to realize the separation of a heavy oil/water/light oil ternary system. When the mixture was poured into a separation funnel, a three-layer structure with heavy oil/water/light oil from the bottom formed (Figure 10.14(a)). After the valve was opened, the mixture flowed downward, and the heavy oil directly penetrated through the material and was collected by a beaker (Figure 10.14(b)), but the middle water (with a low pH) was left in the smart device (Figure 10.14(c)). When the water was transformed to alkaline, it penetrated through the device and was collected by a second beaker (Figure 10.14(d)), leaving the light oil in the smart device (Figure 10.14(e)), thus realizing separation of a ternary oil/water/oil system.[86] This work was also a cooperated result of super-wettability and pH-responsiveness: super-wettability resulted in the penetration of oil or water, and pH-responsiveness led to the controllability of the separation progress.

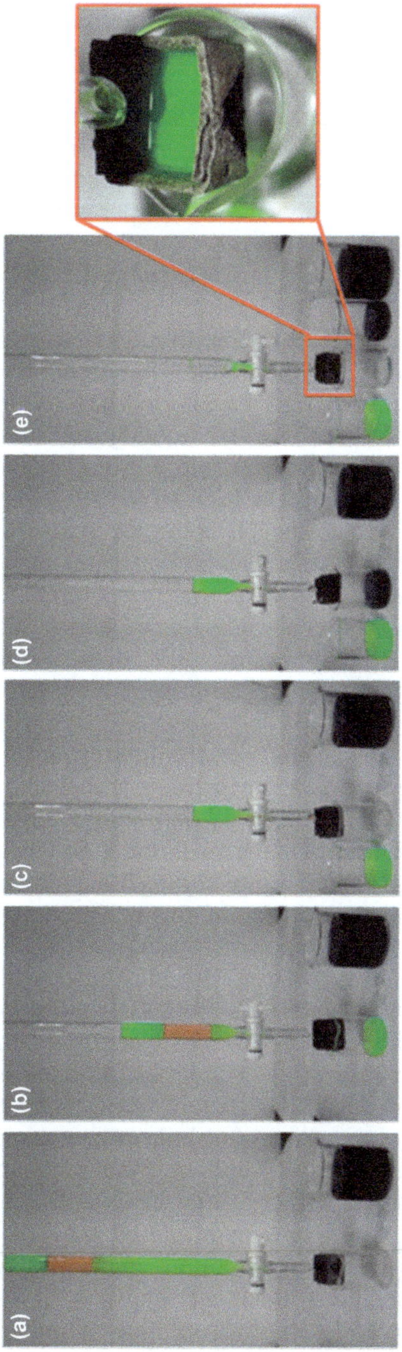

Figure 10.14 pH-Responsiveness-based continuous separation of heavy oil/water/light oil. Reprinted by permission from Macmillan Publishers Ltd: *NPG Asia Mater.* (ref. 86), Copyright 2014.

Using a similar mechanism, Sun and co-workers fabricated a pH-responsive material through the corrosion of a copper mesh and a subsequent modification of mixed thiol molecules ($HS(CH_2)_9CH_3$ and $HS(CH_2)_{10}COOH$). With the magnification of the rough surfaces, the material obtained showed a superhydrophobic property under acidic and neutral conditions and a superhydrophilic property under alkaline conditions. Therefore, when heavy oil/water was poured into the separation device under acidic conditions, the heavy oil wetted the material easily and penetrated through for collection while the upper water layer was left above the material; on the contrary, when the material was treated by a basic solution and a light oil/water solution was poured into the system, water penetrated the materials directly, leaving the light oil above the material (Figure 10.15A).[87] Furthermore, the separation process could be intelligently initiated by regulating the pH of the water. When light oil/water solution was poured into the smart separation device, it stayed above the integrated working film. After the pH of the water was adjusted to a basic value, the material changed from superhydrophobic to superhydrophilic, and water penetrated through the materials easily and was collected by a beaker. When all the water was collected, oil was left above the smart material because of its superoleophobic property, thus realizing oil/water separation with a high efficiency (Figure 10.15B).[88] To sum up, the cooperation of two functions, *i.e.* super-wettability and pH-responsiveness, succeeded in oil/water separation with a smart initiation: the super-wettability of materials resulted in oil/water separation, and the pH-responsiveness enabled the separation process with an intelligent initiation.

Pyridyl is another group that is commonly used to realize pH-responsiveness. In relative work, 2-vinylpyridine was block copolymerized into a polymer brush (PVP) with another kind of monomer on a rough surface. As is shown in the mechanism in Scheme 10.2, under non-acidic conditions, the pyridyl groups were deprotonated and the PVP chains exhibited shrinkage conformation, leaving another block of the polymers with a hydrophobic property in the exterior of the copolymer layer, thus the surface presented a hydrophobic property and was wetted with oil. Under acidic conditions, the pyridyl groups were protonated and the PVP chains exhibited extended conformation, which was induced by the electrostatic repulsion between protonated pyridyl groups, and stretched to the exterior of the copolymer layer, resulting in wetting with an acidic solution. The integrated rough surface magnifies the wettability to a superhydrophilic or superhydrophobic property.

Lu and co-workers modified a graphene foam with an amphiphilic copolymer containing a poly(2-vinylpyridine) block and a polyhexadecyl acrylate block (PVP-*b*-PHA). Due to the hydrophobic/oleophilic property of the PHA block caused by its long hydrophobic chain of alkyl groups, the as-prepared material presented a smart wettability with superhydrophilicity under acidic conditions and superhydrophobicity under basic conditions. Under non-acidic conditions, the PVP chains extended, and the material presented a superhydrophobic/superoleophilic property, thus it selectively

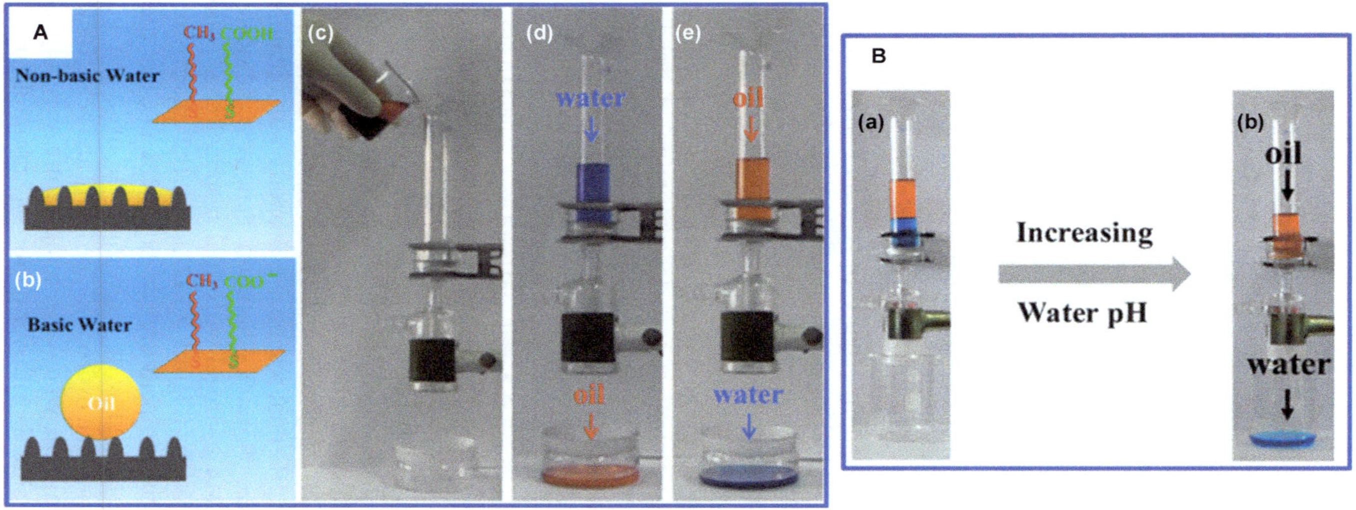

Figure 10.15 Oil/water separation based on the super-wettability originating from mixed thiols. (A) Selective separation of heavy oil or water. Reproduced with permission from Z. J. Cheng, H. Lai, Y. Du, K. W. Fu, R. Hou, C. Li, N. Q. Zhang and K. N. Sun, *ACS Appl. Mater. Interfaces*, 2014, **6**, 636.[87] Copyright (2014) American Chemical Society. (B) Smart separation of light oil/water mixture with a pH-responsive initiation. Reproduced with permission from Z. J. Cheng, J. W. Wang, H. Lai, Y. Du, R. Hou, C. Li, N. Q. Zhang and K. N. Sun, *Langmuir*, 2015, **31**, 1393.[88] Copyright (2015) American Chemical Society.

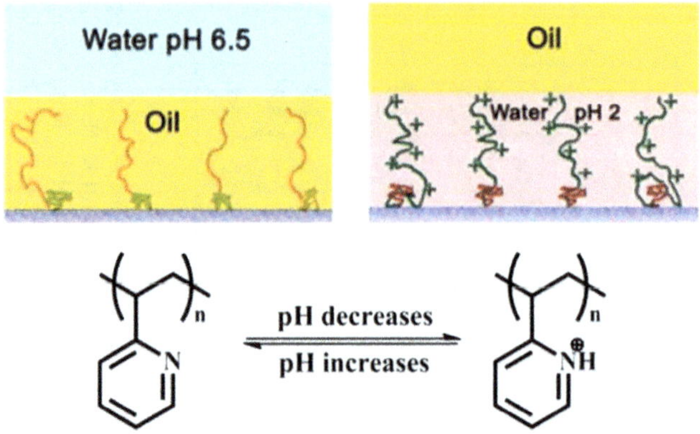

Scheme 10.2 pH-Responsiveness mechanism of PVP copolymers.
Reprinted by permission of Macmillan Publishers Ltd: *NPG Asia Mater.* (ref. 90), Copyright (2014).

adsorbed oil under water rapidly (Figure 10.16A(a)). Furthermore, when the pH value of the water was adjusted to an acidic one, the adsorbed oil was released (Figure 10.16A(b)).[89]

Similarly, Wang *et al.* grafted a pH-responsive poly(2-vinylpyridine) block and an oleophilic/hydrophobic polydimethylsiloxane block (PVP-*b*-PDMS) onto a polyurethane sponge. The PVP chain could alter its chain conformation because of the protonation and deprotonation of pyridine groups responding to the external pH environment and could vary their oily wettability. The results obtained demonstrated that under non-acidic conditions, oil penetrated through the material easily, while leaving the water above; however, when the water was acidic and mixed with heavy oil, the water easily wetted and penetrated through the material, and was collected with a bottom container, leaving the oil above the material, realizing selective adsorption of oil or water and smart oil/water separation (Figure 10.16B).[90]

Besides PVP, poly(dimethylamino)ethyl methacrylate (PDMAEMA) is also used in controlled oil/water separation, not only because of its pH-responsive property induced by the tertiary amine group, but also its thermo-responsiveness (the LCST ranges from 30 to 50 °C with different pH). As a result, a thermo- and pH- dual-responsive property was obtained and respectively used to realize oil/water separation. Feng *et al.* fabricated a mesh coated with a PDMAEMA hydrogel with dual responsiveness and realized smart oil/water separation (Figure 10.17A). In the matter of thermo-responsiveness, when the temperature was below the LCST of the polymer, the polymer chains extended with intermolecular hydrogen bonds with water molecules, resulting in a hydrophilic property and selectively adsorbed water (Figure 10.17B(a)–(c)); when the temperature was adjusted above the LCST, the chains shrank with formation of intramolecular hydrogen bonds, leading to a hydrophobic property and oil penetrated through (Figure 10.17B(d) and (e)). The pH-responsiveness is attributed to the

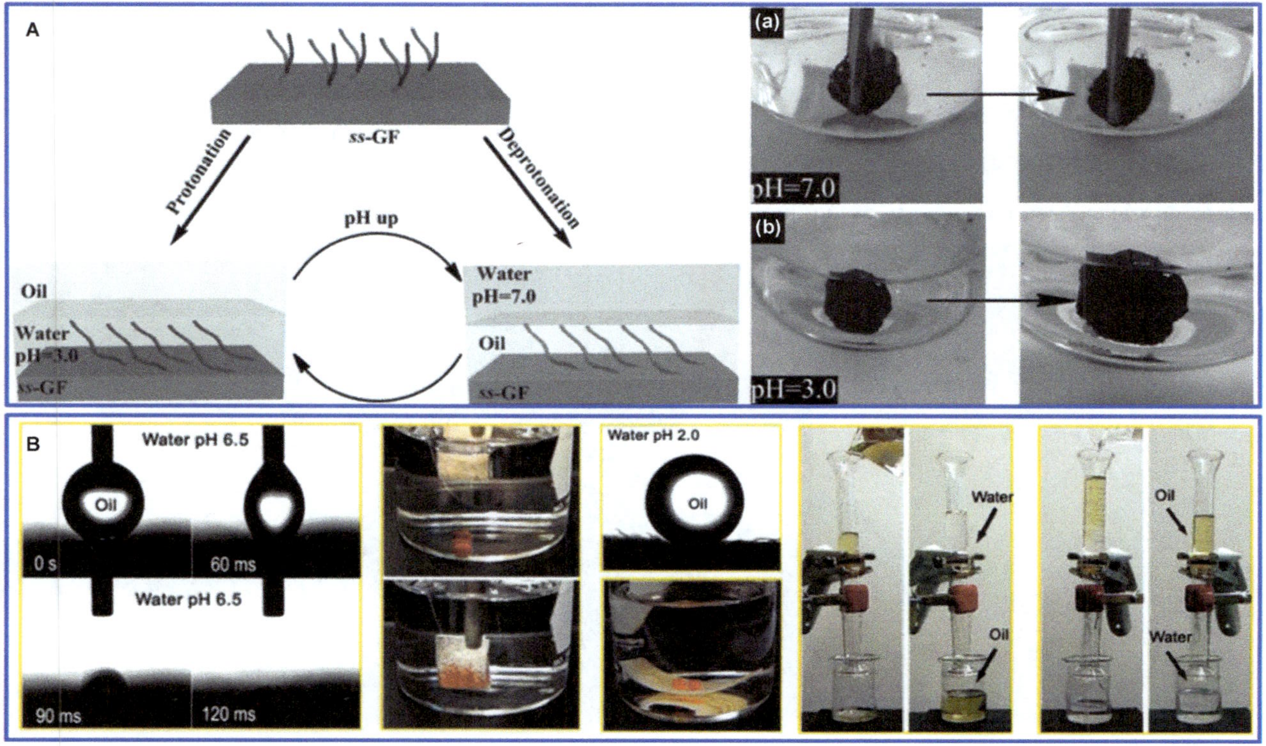

Figure 10.16 Oil/water separation based on the superhydrophobic/superhydrophilic switch that originates from PVP polymers. (A) Reproduced from ref. 89 with permission from John Wiley and Sons. Copyright © 2014 WILEY-VCH Verlag GmbH & Co. KGaA, Weinheim. (B) Reprinted by permission of Macmillan Publishers Ltd: *NPG Asia Mater.* (ref. 90), Copyright (2014).

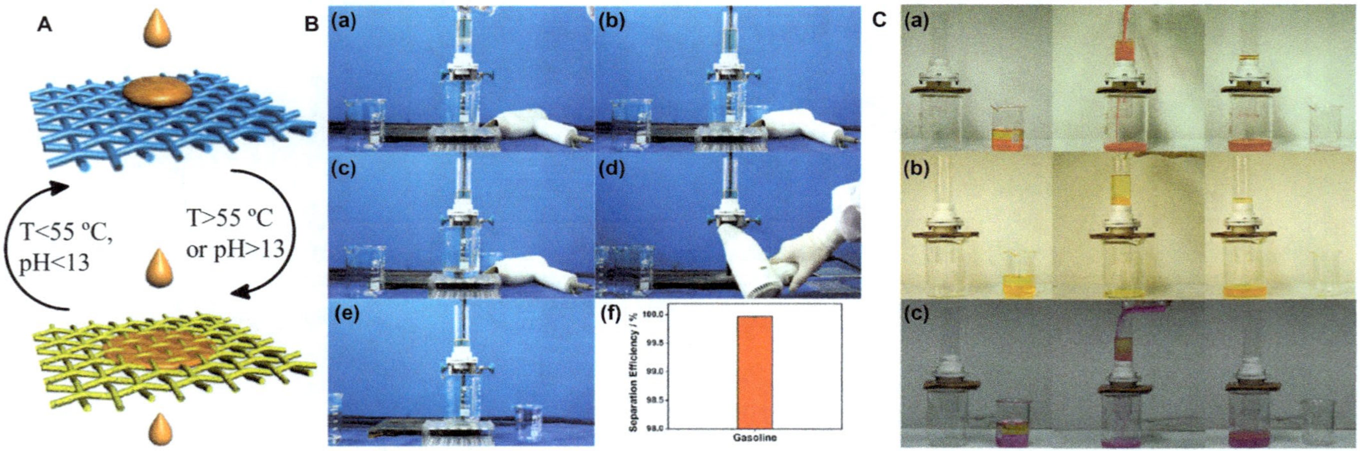

Figure 10.17 Smart oil/water separation responding to temperature and pH based on the thermo- and pH- dual-responsive PDMAEMA polymers.
Reproduced with permission from Y. Z. Cao, N. Liu, C. K. Fu, K. Li, L. Tao, L. Feng and Y. Wei, *ACS Appl. Mater. Interfaces*, 2014, **6**, 2026.[91] Copyright (2014) American Chemical Society.

protonated or deprotonated state of the tertiary amine group under different pH. When the pH was below 8.0, the tertiary amine groups were fully protonated, resulting in a hydrophilic property and water could easily wet the material and flow downward (Figure 10.17C(a) and (b)); when the pH was above 8.0, the tertiary amine groups were deprotonated and the material presented a hydrophobic property, thus leading to the penetration of oil through the material (Figure 10.17C(c)).[91]

10.3.2 Carbon Dioxide-responsive Oil/Water Separation

pH-Responsive materials is the largest subfield in smart oil/water separation; however, the addition of acidic or alkaline solution usually results in inevitable salt accumulation and dilution of solution, which may have an impact on the repeatability of the materials. To solve this problem, Yuan from Tsinghua University fabricated a material with gas responsiveness. The material was made up of a nanostructured block copolymer of poly(*N*,*N*-diethylaminoethyl methacrylate) (PDEAEMA) and a polymethylmethacrylate (PMMA) membrane through electrospinning. When exposed to CO_2 in aqueous media, tertiary amine groups in PDEAEMA were protonated, exhibiting a hydrophilic property, and the material allowed water to penetrate through and be collected by a beaker under the material. After exposure to N_2 to remove CO_2, the material recovered its hydrophobic property because of the deprotonation of the tertiary amine groups. To summarize, using CO_2 as the trigger, the authors realized a novel on–off oil/water separation, and oil easily penetrated through it and was collected. The separation process could be completely reversed without drying the materials (Figure 10.18).[92]

10.3.3 Photoswitchable Oil/Water Separation

Since Fujishima[110] found the water contact angle of polycrystalline anatase TiO_2 was $72 \pm 1^\circ$ and the wetting behaviour could be switched reversibly by regulating UV irradiation, efforts have been devoted to research into the wetting behaviour of inorganic crystals responding to light illumination, among which inorganic oxides such as anatase TiO_2 and ZnO nanorod arrays are the most commonly studied. Through combining photo-responsive wettability conversion with the photocatalytic property of TiO_2, Feng realized oil/water separation and degradation of water-soluble pollutants in one system. They firstly fabricated TiO_2-based mesh films with a micro- and nano-scale hierarchical structure through a hydrothermal method, and then a similar TiO_2-based mesh film modified with octadecylphosphonic acid (ODP) was integrated into one system, resulting in a complex material with a superhydrophobic property. The micro- and nano-scale hierarchical structure amplified the hydrophobic property of ODP and photosensitivity of TiO_2. In the dark, the surface remained stable with a superhydrophobic/superoleophilic property, and it could be used to selectively adsorb oil from an oil/water mixture (Figure 10.19(a) and (b)). Moreover, due to the

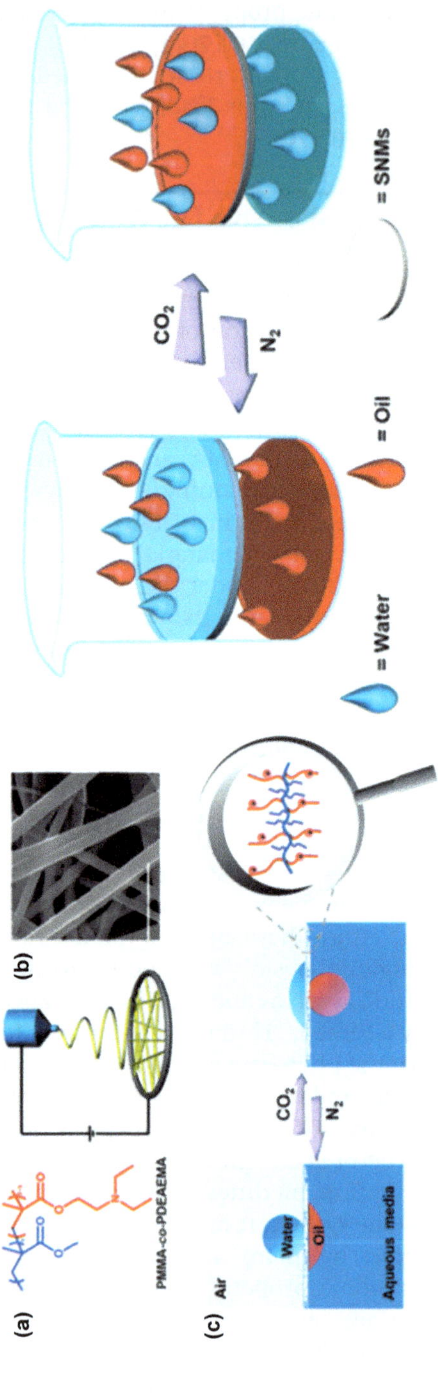

Figure 10.18 Carbon dioxide-responsive oil/water separation based on PDEAEMA nanowires. Reproduced from ref. 92 with permission from John Wiley and Sons. Copyright © 2015 WILEY-VCH Verlag GmbH & Co. KGaA, Weinheim.

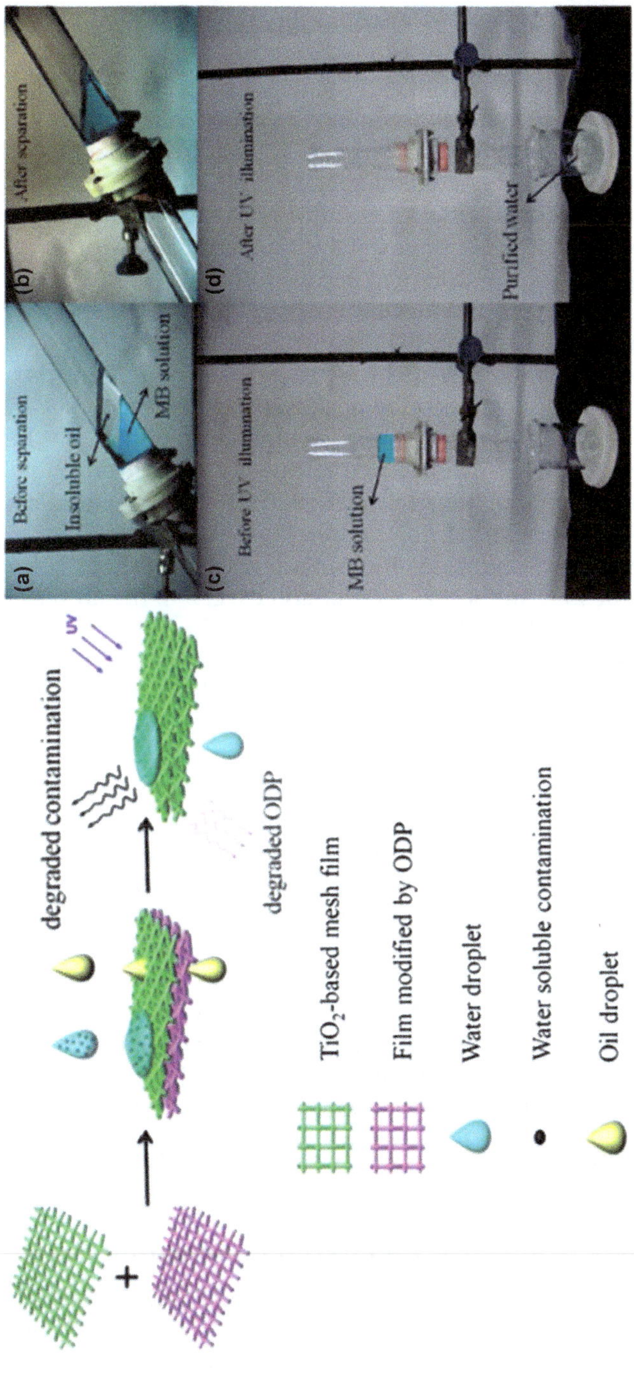

Figure 10.19 Photo-responsive oil/water separation based on the catalysis of TiO_2. Reproduced from ref. 93 with permission from the Royal Society of Chemistry.

photocatalysis of TiO_2, water polluted by organic compounds could be purified when flowing through the TiO_2 film. As can be seen in Figure 10.19(c) and (d), before UV illumination, the material was superhydrophobic and water containing a pollutant was repelled above the smart film; under UV illumination, the ODP gradually degraded and was released, leaving naked substrate and consequently, a superhydrophilic surface, thus water penetrated through the material easily. During the downward flow, the pollutant in the water was photodegraded by TiO_2, thus realizing purification of polluted water. This is an example of the functional cooperation of smart super-wettability and photocatalysis of inorganic oxides, *i.e.* smart super-wettability to control the smart separation of an oil/water mixture, and photocatalysis of inorganic oxides to purify the polluted water.[93]

The working mechanism of ZnO in the oil/water separation process is not exactly the same as that of TiO_2. The wetting conversion of ZnO is based on the variation of intrinsic wettability of the as-prepared materials instead of the degradation of a low-surface-energy compound modified on the surface of inorganic oxides. The intrinsic wettability of ZnO is hydrophobic with a contact angle (CA) of approximately 109°; with the construction of a micro-/nano-scale hierarchical structure, the wettability in the dark is amplified to superhydrophobicity. Under UV illumination, electron–hole pairs will be generated on the ZnO surface, and water and oxygen in the air may compete to incorporate with the holes.[94,95] As a result, the surface hydrophilicity is improved, and is changed to superhydrophilicity with the amplification of the rough surface. Furthermore, the superhydrophobic and superhydrophilic property could be reversibly switched by regulating the UV illumination.

Jiang's group constructed an aligned ZnO nanorod array on a stainless steel mesh through a two-step solution approach. The as-prepared material showed strong control ability on the penetration of water through the material. In the dark, the material showed superhydrophobicity and water was repelled above the material in the separation device; after UV illumination, the material changed to superhydrophilic and water easily wetted the surface and flowed downward (Figure 10.20A).[96] Furthermore, the material was used to separate a crude oil/water mixture. A little water was added to form a water film on the as-prepared material before the oil/water mixture was poured into the separation device. After UV illumination, the material became superhydrophilic and water easily penetrated through to the lower bottle, while the oil was repelled above the material, thus realizing oil/water separation (Figure 10.20B).[97]

10.3.4 Magnetically-driven Oil/Water Separation

Although the oil-collecting ability of superhydrophobic materials has been demonstrated through a large number of reports, realizing oil collection in an enclosed area, which cannot be easily reached, is still a big challenge. A magnetic field was often used to realize long-range control, which is

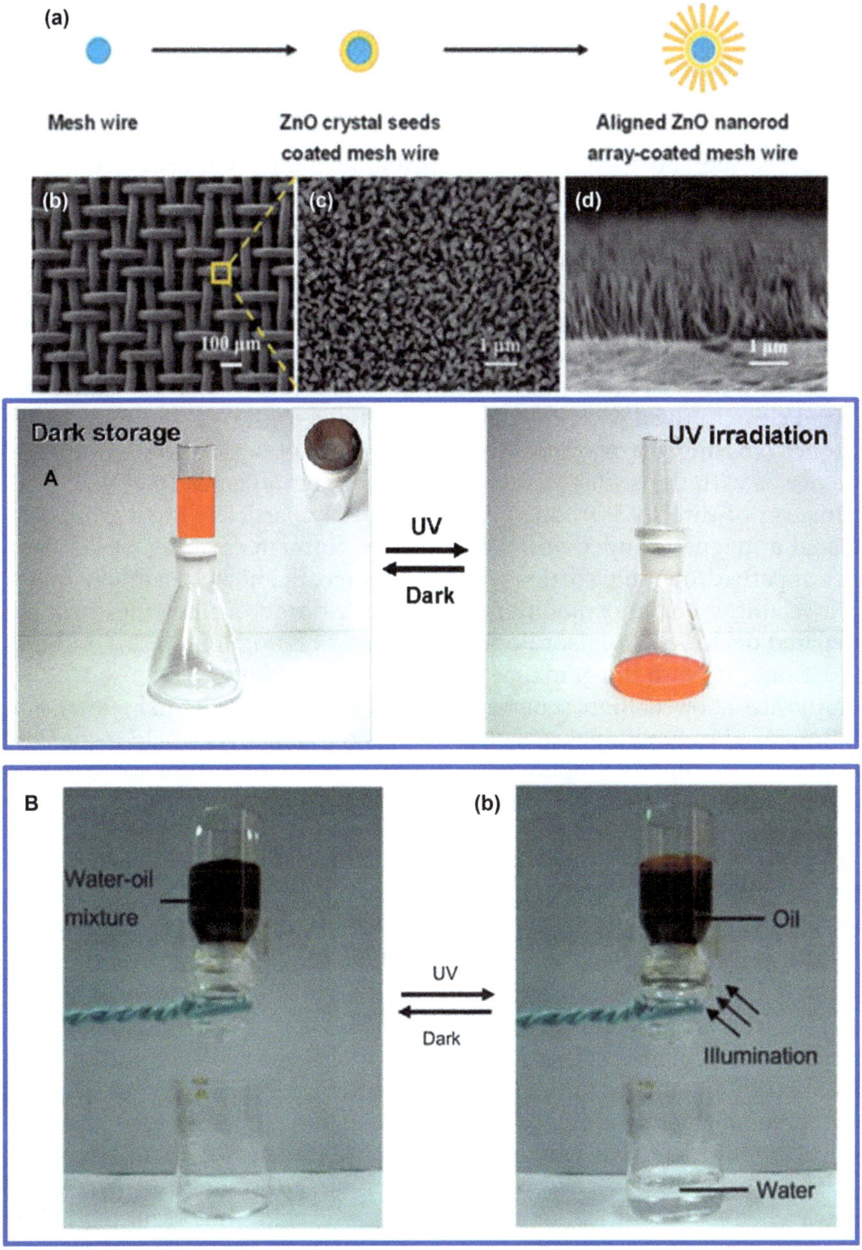

Figure 10.20 Photo-responsive oil/water separation based on ZnO nanorod arrays. (A) Reproduced with permission from D. L. Tian, X. F. Zhang, J. Zhai and L. Jiang, *Langmuir*, 2011, **27**, 4265.[96] Copyright (2011) American Chemical Society; (B) reproduced from ref. 97 with permission from the Royal Society of Chemistry.

thought to show potential in an enclosed area. To realize magnetically driven oil/water separation, the most important elements are the introduction of a paramagnetic property and superhydrophobic character.

Zhang *et al.* constructed a Fe_3O_4@hexadecyltriethoxysilane (HD)-silica/HD-polymer nanocomposite on a piece of polyester film. The material obtained showed superhydrophobicity with a magnetic field-responsive property. Using the as-prepared material in the oil/water system, the magnetic superhydrophobic material was directly guided to the oil and adsorbed it with a high efficiency (Figure 10.21A).[98] Similarly, Calcagnile *et al.* modified polytetrafluoroethylene (PTFE) particles with a hydrophobic property and Fe_3O_4 nanoparticles with a superparamagnetic property on a polyurethane (PU) foam to form a magnetic field-responsive device. Placed on a water/oil system, the device stayed still on the water's surface. When a magnet was placed close to the device, the device was guided to the oil's location by the magnet crossing the air; then the superhydrophobic property empowered the device with the ability to directly adsorb the oil (Figure 10.21B).[99]

Instead of directly introducing magnetic nanoparticles, Shi's group fabricated a magnetic superhydrophobic device through covering nickel foam with superhydrophobic coatings. As a substance, the nickel foam was folded to a seamless box and modified with superhydrophobic coatings. The as-prepared device could adsorb not only floating oil (Figure 10.22A), but also underwater oil with the guidance of a magnet at the bottom (Figure 10.22B). The device showed high separating efficiency at approximately 90%. Furthermore, the functional cooperation of the magnetic field-responsive property and superhydrophobicity provided a bright future for oil/water separation in an enclosed system (Figure 10.22C).[100]

10.4 Smart Ion Channels

Ion channels play an important role in the metabolism process in living organisms. Artificial ion channels in which ions can intelligently pass through have been demonstrated to be of great significance in understanding the working mechanism of natural ion channels, the development of nanofluidics and biosensors *etc.* To achieve these goals, scientists often develop biomimetic asymmetric single or binary responsive nanochannels, and the point of synthesizing such a stimuli-gated nano-device is to selectively introduce stimuli-responsive groups onto the narrow channel surface.

Jiang's group developed a pH-gating ion channel through plasma asymmetric chemical modification of distilled acrylic acid, thus resulting in one side of the hourglass-shaped nanochannel obtained being covered by polyacrylic acid (PAA) (Figure 10.23A). Because of the protonation of the carboxylic groups in PAA below its pK_a of 4.7, the PAA chains shrunk and formed intramolecular hydrogen bonds, resulting in a hydrophobic surface, which repelled flowing water; on the contrary, the PAA chains extended because of the deprotonation of the carboxylic groups and formed intermolecular hydrogen bonds when the pH was above its pK_a of 4.7, and the surface, as a

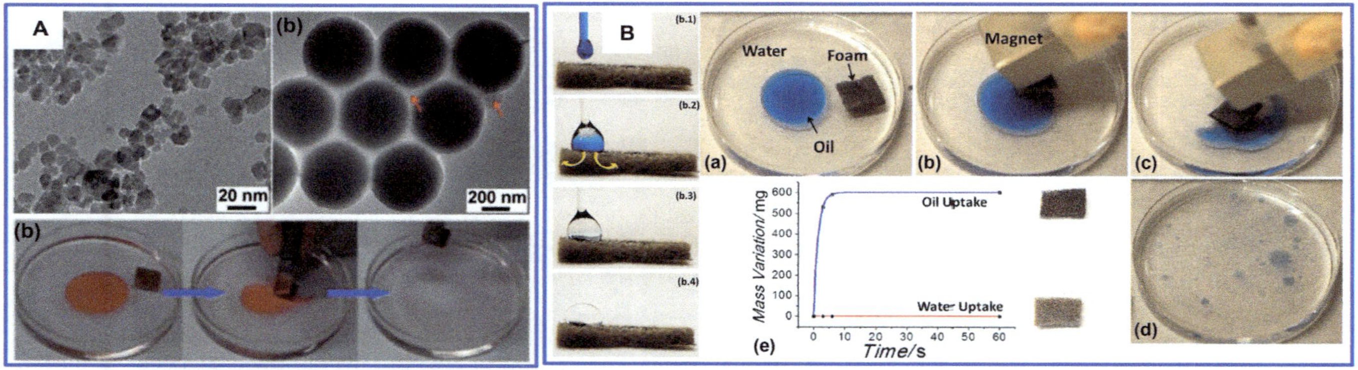

Figure 10.21 Magnetically driven oil/water separation based on ferroferric oxide nanoparticles. (A) Reproduced from ref. 98 with permission from the Royal Society of Chemistry; (B) reproduced with permission from P. Calcagnile, D. Fragouli, I. S. Bayer, G. C. Anyfantis, L. Martiradonna, P. D. Cozzoli, R. Cingolani and A. Athanassiou, *ACS Nano*, 2012, **6**, 5413.[99] Copyright (2012) American Chemical Society.

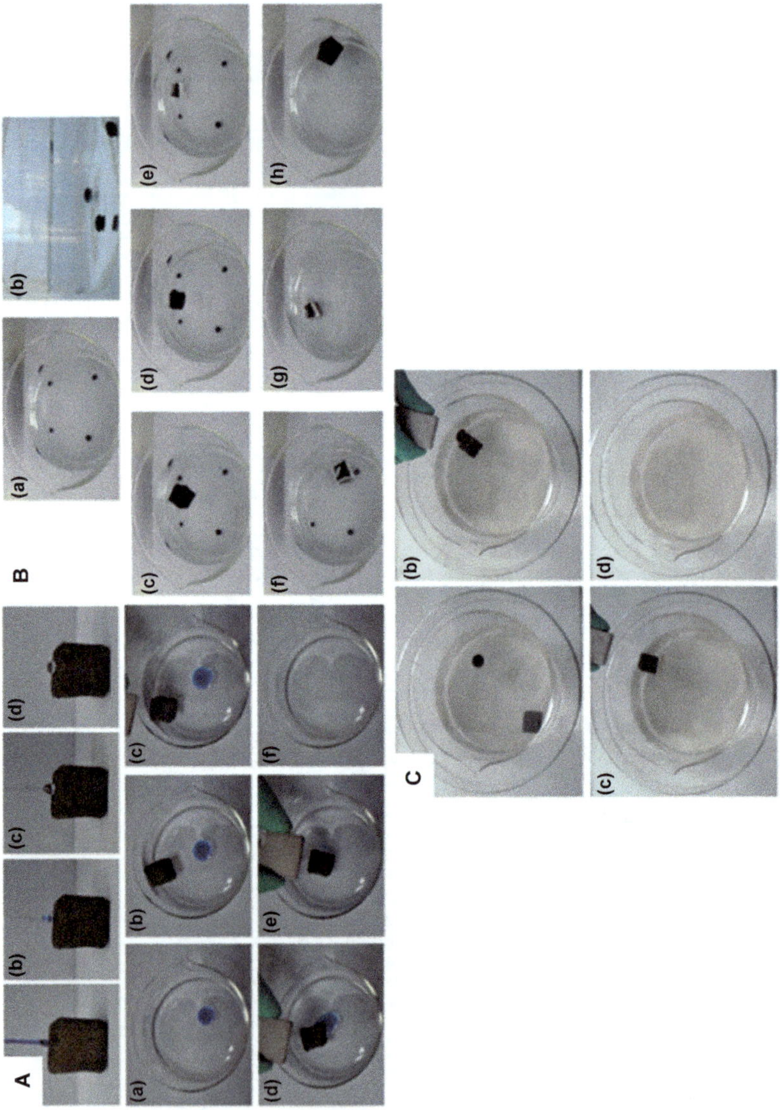

Figure 10.22 Magnetically driven oil/water separation based on nickel foam. Reproduced from ref. 100 with permission from the Royal Society of Chemistry.

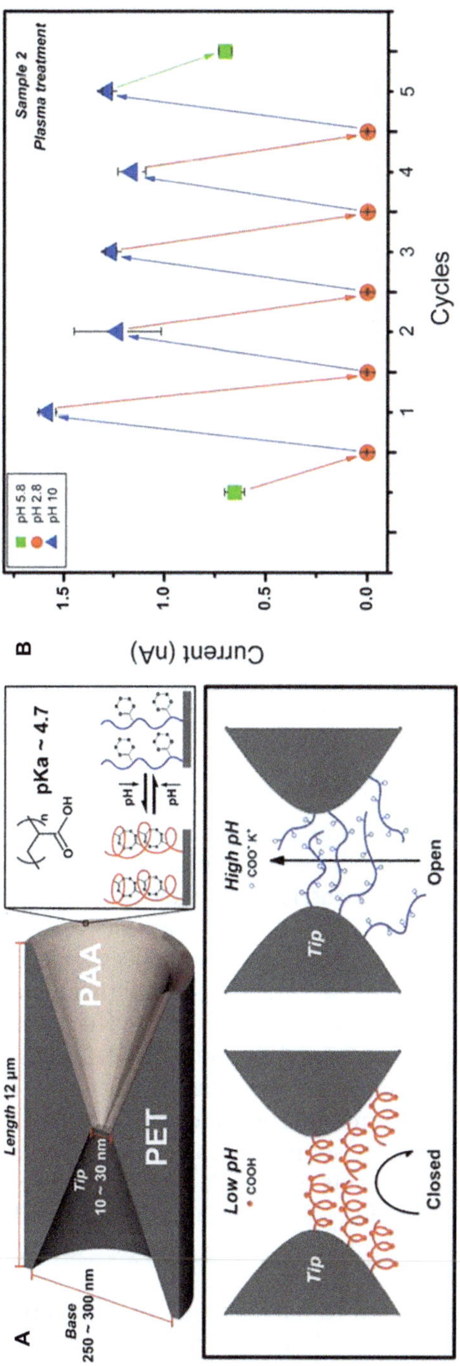

Figure 10.23 pH-Responsive ion nanochannel.
Reproduced from ref. 101 with permission from John Wiley and Sons. Copyright © 2010 WILEY-VCH Verlag GmbH & Co. KGaA, Weinheim.

result, presented hydrophilicity, thus water easily wetted the channel surface and passed through (Figure 10.23B).[101]

When the surface of the other side of the hourglass-shaped nanochannel was modified with PNIPAAm, the exchange of ions through the channels could be regulated through pH and temperature (Figure 10.24A). In this system, the ion transport property was a cooperating result of effective pore size of the nanochannel and the wettability of the inner surface of the nanochannel. Under the best conditions, *i.e.* the environmental temperature was above the LCST of PNIPAAm and the pH was above the pK_a of PAA of 4.7, ions could transport through the channels easily, and through regulating the external temperature and pH, transportation of ions through the nanochannel could be adjusted (Figure 10.24B).[102]

To simulate natural ion channels, the authors developed a nano-capsule with two gates. The two gates of the device were separately modified with polyvinylpyridine (PVP) (pK_a 5.2) and PAA (pK_a 4.7), and these two polymers showed almost opposite wettability under different pH values. As shown in Figure 10.25, below pH 4.7, both polymers were protonated and PVP presented hydrophilicity while PAA was hydrophobic; on the contrary, above pH 5.2, both were deprotonated and PVP showed hydrophobicity while PAA was hydrophilic. When the pH of the PVP side was fixed above 5.2 and that of the PAA side was below 4.7, the two gates were closed, which enabled the nano-capsules to enclose ions inside; on the contrary, if the pH of the PVP side remained below 5.2 and that of the PAA side was above 4.7, the two gates were open for complete penetration of ions.[103]

10.5 Smart Permeability

Since materials with superhydrophobic or superhydrophilic properties have extreme wettability with water, smart materials have potential applications in the selective permeability of water, which may be of significance in solvent separation.

Bergbreiter and co-workers fabricated a solute-responsive film through a layer-by-layer technique. It has been demonstrated that a sulfate radical has a salting-out effect for the internal ingredient, PNIPAAm. Therefore, the cooperation of the salting-out effect of anions on PNIPAAm and its thermo-responsiveness made selective permeability happen. As can be seen in Figure 10.26, under room temperature, water easily wetted the as-prepared material, while 1.2 M Na_2SO_4 solution was repelled with a resulting high contact angle. This was a result of the change of the LCST of PNIPAAm, which was caused by the salt solution; the LCST of PNIPAAm in 1.2 M Na_2SO_4 solution is below room temperature. Because of this effect, the flow rate of water was higher than that of 1.2 M Na_2SO_4 solution at room temperature, which could be used to realize selective permeability.[104]

Soh's group fabricated three kinds of hydrogels: a stretchable (stress-responsive) hydrogel, a pH-responsive hydrogel that expanded in a basic solution, and a thermo-responsive hydrogel that expanded under a high

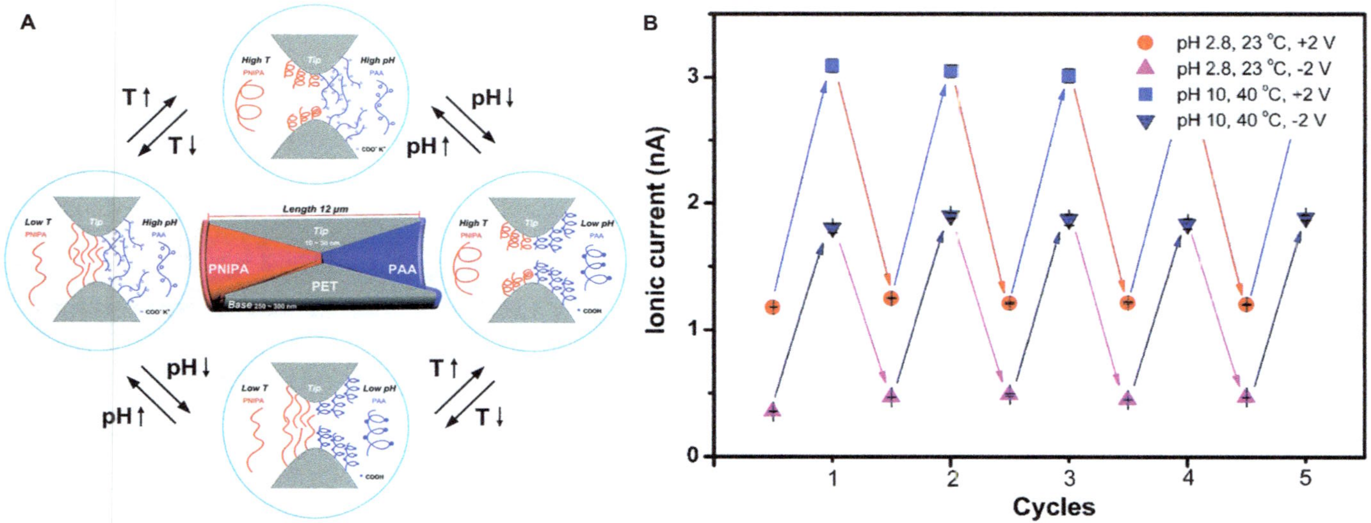

Figure 10.24 Thermo- and pH- dual-responsive ion nanochannel. Reproduced with permission from X. Hou, F. Yang, L. Li, Y. L. Song, L. Jiang and D.B. Zhu, *J. Am. Chem. Soc.*, 2010, **132**, 11736.[102] Copyright (2010) American Chemical Society.

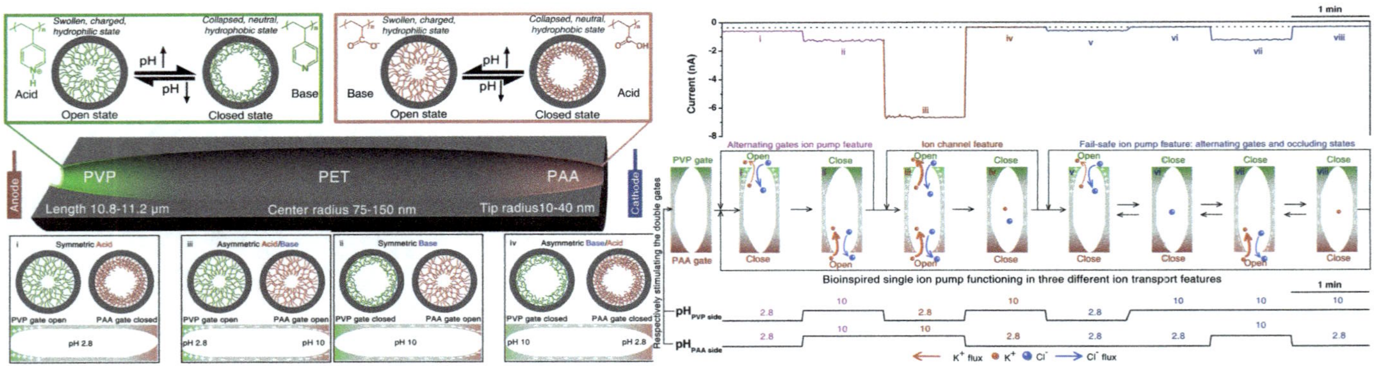

Figure 10.25 Dual pH-responsive ion nanochannel.
Reproduced with permission from H. C. Zhang, X. Hou, L. Zeng, F. Yang, L. Lim D. D. Yan, Y. Tian and L. Jiang, *J. Am. Chem. Soc.*, 2013, **135**, 16102.[103] Copyright (2013) American Chemical Society.

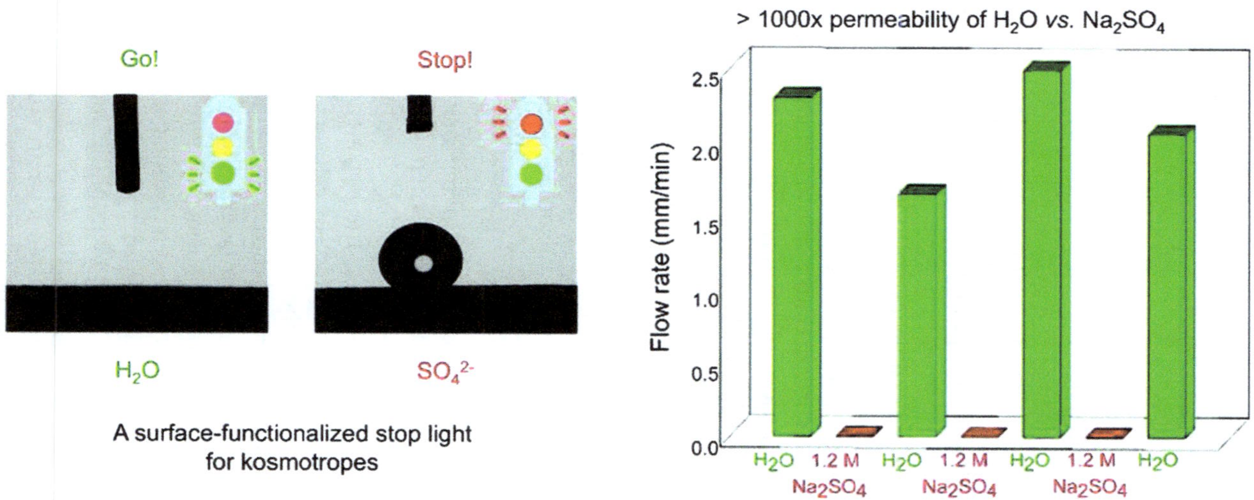

Figure 10.26 Solute- and thermo- dual-responsive smart permeation.
Reproduced with permission from A. L. Allen, K. J. Tan, H. Fu, J. D. Batteas and D. E. Bergbeiter, *Langmuir*, 2012, **28**, 5237.[104] Copyright (2012) American Chemical Society.

temperature (Figure 10.27A). They integrated these three smart materials into one system with four chambers, three of which were for storing solutions and one for mixing solutions (Figure 10.27B(a)). By selectively regulating the conditions, through operating stress, rising temperature or increasing pH, three kinds of hydrogels with different responses could be expanded to allow selective permeation of stored solutions (Figure 10.27B(b)–(e)).[105]

10.6 Bio-applications

Hydrophobic/hydrophilic properties are part of the universal weak interactions in nature, which play a significant role in living organisms, such as the formation of lipid bilayers, protein folding, enzyme–substrate interactions, and endocytosis. However, smart systems are still far from the requirement of controlled reversible capture and release of targeted cells, which is in demand for cell-based cancer diagnostics.

To obtain a thermo-responsive surface for smart capture and release of cells, Wang *et al.* grafted PNIPAAm onto a silicon wafer with nanostructures, followed by the modification of biotinylated epithelial-cell adhesion-molecule antibody (biotin-*anti*-EpCAM) *via* biotin-streptavidin/hydrophobic anchor interactions. When the temperature was fixed above the LCST of PNIPAAm, targeted cancer cells were captured by biotin-*anti*-EpCAM modified on the polymer chains through the hydrophobic interaction between biotin-*anti*-EpCAM and an antigen on the cancer cells' surface. When the temperature was reduced below the LCST, PNIPAAm chains stretched and turned hydrophilic, releasing the adsorbed cancer cells because of the repulsion between the hydrophobic anchor and hydrophilic polymer (Figure 10.28A).[106]

Similarly, Yu *et al.* imported covalently-linking biotin groups into PNIPAAm chains through polymerization, and then strategically introduced streptavidin and biotin-*anti*-EpCAM onto the biotin groups in the PNIPAAm polymers. Above the LCST of PNIPAAm, the polymer chains shrunk with biotin/avidin/biotin-*anti*-EpCAM groups in the external layer, which showed a high adhesion with cancer cells. When the temperature was adjusted below LCST, the polymer chains extended and the consequent chain conformation resulted in the concealment of the biotin/avidin/biotin-*anti*-EpCAM groups; the cancer cells were released from the surface (Figure 10.28B).[107]

Gallium nitride (GaN) is well-known as a photosensitive semiconductor material and performs different wettabilities under different illumination conditions: hydrophobic in the dark and hydrophilic under UV illumination. Based on the photo-responsiveness of the GaN substrate with nanowire structures, Wang and co-workers fabricated a smart surface with reversible superhydrophobic/superhydrophilic properties (Figure 10.29A). In their protein adsorption experiment, in the dark, bovine serum albumin (BSA) was selectively adsorbed onto the superhydrophobic surface because of the hydrophobic interaction between BSA and the surface; while after

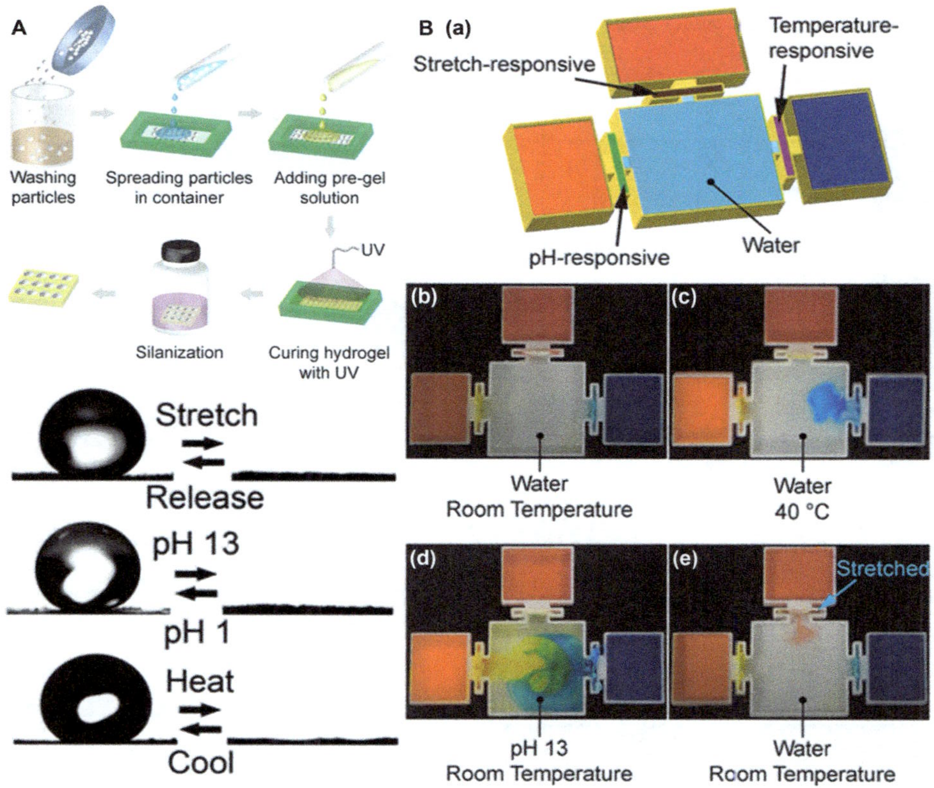

Figure 10.27 Ternary-responsive smart permeation for a designed mixing process. Reproduced from ref. 105 with permission from John Wiley and Sons. Copyright © 2015 WILEY-VCH Verlag GmbH & Co. KGaA, Weinheim.

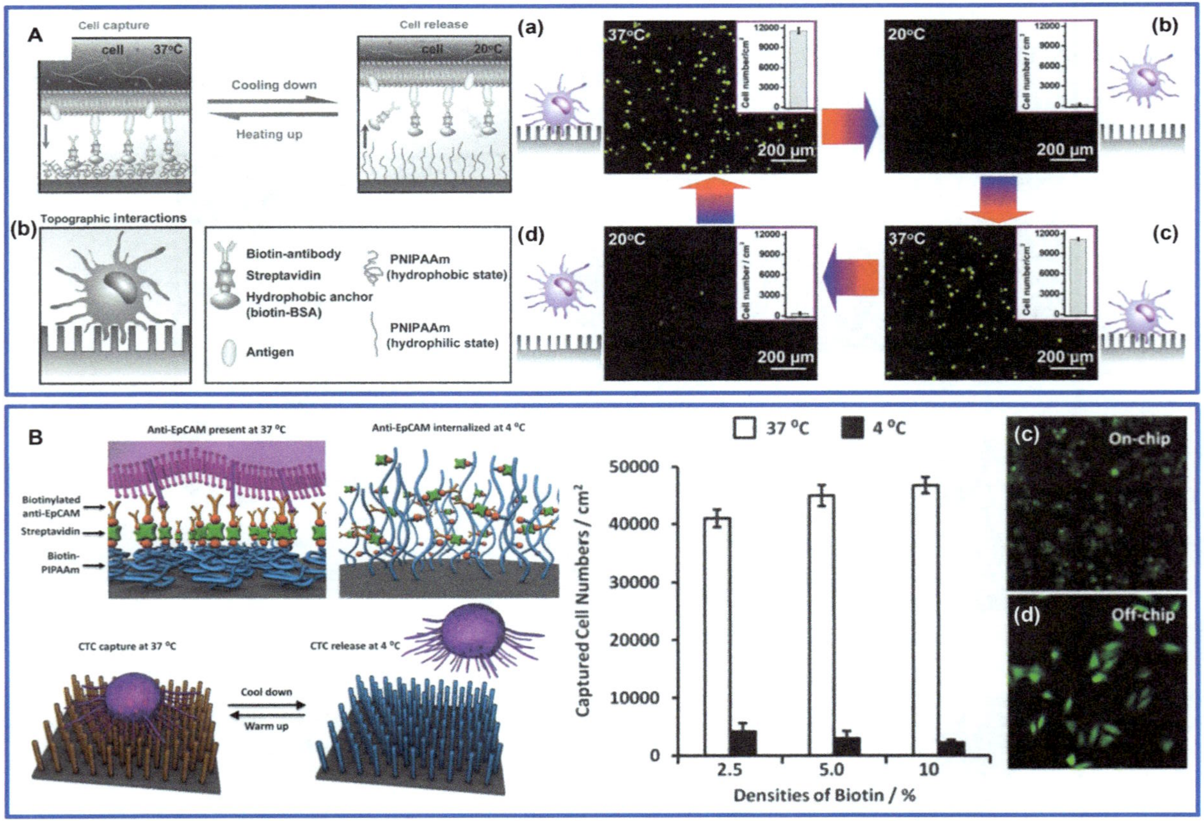

Figure 10.28 Thermo-responsive adhesion of cancer cells.
(A) Reproduced from ref. 106 with permission from John Wiley and Sons. Copyright © 2013 WILEY-VCH Verlag GmbH & Co. KGaA, Weinheim. (B) Reproduced from ref. 107 with permission from John Wiley and Sons. Copyright © 2013 WILEY-VCH Verlag GmbH & Co. KGaA, Weinheim.

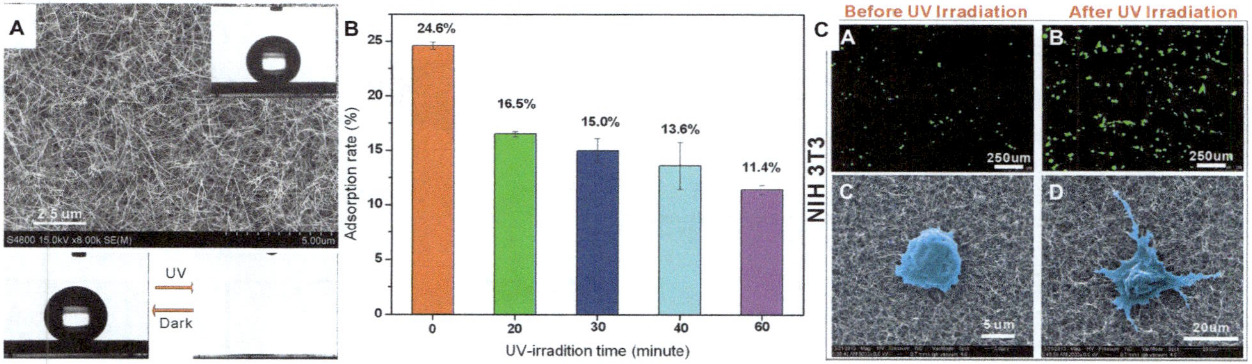

Figure 10.29 Photo-responsive capture/release of cancer cells. Reproduced with permission from J. Y. Li, Q. S. Han, Y. Zhang, W. Zhang, M. D. Dong, F. Besenbacher, R. Yang and C. Wang, *ACS Appl. Mater. Interfaces*, 2013, **5**, 9816.[108] Copyright (2013) American Chemical Society.

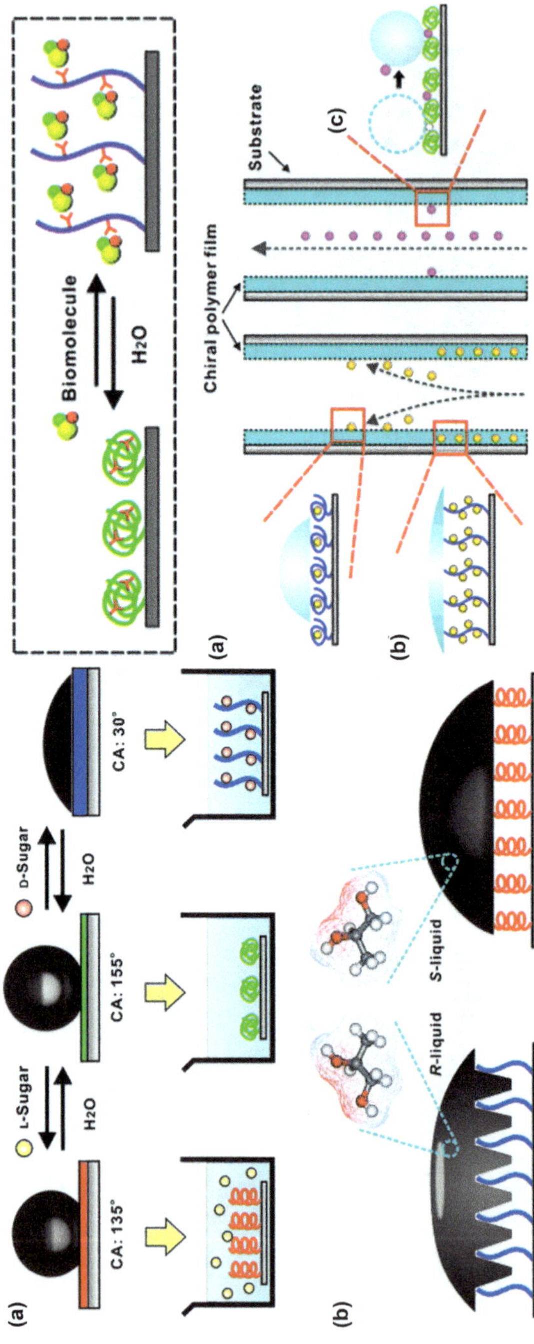

Figure 10.30 Smart separation of biomolecules based on chirality. Reproduced from ref. 109 with permission from John Wiley and Sons. Copyright © 2014 WILEY-VCH Verlag GmbH & Co. KGaA, Weinheim.

UV illumination, BSA was released (Figure 10.29B). Because of the hydrophilic property of the cell surfaces, the as-prepared superhydrophobic surface in the dark showed repulsion towards cells; after UV illumination for 60 min, the surface turned superhydrophilic and cells could be adhered onto it, thus realizing a smart capture/release of targeted cells on the materials (Figure 10.29C).[108]

A large amount of attention has been focused on chirality and relevant studies since the concept of chirality was proposed in 1848. In material science, smart surfaces with chirality have shown great potential in medicines, sensors and biotechnology. Sun's group reported a few smart surfaces that could switch their own wettability in response to chiral molecules. With a chirality-responsive ingredient that could form intermolecular hydrogen bonds with L- or D-molecules integrated into polymer chains, solutions containing L- or D-molecules flowing by the smart surface would be adsorbed onto the surface, which would induce a visual transformation of the surface from hydrophobic to hydrophilic; while solutions containing the other kinds of chiral molecules flowing by the smart surface would stay hydrophobic. This theory showed a potential application of chiral molecule separation (Figure 10.30).[109]

10.7 Outlook

Smart materials have shown great potential in the fields of nanotechnology, biomimetic materials, electronics and fluid mechanics. A reversible superhydrophobic/superhydrophilic switchable property is an important branch in smart materials, which has been demonstrated in the fields of drug delivery, controlled locomotion of small objects, bio-sensors and detectors, *etc.* Integrating smart materials that could switch between superhydrophobic and superhydrophilic with one or more functional material(s) into one system, the given intention could be realized when the integrated functional components work in a designed order. This concept of a functionally cooperated smart device is originally from nature. In this chapter, many successful examples of functionally cooperative smart devices have been discussed; however, research on the devices is just at the beginning, and more attention should be focused on relevant studies, which may bring about a revolution in the development of materials science, biomaterials and nanotechnology.

References

1. Y. G. Jiang, Z. Q. Wang, X. Yu, F. Shi, H. P. Xu and X. Zhang, *Langmuir*, 2005, **21**, 1986.
2. F. Xia, H. Ge, Y. Hou, T. L. Sun, L. Chen, G. Z. Zhang and L. Jiang, *Adv. Mater.*, 2007, **19**, 2520.
3. B. A. Abel, M. B. Sims and C. L. McCormick, *Macromolecules*, 2015, **48**, 5487.

4. L. Tan, M. Y. Yang, H. X. Wu, Z. W. Tan, J. Y. Xiao, C. J. Liu and R. X. Zhuo, *ACS Appl. Mater. Interfaces*, 2015, **7**, 6310.
5. X. X. Chen, J. Gao, B. Song, M. Smet and X. Zhang, *Langmuir*, 2010, **26**, 104.
6. A. Zhuk, R. Mirza and S. Sukhishvili, *ACS Nano*, 2011, **5**, 8790.
7. K. Ichimura, S. K. Oh and M. Nakagawa, *Science*, 2010, **288**, 1624.
8. G. Y. Qing, X. Wang, L. Jiang, H. Fuchs and T. L. Sun, *Soft Matter*, 2009, **5**, 2759.
9. S. P. Wu, X. Q. Zhu, J. L. Yang and J. Nie, *Chem. Commun.*, 2015, **51**, 5649.
10. H. W. Xia, F. Xia, Y. C. Tang, W. Guo, X. Hou, L. Chen, Y. Hou, G. Z. Zhang and L. Jiang, *Soft Matter*, 2011, **7**, 1638.
11. Z. J. Cheng, H. Lai, M. Du, S. C. Zhu, N. Q. Zhang and K. N. Sun, *Soft Matter*, 2012, **8**, 9635.
12. M. M. Ma, L. Guo, D. G. Anderson and R. Langer, *Science*, 2013, **339**, 186.
13. Y. Guo, F. Xia, L. Xu, J. Li, W. S. Yang and L. Jiang, *Langmuir*, 2010, **26**, 1024.
14. T. Hoare, B. P. Timko, J. Santamaria, G. F. Goya, S. Irusta, S. Lau, C. F. Stefanescu, D. Lin, R. Langer and D. S. Kohane, *Nano Lett.*, 2011, **11**, 1395.
15. B. C. K. Tee, C. Wang, R. Allen and Z. N. Bao, *Nat. Nanotechnol.*, 2012, **7**, 825.
16. H. S. Lim, J. T. Han, D. Kwak, M. H. Jin and K. Cho, *J. Am. Chem. Soc.*, 2006, **128**, 14458.
17. G. Y. Qing, X. Wang, H. Fuchs and T. L. Sun, *J. Am. Chem. Soc.*, 2009, **131**, 8370.
18. F. D. Jochum and P. Theato, *Chem. Soc. Rev.*, 2013, **42**, 7468.
19. S. Guragain, B. P. Bastakoti, V. Malgras, K. Nakashima and Y. Yamauchi, *Chem. – Eur. J.*, 2015, **21**, 13164.
20. J. Thévenot, H. Oliveira, O. Sandre and S. Locommandoux, *Chem. Soc. Rev.*, 2013, **42**, 7099.
21. Y. Zhou, D. S. Wang, S. L. Huang, G. Auernahammer, J. J. He, H. J. Butt and S. Wu, *Chem. Commun.*, 2015, **51**, 2725.
22. B. Su, W. Guo and L. Jiang, *Small*, 2015, **11**, 1072.
23. M. Yoshida and J. Lahann, *ACS Nano*, 2008, **2**, 1101.
24. T. L. Sun, G. Y. Qing, B. L. Su and L. Jiang, *Chem. Soc. Rev.*, 2010, **40**, 2909.
25. H. L. Lim, Y. Hwang, M. Kar and S. Varghese, *Biomater. Sci.*, 2014, **2**, 603.
26. F. Xia and L. Jiang, *Adv. Mater.*, 2008, **20**, 2842.
27. E. G. Kelley, J. N. L. Albert, M. O. Sullivan and T. H. Epps, *Chem. Soc. Rev.*, 2013, **42**, 7057.
28. C. A. Lorenzo and A. Concheiro, *Chem. Commun.*, 2014, **50**, 7743.
29. Z. Liu, W. Wang, R. Xie, X. J. Ju and L. Y. Chu, *Chem. Soc. Rev.*, 2016, **45**, 460–475.

30. L. P. Wen, Y. Tian and L. Jiang, *Angew. Chem., Int. Ed.*, 2015, **54**, 3387.
31. X. Yao, Y. L. Song and L. Jiang, *Adv. Mater.*, 2011, **23**, 719.
32. W. Barthlott, *Raster-Elektronenmikroskopie der Epidermis-Oberflachen von Spermatophyten*, Tropische und subtropische Pflanzenwelt (German Edition), 1977.
33. W. Barthlott and C. Neinhuis, *Planta*, 1997, **202**, 1.
34. H. C. von Baeyer, *Science*, 2000, **40**, 12.
35. L. Feng, S. H. Li, Y. S. Li, H. J. Li, L. J. Zhang, J. Zhai, Y. L. Song, B. Q. Liu, L. Jiang and D. B. Zhu, *Adv. Mater.*, 2002, **14**, 1857.
36. S. Yang, J. Ju, Y. C. Qiu, Y. X. He, X. L. Wang, S. X. Dou, K. S. Liu and L. Jiang, *Small*, 2014, **10**, 294.
37. M. Riedel, A. Eichner and R. Jetter, *Planta*, 2003, **218**, 87.
38. I. Scholz, M. Bückins, L. Dolge, T. Erlinghagen, A. Weth, F. Hischen, J. Mayer, S. Hoffmann, M. Riederer, M. Riedel and W. Baumgartner, *J. Exp. Biol.*, 2010, **213**, 1115.
39. E. Gorb, V. Kastner, A. Peressadko, E. Arzt, L. Gaume, N. Rowe and S. Gorb, *J. Exp. Biol.*, 2004, **207**, 2947.
40. L. Feng, Y. N. Zhang, J. M. Xi, Y. Zhu, N. Wang, F. Xia and L. Jiang, *Langmuir*, 2008, **24**, 4114.
41. Y. M. Zheng, X. F. Gao and L. Jiang, *Soft Matter*, 2007, **3**, 178.
42. C. M. Bhadra, V. K. Truong, V. T. H. Pham, M. A. Kobaisi, G. Seniutunas, J. Y. Wang, S. Juodkazis, R. J. Crawford and E. P. Ivanova, *Sci. Rep.*, 2015, **5**, 16817.
43. E. Appel, L. Heepe, C. P. Lin and S. N. Gorb, *J. Anat.*, 2015, **227**, 561.
44. J. Hasan, H. K. Webb, V. K. Truong, G. S. Watson, J. A. Watson, M. J. Tobin, G. Gervinskas, S. Juodkazis, J. Y. Wang, R. J. Crawford and E. P. Ivanova, *Langmuir*, 2012, **28**, 17404.
45. M. X. Sun, G. S. Watson, Y. M. Zheng, J. A. Watson and A. P. Liang, *J. Exp. Biol.*, 2009, **212**, 3148.
46. X. F. Gao and L. Jiang, *Nature*, 2004, **432**, 36.
47. M. Sitti and R. S. Fearing, *J. Adhes. Sci. Technol.*, 2003, **18**, 1055.
48. Q. A. Meng, Q. B. Wang, H. Liu and L. Jiang, *NPG Asia Mater.*, 2014, **6**, e125.
49. Q. B. Wang, B. Su, H. Liu and L. Jiang, *Adv. Mater.*, 2014, **26**, 4889.
50. H. K. Webb, R. J. Crawford and E. P. Ivanova, *Adv. Colloid Interface Sci.*, 2014, **210**, 58.
51. X. M. Li, D. Reinhoudt and M. C. Calama, *Chem. Soc. Rev.*, 2007, **36**, 1350.
52. T. Darmanin and F. Guittard, *J. Mater. Chem. A*, 2014, **2**, 16319.
53. J. L. Yong, Q. Yang, F. Chen, D. S. Zhang, G. Q. Du, J. H. Si, F. Yuan and X. Hou, *J. Micromech. Microeng.*, 2014, **24**, 035006.
54. B. W. Xin and J. C. Hao, *Chem. Soc. Rev.*, 2010, **39**, 769.
55. Y. Wu, X. W. Pei, X. L. Wang, Y. M. Liang, W. M. Liu and F. Zhou, *NPG Asia Mater.*, 2014, **6**, e136.
56. F. Shi, Z. Q. Wang and X. Zhang, *Adv. Mater.*, 2005, **17**, 1005.

57. Y. M. Zheng, H. Bai, Z. B. Huang, X. L. Tian, F. Q. Nie, Y. Zhao, J. Zhai and L. Jiang, *Nature*, 2010, **463**, 640.
58. T. S. Wong, S. H. Kang, S. K. Y. Tang, E. J. Smythe, B. D. Hatton, A. Grinthal and J. Aizenberg, *Nature*, 2011, **477**, 443.
59. M. C. Roco, *AIChE J.*, 2004, **50**, 890.
60. Y. F. Gao, M. J. Cheng, B. L. Wang, Z. G. Feng and F. Shi, *Adv. Mater.*, 2010, **22**, 5125.
61. E. Diller and M. Sitti, *Adv. Funct. Mater.*, 2014, **24**, 4397.
62. Z. Ye and M. Sitti, *Lab Chip*, 2014, **14**, 2177.
63. A. A. Solovev, S. Sanchez, M. Pumera, Y. F. Mei and O. G. Schmidt, *Adv. Funct. Mater.*, 2010, **20**, 2430.
64. S. K. Sailapu and A. Chattopadhyay, *Angew. Chem., Int. Ed.*, 2014, **53**, 1521.
65. Y. Ikezoe, J. Fang, T. L. Wasik, T. Uemura, Y. T. Zheng, S. Kitagawa and H. Matsui, *Adv. Mater.*, 2015, **27**, 288.
66. Z. Li, G. Zhu, R. S. Yang, A. C. Wang and Z. L. Wang, *Adv. Mater.*, 2010, **22**, 2534.
67. J. Bae, J. Lee, S. M. Kim, J. Ha, B. S. Lee, Y. J. Park, C. Choong, J. B. Kim, Z. L. Wang, H. Y. Kim, J. J. Park and U. I. Chung, *Nat. Commun.*, 2014, **5**, 4929.
68. G. Zhu, J. Chen, T. J. Zhang, Q. S. Jing and Z. L. Wang, *Nat. Commun.*, 2014, **5**, 3426.
69. S. Campuzano, D. Kagan, J. Orozco and J. Wang, *Analyst*, 2011, **136**, 4621.
70. J. Wu, S. Balasubramanian, D. Kagan, K. M. Manesh, S. Campuzano and J. Wang, *Nat. Commun.*, 2010, **1**, 36.
71. D. Kagan, P. C. Marzal, S. Balasubramanian, S. Sattayasamitsathit, K. M. Manesh, G. U. Flechsig and J. Wang, *J. Am. Chem. Soc.*, 2009, **131**, 12082.
72. J. Burdick, R. Laocharoensuk, P. M. Wheat, J. D. Posner and J. Wang, *J. Am. Chem. Soc.*, 2008, **130**, 12082.
73. D. Kagan, R. Laocharoensuk, M. Zimmerman, C. Clawson, S. Balasubramanian, D. Kang, D. Bishop, S. Sattayasamitsathit, L. F. Zhang and J. Wang, *Small*, 2010, **6**, 2741.
74. L. Baraban, D. Makarov, R. Streubel, I. Mönch, D. Grimm, S. Sanchez and O. G. Schmidt, *ACS Nano*, 2012, **6**, 3383.
75. R. F. Ismagilov, A. Schwartz, N. Bowden and G. M. Whitesides, *Angew. Chem., Int. Ed.*, 2002, **41**, 652.
76. X. X. Chen, J. Gao, B. Song, M. Smet and X. Zhang, *Langmuir*, 2010, **26**, 104.
77. M. Xiao, X. P. Guo, M. J. Cheng, G. N. Ju, Y. J. Zhang and F. Shi, *Small*, 2014, **10**, 859.
78. L. E. Scriven and C. V. Sternling, *Nature*, 1960, **187**, 186.
79. E. Lauga and A. M. J. Davis, *J. Fluid Mech.*, 2012, **705**, 120.
80. M. Xiao, M. J. Cheng, Y. J. Zhang and F. Shi, *Small*, 2013, **9**, 2509.
81. M. Xiao, C. Jiang and F. Shi, *NPG Asia Mater.*, 2014, **6**, e128.

82. G. N. Ju, M. J. Cheng, M. Xiao, J. M. Xu, K. Pan, X. Wang, Y. J. Zhang and F. Shi, *Adv. Mater.*, 2013, **25**, 2915.
83. M. M. Song, M. Xiao, L. N. Zhang, D. Q. Zhang, Y. T. Liu, F. Wang and F. Shi, *Nano Energy*, 2016, **20**, 233.
84. M. M. Song, M. J. Cheng, G. N. Ju, Y. J. Zhang and F. Shi, *Adv. Mater.*, 2014, **26**, 7059.
85. L. N. Zhang, M. M. Song, M. Xiao and F. Shi, *Adv. Funct. Mater.*, 2016, **26**, 851–856.
86. G. N. Ju, M. J. Cheng and F. Shi, *NPG Asia Mater.*, 2014, **6**, e111.
87. Z. J. Cheng, H. Lai, Y. Du, K. W. Fu, R. Hou, C. Li, N. Q. Zhang and K. N. Sun, *ACS Appl. Mater. Interfaces*, 2014, **6**, 636.
88. Z. J. Cheng, J. W. Wang, H. Lai, Y. Du, R. Hou, C. Li, N. Q. Zhang and K. N. Sun, *Langmuir*, 2015, **31**, 1393.
89. H. G. Zhu, D. Y. Chen, N. J. Li, Q. F. Xu, H. Li, J. H. He and J. M. Lu, *Adv. Funct. Mater.*, 2015, **25**, 597.
90. L. B. Zhang, Z. H. Zhang and P. Wang, *NPG Asia Mater.*, 2012, **4**, e8.
91. Y. Z. Cao, N. Liu, C. K. Fu, K. Li, L. Tao, L. Feng and Y. Wei, *ACS Appl. Mater. Interfaces*, 2014, **6**, 2026.
92. H. L. Che, M. Huo, L. Peng, T. Fang, N. Liu, L. Feng, Y. Wei and J. Y. Yuan, *Angew. Chem., Int. Ed.*, 2015, **54**, 8934.
93. C. R. Gao, Z. X. Sun, K. Li, Y. N. Chen, Y. Z. Cao, S. Y. Zhang and L. Feng, *Energy Environ. Sci.*, 2013, **6**, 1147.
94. X. J. Feng, L. Feng, M. H. Jin, J. Zhai, L. Jiang and D. B. Zhu, *J. Am. Chem. Soc.*, 2004, **126**, 62.
95. R. D. Sun, A. Nakajima, A. Fujishima, T. Wantanabe and K. Hashimoto, *J. Phys. Chem. B*, 2001, **105**, 1984.
96. D. L. Tian, X. F. Zhang, J. Zhai and L. Jiang, *Langmuir*, 2011, **27**, 4265.
97. D. L. Tian, X. F. Zhang, Y. Tian, Y. Wu, X. Wang, J. Zhai and L. Jiang, *J. Mater. Chem.*, 2012, **22**, 19652.
98. L. Wu, J. P. Zhang, B. C. Li and A. Q. Wang, *Polym. Chem.*, 2014, **5**, 2382.
99. P. Calcagnile, D. Fragouli, I. S. Bayer, G. C. Anyfantis, L. Martiradonna, P. D. Cozzoli, R. Cingolani and A. Athanassiou, *ACS Nano*, 2012, **6**, 5413.
100. M. J. Cheng, G. N. Ju, C. Jiang, Y. J. Zhang and F. Shi, *J. Mater. Chem. A*, 2013, **1**, 13411.
101. X. Hou, Y. J. Liu, H. Dong, F. Yang, L. Li and L. Jiang, *Adv. Mater.*, 2010, **22**, 2440.
102. X. Hou, F. Yang, L. Li, Y. L. Song, L. Jiang and D. B. Zhu, *J. Am. Chem. Soc.*, 2010, **132**, 11736.
103. H. C. Zhang, X. Hou, L. Zeng, F. Yang, L. Lim, D. D. Yan, Y. Tian and L. Jiang, *J. Am. Chem. Soc.*, 2013, **135**, 16102.
104. A. L. Allen, K. J. Tan, H. Fu, J. D. Batteas and D. E. Bergbreiter, *Langmuir*, 2012, **28**, 5237.
105. X. Huang, Y. J. Sun and S. Soh, *Adv. Mater.*, 2015, **27**, 4062.
106. H. L. Liu, X. L. Liu, J. X. Meng, P. C. Zhang, G. Yang, B. Su, K. Sun, L. Chen, D. Han, S. T. Wang and L. Jiang, *Adv. Mater.*, 2013, **25**, 922.

107. S. Hou, H. C. Zhao, L. B. Zhao, Q. L. Shen, K. S. Wei, D. Y. Suh, A. Nakao, M. A. Carcia, M. Song, T. Lee, B. Xiong, S. C. Luo, H. R. Tseng and H. H. Yu, *Adv. Mater.*, 2013, **25**, 1547.
108. J. Y. Li, Q. S. Han, Y. Zhang, W. Zhang, M. D. Dong, F. Besenbacher, R. Yang and C. Wang, *ACS Appl. Mater. Interfaces*, 2013, **5**, 9816.
109. G. Y. Qing and T. L. Sun, *Angew. Chem., Int. Ed.*, 2014, **53**, 930.
110. A. Fujishima, *Nature*, 1972, **238**, 37.

Subject Index